Numbers and nationhood explores the Italian inflection of a Europe-wide phenomenon in the nineteenth century: the rise of statistics as a mode of representation of modern society. Silvana Patriarca examines the ideologies which informed the copious statistical literature produced between the 1820s, when statistical publications began to proliferate in the Italian states, and the 1870s, when a unified Italy entered a fully positivistic era. Her innovative study illuminates the meanings that the science of statistics had for its mostly liberal promoters and explores the uses of statistics in the process of nation-building. She shows how statistics helped to generate a long-lasting image of Italy that accentuates its internal territorial divisions. By examining the power of numerical representations, *Numbers and nationhood* provides a fresh approach to the history of Risorgimento Italy and of positivism, bringing to the fore issues of science, politics, and representation.

CAMBRIDGE STUDIES IN ITALIAN HISTORY AND CULTURE

NUMBERS AND NATIONHOOD

CAMBRIDGE STUDIES IN ITALIAN HISTORY AND CULTURE

This series comprises monographs and a variety of collaborative volumes, including translated works, which will concentrate on the period of Italian history from late medieval times up to the Risorgimento. The editors aim to stimulate scholarly debate over a range of issues which have not hitherto received, in English, the attention they deserve. As it develops, the series will emphasize the interest and vigor of current international debates on this central period of Italian history and the persistent influence of Italian culture on the rest of Europe.

Titles in the series

Family and public life in Brescia, 1580–1650
The foundations of power in the Venetian state
JOANNE M. FERRARO

Church and politics in Renaissance Italy
The life and career of Cardinal Francesco Soderini, 1453–1524
K. J. P. LOWE

Crime, disorder, and the Risorgimento
The politics of policing in Bologna
STEVEN C. HUGHES

Liturgy, sanctity and history in Tridentine Italy
Pietro Maria Campi and the preservation of the particular
SIMON DITCHFIELD

Lay confraternities and civic religion in Renaissance Bologna
NICHOLAS TERPSTRA

Society and the professions in Italy, 1860–1914
Edited by MARIA MALATESTA
Translated by Adrian Belton

Numbers and nationhood
Writing statistics in nineteenth-century Italy
SILVANA PATRIARCA

Other titles are in preparation

NUMBERS AND NATIONHOOD

WRITING STATISTICS IN NINETEENTH-CENTURY ITALY

SILVANA PATRIARCA

Columbia University

CAMBRIDGE
UNIVERSITY PRESS

Published by the Press Syndicate of the University of Cambridge
The Pitt Building, Trumpington Street, Cambridge CB2 1RP
40 West 20th Street, New York, NY 10011–4211, USA
10 Stamford Road, Oakleigh, Melbourne 3166, Australia

First published 1996

Printed in Great Britain at the University Press, Cambridge

A catalogue record for this book is available from the British Library

Library of Congress cataloguing in publication data

Patriarca, Silvana.
Numbers and nationhood: writing statistics in nineteenth-century Italy / Silvana Patriarca.
p. cm. – (Cambridge studies in Italian history and culture)
Includes bibliographical references.
ISBN 0 521 46296 7
1. Statistics – Italy – History – 19th century. I. Title.
II. Series.
HA19.P37 1996
314.5′09′034 – dc20 95–4328 CIP

ISBN 0 521 46296 7 hardback

CE

For my parents

CONTENTS

ILLUSTRATIONS

ACKNOWLEDGMENTS

This book began a few years ago as a dissertation project at the Johns Hopkins University. The generosity of several institutions helped me throughout my research and writing. The Charles Singleton Center for Italian Studies of the Johns Hopkins University awarded me a fellowship at the initial stage of my work. I received a doctoral fellowship from the Joint Committee on Western Europe of the American Council of Learned Societies and the Social Science Research Council. The Councils for Research in the Humanities and Social Sciences of Columbia University provided further support through the award of two summer fellowships, and I also profited from the research and study leave that Columbia University awards to its junior faculty. My thanks to all these institutions.

While researching and writing I have benefited from the scholarship and encouragement of many people. I wish to thank Mack Walker, my adviser at the Johns Hopkins University, who has given me invaluable help and encouragement in all stages of this work and has provided an attentive and sensitive reading of the manuscript. I am grateful to Giovanni Levi for being an inspiring mentor when I was at the University of Turin, and a challenging intellectual presence afterwards; throughout the years he has never failed as a source of moral support and hospitality. My thanks go to Cesare Mozzarelli who spurred me on with constructive criticism and helped me find not only some relevant sources, but also a place to stay in Milan while I was doing research at the local Archivio di Stato. I want to express my gratitude to Lucy Riall and Marta Petrusewicz who read the whole manuscript and generously provided valuable criticisms and suggestions. Of course the responsibility for the final product is entirely mine.

For the advice and criticism offered at different stages of this project I wish to thank Carlo Corsini, Richard Goldthwaite, Raymond Grew, Maurizio Gribaudi, Patrizia Guarnieri, Edward Malefakis, Alessandro

Pizzorno, Mary Poovey, Theodore M. Porter, Raffaele Romanelli, Nancy Stepan, Geoffrey Symcox, Stuart J. Woolf. I also would like to thank Pina Catalonotto in Palermo, Simonetta Soldani in Florence and Nerina Astarita in Naples for the generous hospitality they gave me during visits to the local archives and libraries, and Benedetto La Rosa in Palermo who helped me locate material at the Archivio di Stato of that city. My thanks also to the staffs of the Biblioteca dell'Accademia delle Scienze in Turin and of the Gabinetto Vieusseux in Florence who photocopied a great quantity of material for me.

I am grateful to several friends who have been patient enough to listen to my frequent complaints and have also helped me in this project in many significant ways: Tommaso Astarita, Lanfranco Blanchetti-Revelli, Francesca Bonazzi, Luigi Brossa, Caterina Budano, Kathleen Canning, Carla Cappetti, Eric Firstenberg, William V. Harris, Deborah Levenson-Estrada, Gregory Mefferd, Francesca Nitti, Marisa Patriarca, Michael Schaffer, Scott Spector, Michael Stolberg, Margherita Zanasi. This book is dedicated to my parents, Elda Sodano and Bruno Patriarca, who taught me the virtue of perseverance.

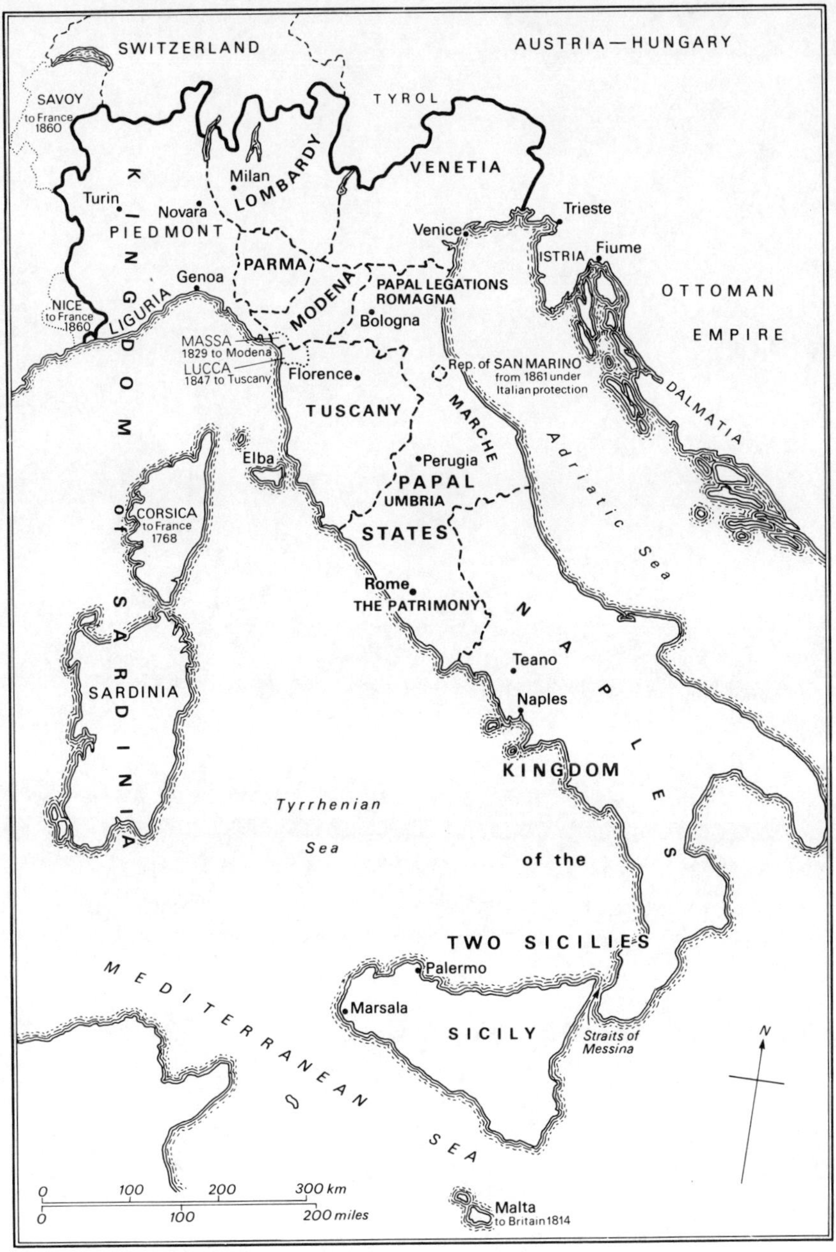

Map 1 Italy during the process of unification.

CHAPTER I

INTRODUCTION

When we think about modern states we often envision statistical aggregates. Area, population, size of the gross national product, the level of the national debt, these are identifying features that we take for granted, as an obvious way of representing territorial entities. Yet numbers have not always enjoyed this position, and in fact their rise as a fundamental mode of representation is a relatively recent process. Although the first systematic attempts to provide numerical evaluation of the population and wealth of states can be traced back to the consolidation of modern states in the seventeenth and eighteenth centuries, it is particularly when the form of the nation-state imposed itself in Europe and America, in the late eighteenth and during the nineteenth century, that statistics became a widespread practice and attracted the solicitous attention of ruling elites and reformers alike.

Since Harald Westergaard's studies on the history of statistics, historians have been aware of the "enthusiasm" for social counting that gripped several European countries in the 1830s–1840s.[1] This enthusiasm led to the launching of specialized journals, the formation of statistical societies for the purpose of collecting numerical information on the most diverse social facts and phenomena, and the establishment and consolidation of state bureaus for the collection, classification, and analysis of increasing quantities of data which administrators and political leaders deemed essential for the guidance of political decisions and for the governing of society. It is well known for example that the business of parliamentary government in liberal England was largely based on a vast amount of statistical information collected and diffused by both state agencies and private "statistical societies" – the great collectors of "Facts" which Charles Dickens satirizes in his novel *Hard*

1 *Contributions to the History of Statistics* (London, 1932). Westergaard uses the expression "era of enthusiasm" to refer to the period 1830–1849 (see pp. 136–171).

Times (1854).[2] In France, statistics flourished particularly during the Napoleonic regime which vigorously promoted the collection of information for administrative purposes;[3] interest in the numerical description and analysis of society powerfully re-emerged among the French elites in the 1820s and 1830s, giving rise to authoritative institutions and publications.[4] In the United States, from the very beginning of the nation's history, constitutional provisions linking the apportionment mechanism to census-taking brought statistics to the very center of political and social life.[5] If perhaps not to the same extent as in these countries, a similar trend occurred in other states.

Besides being linked to the consolidation of modern states, the development of statistics in the nineteenth century paralleled the growth of a reading public and other historical processes to which we usually refer through the short-hand labels of industrialization, urbanization, and the emergence of a mass society. Almost by definition, since statistics is the domain of large numbers, it may seem to be an ideal way of representing these large-scale processes; some have even maintained that statistics is "a perspective genuinely in harmony with the statistical structure of modern society."[6] Yet the reason why we think about modern societies the way we do, namely as aggregates of large numbers, is in turn the result of the very rise of statistics to a hegemonic status as a mode of representation – a process which did not

2 On the parliamentary uses of statistics see D. Eastwood, "Amplifying the Province of Legislature: the Flow of Information and the English State in the Early Nineteenth Century," *Historical Research* 62 (1989), pp. 276–294; on British statistical societies see M. J. Cullen, *The Statistical Movement in Early Victorian Britain. The Foundations of Empirical Social Research* (New York, 1975); V. L. Hilts, "*Aliis exterendum*, or, the Origins of the Statistical Society of London," *Isis* 69 (1978), pp. 21–43.

3 See *La Statistique en France à l'époque napoléonienne*, Journée d'étude, Paris, 14 février 1980 (Brussels, 1981); J.-C. Perrot and S. J. Woolf, *State and Statistics in France 1789–1815* (London, 1984); M.-N. Bourguet, *Déchiffrer la France. La statistique départementale à l'époque napoléonienne* (Paris, 1988).

4 On the various aspects of nineteenth-century statistical investigation in France (crime statistics, public hygiene, etc.) see the essays in the collective volume *Pour une histoire de la statistique* (Paris, 1977). For a general overview see H. Le Bras, "La Statistique Générale de la France," in P. Nora, ed., *Les Lieux de la mémoire*, vol. II, *La Nation* (Paris, 1986), pp. 317–353. See also S. J. Woolf, "Statistics and the Modern State," *Comparative Studies in Society and History* 31 (1989), pp. 588–603.

5 See P. C. Cohen, *A Calculating People: The Spread of Numeracy in Early America* (Chicago and London, 1982); M. J. Anderson, *The American Census. A Social History* (New Haven and London, 1988).

6 K. H. Metz, "Paupers and Numbers: The Statistical Argument for Social Reform in Britain during the Period of Industrialization," in L. Krüger, L. J. Daston, and M. Heidelberger, eds., *The Probabilistic Revolution*, vol. I, *Ideas in History* (Cambridge, Mass., and London, 1987), p. 338.

arise automatically from social transformations, but has in fact a complex history. It is the story of this rise in a specific historical setting that I tell in this book by focusing on the case of nineteenth-century Italy.

In the histories of positivism and the social sciences, nineteenth-century Italy stands out almost exclusively as the country which produced Cesare Lombroso and his school of criminal anthropology, the epitome of a stultifying application of measurements and numbers to the understanding of human differences. Although Lombroso will make an appearance towards the end of this story, this book does not deal with him or his followers, but focuses on an earlier period, the years between the Restoration of 1815 and the early 1870s which were characterized by the movement for Italian independence and unification and by the first efforts at national integration. In the period before the proclamation of the Kingdom in 1861, statistical investigations were strongly promoted and developed by "opposition intellectuals" in their struggle against domestic "autocratic" regimes and foreign domination.[7] Journals largely devoted to statistics were founded in places as diverse as Milan and Palermo. From the 1830s on several private researchers undertook to describe not only single communities and provinces, but also the condition of the peninsula as a whole, as if Italy were already a unified country. During the same period, the absolute governments re-established by the Congress of Vienna collected statistics for diverse administrative purposes, at first keeping their results secret but later increasingly making them public.

Historians of Italy have often used this literature in their reconstructions of the economic and social conditions of the Italian states before unification, and to portray the condition of the new state at the beginning of its existence. The debates on statistics which took place in the 1820s and 1830s have been studied by historians of economic thought as episodes in the formation of the specific inflection of political economy that prevailed in the Italian states of the first half of the nineteenth century. Aurelio Macchioro has coined the word *statisticismo* to indicate a political economy which clung to its old identity of

7 "Opposition intellectuals" is an expression used by M. Berengo, "Intellettuali e organizzazione della cultura nell'età della Restaurazione," in *La Restaurazione in Italia. Struttura e ideologie. Atti del XLVII Congresso di Storia del Risorgimento italiano* (Rome, 1976), pp. 297–307. As for "intellectuals," I find it useful to think about this category in the way that Katherine Verdery, following the lead of Zygmunt Bauman, does: "sometime occupants of a site that is privileged in forming and transmitting discourses, in constituting thereby the means through which society is 'thought' by its members" (*National Ideology under Socialism. Identity and Cultural Politics in Ceauçescu's Romania* [Berkeley, Los Angeles, and Oxford, 1991], p. 17).

empirical discipline;[8] Roberto Romani has recently emphasized the long-lasting presence of a "culture of observation" among nineteenth-century Italian economists.[9] The importance that statistics had for the new ruling elites after the proclamation of the Kingdom of Italy (1861) has been underlined in studies which have stressed how it became the self-legitimating ideology of the new state apparatus engaged in the process of nation-building.[10] During the first decade of existence of the new state, the publication of grand statistical summaries actually constituted, as Raffaele Romanelli has observed, one of the few truly "national" manifestations of the country.[11]

But if historians of Italy have thus often noticed that the ruling elites' first knowledge of the country was of a statistical nature,[12] and have written extensively on the politics of Italian positivism;[13] yet they have not investigated the cognitive implications and the "reality effects" of this practice of knowledge. Nor have they investigated how it contributed to the understanding of what Italy was about, or how it shaped the image of this entity both before and after its existence as an independent state. This study will explore precisely these issues and will contend that in Italy statistics not only performed a work of ideological and political legitimation, but also contributed to the creation, the "production" as it were, of the Italian nation, that is of the very entity that they were supposed to describe.

What do I mean by this statement? It is important to explain here how I use the term "statistics" since it has several and historically changing meanings. In today's ordinary use, the word "statistics" refers not only to the methodology for the treatment of quantitative data, but

8 A. Macchioro, "La Raccolta Custodi di 'Scrittori Classici di Economia' fra la statistica e l'economia politica," in D. Rota, ed., *Pietro Custodi tra Rivoluzione e Restaurazione. Atti del Primo Convegno Nazionale* (Lecco, 1989), pp. 139–164.

9 See R. Romani, *L'economia politica del Risorgimento italiano* (Turin, 1994), p. 26.

10 See R. Romanelli, "La nuova Italia e la misurazione dei fatti sociali. Una premessa," *Quaderni storici* 15 (1980), pp. 765–778. This essay is the introduction to a monographic issue of the journal devoted mainly to the study of the political context of the production, and the actual outcome, of several statistical investigations in post-unification Italy.

11 R. Romanelli, *L'Italia liberale (1861–1900)* (Bologna, 1979), pp. 8–9.

12 Besides the already mentioned works by Romanelli, see E. Ragionieri, "La storia politica e sociale," in *Storia d'Italia*, vol. IV, *Dall'Unità a oggi* (Turin, 1976), pp. 1714–1716.

13 On the period before unification see for example G. C. Marino, *La formazione dello spirito borghese in Italia* (Florence, 1974); among recent contributions on the post-unification period see G. Sola, "Sviluppi e scenari della sociologia italiana: 1861–1890," in G. Barbano and G. Sola, *Sociologia e scienze sociali in Italia 1861–1890. Introduzioni critiche e repertorio bibliografico* (Milan, 1985), and R. Camurri, ed., *La scienza moderata. Fedele Lampertico e l'Italia liberale* (Milan, 1992).

also to the data themselves and to the practice of collecting these data performed by the state or by private agencies or individuals. In this book I will also use the term "statistics" in its nineteenth-century meanings to refer both to a specific governmental science and to a genre of writing which employed numbers to describe territorial entities and collectivities – a mode of representation which first emerged in the mid-eighteenth century and became very popular throughout the following century.[14] When I say that statistics contributed to the "production" of the Italian nation I mean to refer, on the one hand, to the place that statistics occupied in the formation of a liberal and nationalist outlook and, on the other hand, to the contribution it gave to the very imagining and shaping of a national space.[15]

Risorgimento nationalism is a topic from which recent historiography, especially in Italy, has shied away. Both general historiographical trends of the past few decades (the rise of social history) and the troubled relation of Italians with their national identity, especially after the disastrous experience of fascism and of World War II, have made the issue of Italian nationalism an altogether unpopular topic of investigation. There is of course an old tradition of studies on the ideas of nationalist thinkers – Giuseppe Mazzini primarily, and the moderates such as Vincenzo Gioberti and Cesare Balbo – but little innovative research from new perspectives has appeared in recent years.[16] In this book I do not deal directly with nationalist ideology, with the ideas of major or secondary thinkers of nationalist sentiment, but I look at the making of the nation which took place in the practices of description

[14] When I use the term in the plural I mean the actual numbers and the texts in which they are collected, in the singular all the other meanings of the word. Necessarily in several cases it will be the reader's task to elicit from the context the sense in which the term is being used.

[15] I use the term "production" to stress that statistics, like any representation, do not merely "reflect," but "supplement" reality and contribute to its making: see D. LaCapra, "Rethinking Intellectual History and Reading Texts," in D. LaCapra and S. L. Kaplan, eds., *Modern European Intellectual History: Reappraisals and New Perspectives* (Ithaca, N.Y., 1982), pp. 47–85. On representations as social practices see R. Chartier, *Cultural History: Between Practices and Representations* (London, 1988).

[16] For an updated discussion of these issues see L. Riall, *The Italian Risorgimento. State, Society and National Unification* (London, 1994), ch. 5. For a recent overview of nineteenth-century Italian nationalism see A. Lyttelton, "The National Question in Italy," in M. Teich and R. Porter, eds., *The National Question in Europe in Historical Context* (Cambridge, 1993), pp. 63–105. F. Chabod's 1943–1944 lectures on ideas of nationality in Italy and Europe (*L'idea di nazione*, ed. A. Saitta and E. Sestan [Bari, 1961]) are still a useful introduction to Risorgimento nationalism, as well as a touching document of Chabod's attempt to affirm a democratic idea of nation amidst the tragedies brought about by the extreme nationalisms of this century.

of a "literature of facts" that aimed at producing an authoritative image of Italy based on "positive," that is numerical, knowledge.

While nationalism became a central political issue on the agenda of the Italian moderates only after 1848, already before then debates, projects and writings about statistics constituted an organic component of the epistemological and ideological battle which liberal writers conducted in order to create a public opinion favorable to the cause of economic progress and the reform of government structures. As the dialogue and the exchanges taking place between intellectuals across the various Italian states show, this public opinion was increasingly national. In these states, I will argue, the practice of statistical writing was the expression of the awareness, widespread among the educated elites, of the relative backwardness of Italian society; it was part of their attempt to modify that condition and had an important pedagogical value. From the 1820s and throughout the 1850s authors of statistical works engaged in a work of description of individual localities or of the whole of Italy which embodied an aspiration to re-ordering and rationalizing society and to the exercise of power. They established new taxonomies reflecting bourgeois values and aspirations. The liberal reformers' projects of modernization of society and the state were predicated on an interest in economic development and the introduction of representative institutions, and on their anxiety and fears about the consequences of uncontrolled economic change and popular unrest. A more effective government of society meant for them the capacity to check all possible threats to the social order. But making government more effective was a goal to which Restoration rulers too could not be indifferent since they faced severe difficulties in governing societies increasingly disrupted by economic change and political turmoil.[17] They could not ignore the advantages offered by a statistical surveillance, as it were, over the societies they ruled, and therefore did not dismantle the structures for data collection which in most cases they had inherited from the Napoleonic regimes. In fact they even created new ones, with uneven results.

After 1848, some nationalist liberals explicitly enrolled statistics in the cause of national independence as one of their tools for the creation of a new "fatherland." Statistics, which liberals considered a fundamental institution of representative government, destined to spread with the diffusion of liberal institutions, then became part of the project of a minority – but a very vocal minority – which aimed at the

[17] For a detailed overview of the problems of "public order" faced by Restoration governments see J. A. Davis, *Conflict and Control: Law and Order in Nineteenth-Century Italy* (London and Atlantic Highlands, N. J., 1988).

building of an Italian nation-state capable of becoming a significant presence in the arena of the advanced European nations. In that context, I argue, descriptive statistics had a "constitutive" function which its proponents did not fail to deploy as a tool in their political struggle. In the hands of supporters of the national cause such as Pietro Maestri and Cesare Correnti they truly became "patriotic" statistics,[18] instruments of nationalist propaganda aiming to show the viability of the future nation. In the Kingdom of Italy, I further argue, statistics continued to perform a constitutive task by offering the new ruling elites the help of "scientific" observation in order to address and solve contentious issues facing the new state. The proliferation of statistical investigations in the 1860s was directly linked to problems of governance and aimed at mapping the distribution of people, resources and institutions on the national territory; statisticians monitored the development of the population and attempted to check the performance of local government in the context of the new liberal institutions.

Statistical investigations and descriptions were predicated on an idea of the nation as an object to be known, measured, compared, and governed. Variously conceived in terms of territorial boundaries, Italy appeared essentially as an ensemble of resources; the "people," understood by romantic nationalism as the bearers of a common language and culture, and especially of a collective will, were absent from statistical descriptions except as population, namely as a fundamental resource to be governed. They made up a society of which the state was the "tutor," its paternal and paternalistic educator and the enforcer of social discipline. This vision fitted well the ideology of the Italian moderate liberals whose interest in change and reform coincided to a large extent with what they considered useful and necessary to prevent greater and uncontrollable change.

But statistics also created a particular image of the national space, they gave a body to an abstract entity. Of course, they were not the only genre of writing that tried to give a body to this abstract entity; other discourses, notably history, a favored genre of patriotic authors, and geography, did the same. What was specific about the image of Italy created by statistics? What distinguished and distinguishes statistics as a mode of representation? This book tries to answer these questions by tracing the projects, expectations, and conflicts that surrounded the writing of statistics at the moment of its first rise to prominence and by analyzing the logics of statistical descriptions. From a conceptual stand-

[18] This role of statistics had already been noticed by C. Pazzagli, "Statistica 'investigatrice' e scienze 'positive' nell'Italia dei primi decenni unitari," *Quaderni storici* 15 (1980), p. 797.

point the three genres I just mentioned, were, at the beginning of the nineteenth century, much closer than they are today. Many of the works defined by early nineteenth-century writers as statistical were largely qualitative and difficult to distinguish from works of political geography. Statistical works moreover sometimes incorporated historical narratives; and the famous phrase coined by August Ludwig von Schlözer in the early 1800s – "History is a statistics in movement, statistics is stationary history"[19] – which pointed to the common object of both, the state, continued to make sense for some time in the nineteenth century.

This proximity was partly due to the fact that the three genres shared a similar genealogy: in one form or another these modes of writing were all actively engaged in the service of the modern state. Since the sixteenth century at least, as Michel De Certeau has observed, historical discourse "provides [power] with a familial, political or moral genealogy."[20] As an academic discipline history was promoted in the nineteenth century to glorify the national state; where national states did not yet exist national elites enrolled history in their cause.[21] As the knowledge of the king's lands, geography too developed in an organic relation with the growth of state institutions; historically it had served the military needs of state elites, the practice of military reconnaissance providing in turn a model for the geographer's observation. In the nineteenth century, with the growth of nationalism and the nation-state, geography acquired new tasks and functions, and was taught in schools, along with history, to arouse "love of fatherland" in the hearts of prospective citizens.[22]

Recalling the proximity among these genres does not mean to deny the existence of obvious differences, which actually increased during the nineteenth century. While old descriptive geography was being questioned by a new geography of Humboldtian inspiration, statistical writings became more numerically oriented. This transformation was not without important epistemological and ideological consequences

19 *Theorie der Statistik nebst Ideen über das Studium der Politik überhaupt* (Göttingen, 1804), p. 86.

20 *The Writing of History* (New York, 1988), p. 7.

21 Many decades ago Benedetto Croce pointed out the relation between the development of historical writing and the formation of a national consciousness in nineteenth-century Italy in his *Storia della storiografia italiana nel secolo decimonono*, 2nd edn. (Bari, 1930), especially ch. 5 in vol. 1.

22 By saying this, I do not intend to rule out other matrices which shaped the development of geography. For a general overview of the history of geography see G. Dematteis, *Le metafore della Terra. La geografia umana tra mito e scienza* (Milan, 1985).

which we usually tend to take for granted. The way in which numbers construct and visualize entities of various kinds is distinctive and has consequences that go beyond their capacity for manipulation. Even more strongly than other forms of inscription, numbers insure the mobility of things by "immobilizing" them.[23] The use of numbers in the representation of the world is predicated upon procedures of classification and separation of the "identical" and the "different" which result in the building of a rigid perception of reality.[24] Applied to society, quantifying procedures, while making comparison easier (if not providing its only basis), give a concrete body to abstract entities and ideas, and make concrete things more abstract. The full sense of these observations will become apparent in the course of our account of the use of statistics in the representation of Italy.

STATISTICS AND OBJECTIFICATION

To look at statistics means to look at a particular component of elite culture. In the native land of Croce and Gramsci, where the traditions both of idealism and of historical materialism have exercised, and still exercise, a great influence, the history of the Italian elites' culture represents a prominent field of investigation. Scholars of diverse ideological orientation have worked predominantly along the traditional lines of intellectual history.[25] The conscious dimensions of ideas and systems of thought, seen either as the autonomous creation of intellectuals, or as the more or less mediated reflection of class interests, have been the major focus of investigation. In contrast little attention has been paid to the analysis of discursive practices and systems of representation from the standpoint of their own internal logic and specificity, of the "very devices, rhetorical articulations, and argumentative strate-

[23] Bruno Latour has called attention to this property possessed by all forms of inscription in "Visualization and Cognition: Thinking with Eyes and Hands," *Knowledge and Society: Studies in the Sociology of Culture Past and Present* 6 (1986), pp. 1–40.

[24] For important reflections on the characteristics of a knowledge based on procedures of classification and the search for the "identical" as it emerged in early modern culture see M. Foucault, *The Order of Things. An Archeology of the Human Sciences* (New York, 1970), esp. chs. 3 and 5.

[25] The essays on "culture" in the volumes of the *Storia d'Italia* published by Einaudi in the 1970s are paradigmatic in this respect. Things began to change slowly, however, in the 1980s; and the essays in the *Annali* of the same *Storia d'Italia* are indicative of an attention to some new perspectives and new subjects. For a critique of the study of culture in Italian historiography see also the introduction to the recent volume by S. Soldani and G. Turi, eds., *Fare gli italiani. Scuola e cultura nell'Italia contemporanea*, vol. 1, *La nascita dello stato nazionale* (Bologna, 1993), pp. 9–33.

gies"[26] which characterize them. In this book I try to combine a textual analysis of descriptive practices with an attention to the conscious projects and claims of individuals and groups. I am convinced that the adoption of this kind of approach – one that is partly inspired by the so-called "new cultural history"[27] – can bring a fresh perspective on the history of nineteenth-century Italian culture. By increasing an awareness of the complex interplay between ideologies and representations,[28] it can further a less ideologically constrained reading of the period and open new avenues of investigation.

Besides being in dialogue with the historiography on modern Italy, this book is also conversant with the growing body of literature which in recent years has been investigating the history of statistics in both its administrative and scientific dimensions – as a technology of power and knowledge, to employ a Foucauldian expression. In fact the original impulse to undertake the study that then evolved into this book came to me from a need to question assumptions and implications about the use of quantitative methodologies in history and the social sciences. Like many other social historians employing quantitative sources and methods in their work, in the mid- to late 1980s – partly in connection with a more general epistemological shift across the human sciences, which has come to be known as the "linguistic turn," and partly out of a personal dissatisfaction with quantifying procedures – I began to feel a desire for a better understanding of the epistemological and rhetorical foundations of my work.[29] The new sensitivity for

26 The expression is from R. Chartier's rich introduction to his *Cultural History*, p. 10.

27 See the introduction by L. Hunt to Hunt, ed., *The New Cultural History* (Berkeley and Los Angeles, 1989).

28 I tend to use the term ideology to mean a "system" or a "set" of beliefs (without implying unity and coherence), those "symbolic practices" which pertain to social action or political projects. There is a useful discussion in J. B. Thompson, *Studies in the Theory of Ideology* (Berkeley and Los Angeles, 1984), although the author makes a plea for what he calls a "critical conception" of ideology on which I have many reservations. The notion of representation is not easy to pinpoint especially in its relation to ideology: representations could be seen as parts of ideologies, but they can also in turn "contain" the latter: see H. Lefebvre, *La Présence et l'absence. Contribution à une théorie des représentations* (Paris, 1980). I find particularly insightful M. Poovey's work on gender representations, where these are defined as "part of the system of interdependent images through which ideologies be[come] accessible to individual men and women" and as "sites on which ideological systems were simultaneously constructed and contested" (*Uneven Developments. The Ideological Work of Gender in Victorian England* [Chicago, 1988], p. 2). I believe, however, as I will try to make clear in this book, that existing "technologies of knowledge" – statistics being one of them – play a role in shaping social representations.

29 The literature on the linguistic turn is becoming immense. Still useful is J. E. Toews, "Intellectual History after the Linguistic Turn: The Autonomy of

the role played by language in the shaping of reality has translated since then into a greater awareness of the linguistic nature of the sources that historians have been employing primarily in a documentary fashion – censuses and statistical surveys being the case in point. It has also translated into an investigation of the discursive production of categories which give meaning to and shape people's lives. In this context from instruments of historical investigation and analysis, statistics have increasingly become objects of historical analysis and critical deconstruction.

The new literature on the history and sociology of statistics has greatly contributed to our understanding of the role that statistical practices play in contemporary societies. Not only are statistics shaped by ideologies, as Joan Scott, to cite a well-known example, has shown in her study of the statistical representations of industry in mid-nineteenth-century Paris;[30] in their turn, through their "power of naming,"[31] they also tend to shape perceptions and ideologies. Several scholars, most notably Benedict Anderson and Bernard I. Cohn, have argued that censuses, those statistical operations *par excellence*, have had and continue to have an important cultural and political impact as tools for the construction of national populations and the shaping of social and ethnic identities.[32] Through an array of state rituals and social practices, people consciously or unconsciously conform to the categories set up in official statistics. Census-takers and statisticians, however, have little control over the life of the categories and classifications that they establish. "Counting" – as Ian Hacking observes – "generates its own subdivisions and rearrangements."[33] Through practices aiming to shape national populations, it is in some cases the identity of minority groups themselves that is, if not created,

Meaning and the Irreducibility of Experience," *American Historical Review* 92 (1987), pp. 879–907, and J. W. Scott's essays in her *Gender and the Politics of History* (New York, 1988).

30 "A Statistical Representation of Work. *La Statistique de l'industrie à Paris, 1847–48*," in *Gender and the Politics of History*, p. 115 (an earlier version of this essay is in S. L. Kaplan and C. J. Koepp, eds., *Work in France. Representations, Meaning, Organization, and Practice* [Ithaca and London, 1986], pp. 335–363).

31 P. Bourdieu, "The Social Space and the Genesis of Groups," *Theory and Society* 14 (1985), pp. 723–744.

32 B. Anderson, *Imagined Communities. Reflections on the Origins and Spread of Nationalism*, 2nd edn. (London and New York, 1991), ch. 10; B. I. Cohn, "The Census, Social Structure and Objectification in South Asia," in Cohn, *An Anthropologist among the Historians and Other Essays* (Delhi and Oxford, 1987), pp. 224–254.

33 "Making up People," in T. C. Heller, M. Sosna, and D. E. Wellbery, eds., *Reconstructing Individualism: Autonomy, Individuality, and the Self in Western Thought* (Stanford, Calif., 1986), p. 223.

certainly reinforced.[34] Unintended or subversive effects, in contrast to those sought by the makers of statistics, are often the most enduring outcome of the processes of objectification initiated by statistical practices.[35] To understand that practices of quantification have also their own logics is perhaps the most important outcome of the recent flourishing of studies on statistics.

As the establishment of modern statistical practices is at the origin of powerful processes of objectification, in-depth studies of these processes are needed in order to illuminate the genesis of many of the social and spatial categories we use in our own investigative work. Only in this manner can we hope to free ourselves from some of the thick lenses that the growth of administrative apparatuses has imposed on our way of studying the social world. This book is intended to be a contribution to this kind of enterprise. It should therefore not be read exclusively as a contribution to the history of nineteenth-century Italy and the culture of its elites. By reconstructing the emergence and consolidation of a mode of representing Italy which still conditions our reading of the Italian past, as well as of its present, this study is intended as a stimulus to the critical scrutiny and interrogation of the categories and units of analysis employed by social scientists and historians.

HISTORICAL PRECEDENTS: THE LEGACY OF THE ENLIGHTENMENT AND THE NAPOLEONIC ERA

Although my purpose is not that of writing a history of the science of statistics or a history of statistics as an administrative practice, but that of exploring the working of statistics as a mode of representation in a specific historical setting, both the scientific and (to a lesser extent) the administrative dimension of statistics are relevant to my investigation and are thus part of my story of numbers and nationhood in nineteenth-century Italy. To tell this story requires taking into consideration the authors and the actual sites of production of statistics, the conceptualization of the discipline and of its position in the system of the sciences, the theories, methodologies and epistemologies that

[34] For an American example see W. Petersen, "Politics and the Measurement of Ethnicity," in W. Alonso and P. Starr, eds., *The Politics of Numbers* (New York, 1987), pp. 187–233.

[35] I. Hacking, "Bio-power and the Avalanche of Printed Numbers," *Humanities in Society* 5 (1982), pp. 279–295. See also by the same author *The Taming of Chance* (Cambridge, 1990) which more specifically deals with the history of probability theory and the rise of indeterminism.

informed statistical texts in relation to the politics and ideologies of their authors, the actual practices of description and analysis and the subjects they made visible and objectified, along with those that they ignored or suppressed.

But before I can start to tell this story it is necessary to take a look at the larger context of investigative traditions and practices in which Risorgimento statistics were located. By making the Restoration years the moment of departure for this study I do not intend to play down the relevance of earlier experiences and traditions in the formation of nineteenth-century Italian statistics, nor to claim any originality or primacy for the Italians. It cannot be disputed that any proper history of Risorgimento Italy should start with the French era. This era brought into being the "institutional unity" of the peninsula and gave origin to a ruling class "endowed with common qualifications [*requisiti*], which faced with common intensity the problem of participation in the life of the state."[36] My study is however not interested in origins or in offering a "complete" history of statistics in Risorgimento Italy. It has the more limited, and more focused, goal of investigating the uses of statistics on the part of liberals and nationalists in the central years of the Risorgimento and of offering elements of reflection on the instruments that modernizers and supporters of the national cause possessed for knowing and representing the country in which they lived. In order to provide a background for a better understanding of the characteristics of the statistical literature examined in this book, I will briefly recall in this section the distinguishing traits of these traditions and in the next one I will outline some of the new conceptions of statistics that emerged in the nineteenth century.

Several studies have already mapped out the experiences and traditions that preceded the period I will examine and that were influential in shaping some of the later output: the traditions of so-called political arithmetic and of German academic statistics (known as *Staatenkunde* or *Statistik*) which emerged in the late seventeenth century and flourished in the eighteenth century, and especially the experience of the governments established during the French occupation which provided important models for the development of statistics in the Italian states of the first half of the nineteenth century.

The expression "political arithmetic" was coined by the English army physician and land surveyor, William Petty, in the 1670s. Emerging in a period of civil strife, the main objects of investigation of political arithmetic were population and government revenues, the

[36] See A. Scirocco, *L'Italia del Risorgimento 1800–1871* (Bologna, 1990), p. 7.

knowledge of which had the aim of strengthening the power of central authority and of the state.[37] Following the developments and vicissitudes of the modern state on the continent, studies of a similar kind were increasingly carried out in France and in the German states in the eighteenth century.[38] In the absence of complete enumerations and reliable numbers, political arithmeticians produced estimates of the resources – both human (population) and material (wealth and revenues) – of states. The findings of political arithmeticians must also be located within the framework of the development of insurance schemes in early modern Europe. One of the most important constructs of political arithmetic, the mortality table, was devised and perfected in connection with the problems of annuities and life insurance, even though, as Lorraine Daston has observed, the actual practice of insurance had little to do with these studies before the nineteenth century.[39] It is this development of demographic statistics in eighteenth-century Europe that in his influential work Michel Foucault has interpreted as pointing to the formation of a new kind of power – a "bio-power" – targeting human beings considered as living beings, as a species which has to be regulated in order to be preserved and increased.[40]

What conventional historians of statistics have usually presented as the alternative to the quantifying methods of political arithmetic (and Foucault sees as an instance of the "governmentalization" of the

[37] For an outline of the development of political arithmetic see Westergaard, *Contributions*, pp. 16–100; see also for a more recent assessment J. and M. Dupâquier, *Histoire de la démographie. La statistique de la population des origines à 1914* (Paris, 1985), pp. 129–198, and T. M. Porter, *The Rise of Statistical Thinking 1820–1900* (Princeton, 1986), pp. 18–23. On English political arithmetic in particular see P. Buck, "Seventeenth-Century Political Arithmetic: Civil Strife and Vital Statistics," *Isis* 68 (1977), pp. 67–84, and "People Who Counted: Political Arithmetic in the Eighteenth Century," *Isis* 73 (1982), pp. 28–45.

[38] There is some general information in the works of Westergaard and of Dupâquier cited in the previous note. On political arithmetic in France see also J. Hecht, "L'Idée de dénombrement jusqu'à la Révolution," in *Pour une histoire de la statistique*, pp. 21–82; M.-N. Bourguet, *Déchiffrer la France*, pp. 40–44. On German demographic studies see also J. Hecht, "Johann Peter Süssmilch point alpha ou omega de la science démographique naïve?," *Annales de démographie historique* (1979), pp. 101–144.

[39] See L. Daston, "The Domestication of Risk: Mathematical Probability and Insurance 1650–1830," in Krüger *et al.*, eds., *The Probabilistic Revolution*, vol. 1, pp. 237–260. On the mortality table see Dupâquier and Dupâquier, *Histoire de la démographie*, ch. 6.

[40] Foucault defines "bio-power" as "what brought life and its mechanisms into the realm of explicit calculations and made knowledge-power an agent of transformation of human life" (*History of Sexuality*, vol. 1, *An Introduction* [New York, 1980], p. 143).

state),[41] namely German academic statistics, was first introduced at the University of Helmstadt by Hermann Conring in the late seventeenth century. The major systematizer of this discipline was Gottfried Achenwall, professor at Göttingen between 1748 and 1772.[42] Academic statistics developed as a descriptive discipline whose aim was the knowledge of the power of a state through the accurate and orderly description of its components. This discipline had a political purpose analogous to that of political arithmetic; yet it distinguished itself from the latter by resorting to numbers only for purposes of description of states and comparison between states, and by refraining from the calculations and estimates of political arithmeticians. Moreover, while political arithmetic implied a conception of the autonomy of society, this was clearly absent from *Staatenkunde*, which has been aptly described as a "morphology of the state."[43]

In her study of Napoleonic statistics Marie-Noëlle Bourguet has suggested that it was not by accident that this latter tradition was preferred in revolutionary France. Under the absolute monarchy, Enlightenment scholars had often to resort to the calculations of political arithmetic in order to make up for the lack of official data. Republican administrators, on the contrary, intended to mark the beginning of a new era by producing a complete statistical description of all the French *départements* which could serve the state and inform the nation. While relying on the French monarchical traditions of investigation and of erudite topographical descriptions – which had involved physicians and savants of various kinds – the statisticians of the Napoleonic era also looked with interest at the descriptive models set up by German scholars.[44] Around this enterprise many people, both within

41 "Governmentality," in G. Burchell, C. Gordon and P. Miller, eds., *The Foucault Effect. Studies in Governmentality* (Chicago, 1991), pp. 87–104 (this is the text of a lecture originally given at the Collège de France in 1978).

42 On German academic statistics see, besides Westergaard, *Contributions*, pp. 4–15, and Dupâquier and Dupâquier, *Histoire de la démographie*, pp. 114–128, P. F. Lazarsfeld, "Notes on the History of Quantification in Sociology. Trends, Sources and Problems," *Isis* 52 (1961), pp. 283ff.; J. Hoock, "D'Aristote à Adam Smith: quelques étapes de la statistique allemande entre le XVIIe et le XIXe siècle," in *Pour une histoire de la statistique*, pp. 477–491; A. Seifert, "Staatenkunde: eine neue Disziplin und ihr wissenschaftstheoretischer Ort," in M. Rassem, ed., *Statistik und Staatsbeschreibung in der Neuzeit* (Paderborn, 1980), pp. 217–248.

43 See Bourguet, *Déchiffrer la France*, pp. 49–50. For a comparison of the two "traditions" see also Lazarsfeld, "Notes on the History of Quantification," pp. 279–292.

44 Bourguet, *Déchiffrer la France*, pp. 22–52; F. Sofia, *Una scienza per l'amministrazione. Statistica e pubblici apparati tra età rivoluzionaria e restaurazione* (Rome, 1988) has concentrated her analysis on the French theoretical debates of the same period to show the linkage with German statistics (ch. 1).

the state apparatus and close to it, were mobilized; new theoretical works made their appearance along with a large number of empirical descriptions.

The eighteenth-century Italian states as well saw a flourishing of political arithmetic studies among Enlightenment intellectuals. While sometimes helped by enlightened princes, these investigators, like their French colleagues, had often to gather information on their own in the face of difficulties and disappointments. This was the case with the Piedmontese political arithmeticians Giambattista Vasco and Carlo Ludovico Morozzo, and with the Tuscan Marco Lastri.[45] In Austrian Lombardy, in contrast, the monarch took the initiative. In 1750 Maria Theresa launched an impressive work of investigation into the state of land and agriculture with the aim of drawing up a land register. This was the basis for a major reform of the taxation system and received the praise of intellectuals such as Pietro Verri.[46] In the Kingdom of the Two Sicilies the reforming zeal of enlightened intellectuals, although not employed in political arithmetic studies, generated remarkable descriptive material: in the 1780s the *visitatore* (inspector) Giuseppe Maria Galanti, a student of the Neapolitan economist Antonio Genovesi and the most illustrious exponent of the second generation of the southern Enlightenment, was allowed by the king to tour the provinces of the Kingdom in order to report to him about their condition. The result was a work that Vincenzo Cuoco, a political and intellectual protagonist of the Napoleonic period and a student of Galanti, would define as the first "statistical" work produced in Italy.[47]

It was thanks to these eighteenth-century experiences that, on their

[45] On political arithmetic in Piedmont see G. Levi, "Gli aritmetici politici e la demografia piemontese negli ultimi anni del '700," *Rivista Storica Italiana* 86 (1974), pp. 201–261; on Tuscany see M. P. Paoli and R. Graglia, "Marco Lastri: aritmetica politica e statistica demografica nella Toscana del '700," *Annali della Fondazione Luigi Einaudi* 12 (1978), pp. 117–215. On the quantitative approaches of eighteenth-century Italian economists see M. Bianchini, *Alle origini della scienza economica. Felicità pubblica e matematica sociale negli economisti italiani del Settecento* (Parma, 1982).

[46] For a general account see D. Sella and C. Capra, *Il Ducato di Milano dal 1535 al 1796* (Turin, 1984), pp. 310–323, 350ff.

[47] See the introduction by D. Demarco to G. M. Galanti, *Della descrizione geografica e politica delle Sicilie*, 2 vols. (Naples, 1969), and Sofia, *Una scienza per l'amministrazione*, pp. 144–155. On statistical investigations in the South see also R. De Lorenzo, "Strategie del territorio e indagini statistiche nel Mezzogiorno fra Settecento e Ottocento," in R. De Lorenzo, ed., *L'organizzazione dello stato al tramonto dell'Antico Regime* (Naples, 1990), pp. 129–185. Reference should also be made to the survey activity (enumerations of goods and souls, for example) of the state: but for this see ch. 4 below.

arrival in the Italian peninsula, the French found an intellectual milieu which was very receptive to their will to accumulate knowledge about the new territories.[48] While in the territories directly annexed as *départements* to the Empire the French pursued their grand program of departmental or prefects' statistics, in the northern Italian Republic, which became the Kingdom of Italy in 1805, and in the Napoleonic Kingdom of Naples, independent offices were established and programs of statistical investigations were initiated, with various outcomes.[49] It was in the service of the Napoleonic state apparatuses that some of the protagonists of the debate on statistics during the Restoration, notably Melchiorre Gioia and Luca de Samuele Cagnazzi, began their statistical apprenticeship. In the effort to recover the specific meaning that the "science" of statistics acquired within the institutional and constitutional arrangements and rearrangements of Napoleonic Italy, Francesca Sofia has recently defined this statistical enterprise as the "last and fascinating chapter of the eighteenth-century 'police science.' "[50] Through Sofia's strongly juridico-political lens, the debates about statistical research during the Napoleonic era emerge as episodes embodying the institutional conflicts of those years, in particular the conflict between the attempt of the French authorities to centralize the organs of government and the resistance of the local notables, in other words between different modes of conceiving the nature and tasks of a public administration.

But there was more to the proliferation of statistics than institutional and administrative transformations, however important these are for an understanding of the debates and conceptualizations of statistics taking place at the time. Already during the Napoleonic period statistics began to escape a purely administrative dimension and to flow out into an emerging public sphere. The appropriation by the "public" of this language of power is perhaps the most distinctive trait of statistics in the nineteenth century, along with the definitive institutionalization of statistics within the states' apparatuses, and the conceptual transformations which concerned the scientific dimension of statistics. It is necessary at this point to introduce these conceptual transformations since their occurrence inevitably had an impact on the empirical works undertaken by administrations and by private statisticians alike.

[48] As Sofia observes in *Una scienza per l'amministrazione*, ch. 2.

[49] Besides Sofia, on the Kingdom of Italy see A. Cova, "Osservazioni sulla origine delle statistiche del Regno italico," *Annali dell'Istituto Italo-Germanico in Trento* 5 (1979), pp. 117–141. On the Kingdom of Naples see S. Martuscelli, ed., *La popolazione del Mezzogiorno nella statistica di Re Murat* (Naples, 1979).

[50] *Una scienza per l'amministrazione*, p. 10.

STATISTICS IN THE NINETEENTH CENTURY: OLD AND NEW PRACTICES AND CONCEPTIONS

In the 1820s and 1830s a new kind of statistical science emerged in some European countries – in France, England, and Belgium in particular: Theodore M. Porter has defined it a "numerical social science of facts." It quantified the dimensions of and studied the relations between particular aspects or "problems" of a society: crime, the primary object of "moral statistics," education, public health, the condition of the working classes, and so on.[51] This practice of statistics was closely linked to urban realities and to the tensions of industrializing societies. Most of its practitioners – such as the members of the British statistical societies – engaged primarily in the collection and publication of "facts." There were also others, such as the Belgian astronomer and mathematician Adolphe Quételet and the Frenchman A.-M. Guerry, who, on the basis of numbers most often collected by state bureaucracies, searched for the regular trends present in averages and rates over time and across countries, and for the patterns of correlation among different kinds of social facts. They tried to make generalizations and looked for the "laws" of society. Quételet was actually the principal inventor of a "numerical social science of laws,"[52] to use again Porter's definition, which he called "social physics." He attempted to apply methods and notions derived from the calculus of probability in the analysis of social numbers and in 1835 invented a construct – the *homme moyen* or "average man" – which was destined to have considerable success in the years to come while at the same time eliciting a lot of controversy. The Belgian savant contributed greatly to making society an entity in its own terms. In many respects the social world that his social physics conjured up is the world which we inhabit today.

Late-nineteenth-century and even twentieth-century statisticians who aimed at establishing a proper genealogy for their own idea and practice of statistics have usually contrasted the "old" science of the power of the state, descriptive statistics, with Quételet's more mathematically oriented social physics. The former did not survive as an independent science and was eventually superseded by the other, more "investigative" and mathematical in kind.[53] In fact things were more

51 For this account see Porter, *Rise of Statistical Thinking*, chs. 1–2.

52 Ibid., p. 41. The most complete treatment of Quételet's work remains J. Lottin, *Quételet: statisticien et sociologue* (Louvain, 1912).

53 See for example A. Gabaglio, *Teoria generale della statistica*, 2nd edn. (Milan, 1888); V. John, *Geschichte der Statistik* (Stuttgart, 1884); Westergaard, *Contributions*.

complicated than that. Throughout the whole nineteenth century many "schools" or different modes of understanding and practicing statistics coexisted, articulated as they were with different institutions and social figures within each country. The coexistence was not always peaceful, to be sure. Yet there was no linear development and superseding of one school by another, but a complex process of change and exchange.[54] Quételet himself did not see his social physics as an alternative to a practice of statistics more oriented towards description, but as its necessary complement.[55] Recent research in the history of science and probability has also tended to limit the significance of the role played by Quételet in the making of modern statistics and has emphasized his "failures" *vis-à-vis* the latter, ranging from his inability to distinguish the influence of different causes,[56] to his very limited use of mathematics.[57]

Quételet's work, however, if seen, as it should be, in the context of the practices of his own time, certainly exhibited many innovative features. In that context, as Joshua Cole has recently observed, by "eliminating the one to one correspondence between objects in the world and digits in the statistical table," Quételet's work on averages represented an important innovation in the treatment of numbers not only at a conceptual but also at a rhetorical level; his average man was an "allegory of difference, an avowedly fictional representation for multiplicity and variety in a single object."[58] But Quételet's impact was to be felt at different times in different countries. As a matter of fact, each European country "went statistical" in its own way.[59] This meant not only that who counted and what was counted differed in each country (as Lorraine Daston has humorously observed, in the nineteenth century "the French counted criminals, the British paupers,

54 M. Armatte makes a strong argument in this sense: "Une discipline dans tous ses états: la statistique à travers ses traités," *Revue de synthèse*, n. s., 4 (1991), pp. 161–206. Interestingly, Fedele Lampertico, a late-nineteenth-century Italian statesman and statistician, also observed that there is no neat sequence of phases and periods in the history of statistics: "Sulla statistica teorica in generale e su Melchiorre Gioia in particolare," *Annali di statistica*, 2nd s., 7 (1879), p. 130.

55 This is particularly clear in his *Lettres à S. A. R. le Duc Régnant de Saxe-Coburg et Gotha sur la Théorie des probabilités appliquée aux sciences morales et politiques* (Brussels, 1846), a work divided between a large section on probability theory and social physics and another on statistics for administrative uses.

56 S. M. Stigler, *The History of Statistics. The Measurement of Uncertainty before 1900* (Cambridge, Mass., and London, 1986), pp. 174ff.

57 Porter, *The Rise of Statistical Thinking*, p. 46.

58 "The Chaos of Particular Facts: Statistics, Medicine, and the Social Body in Early Nineteenth-Century France," *History of the Human Sciences* 7 (1994), p. 13.

59 I am borrowing the expression from Hacking, *The Taming of Chance*, p. 17.

and the Prussians foreigners"),[60] but also that interpretations of what statistics was about varied and the diffusion of statistical innovations, of new methods and languages, followed very different paths in different societies.

In the Italian states of the 1820s–1850s, the continuities with the Napoleonic tradition of descriptive statistics were remarkable. Not only was interest in statistics kept alive by former functionaries of the Napoleonic regimes who animated an expanding publishing sector, but the works defined as statistical were closely linked to that tradition of research. They were most often descriptions of given territorial or administrative entities (cities, communes, provinces, states) that made use of numbers but also contained extensive non-quantitative material. Less numerous, but increasing over time, were the works devoted to the study of a single category of phenomena, such as crime or public health. Statistics continued to be conceived predominantly as an administrative or governmental science and enjoyed a status similar to that of political economy.

The Italian promoters of statistics were aware of the new developments in quantitative investigation that were taking place in England, France, and Belgium, but until the 1860s their contribution to the new statistics was negligible. They lacked the conditions for developing the kind of investigations on which the new statistics relied: there was no unified state with its own statistical apparatus to produce the "large numbers" needed by a consistently numerical social science and necessary to further disciplinary specialization. Moreover, practitioners of statistics usually did not have any advanced knowledge of those disciplines such as mathematics that could have stimulated a different kind of quantification. They were mostly humanists with a background in legal studies and political economy, or physicians and naturalists. The impact of the new Quételetian statistics would be felt in Italy only after the unification of the country in the 1860s and 1870s with the start of the debate on positivism.

As a similar pattern of development, characterized by a long persistence of the tradition of descriptive statistics, was to be found also in the German states, one could be tempted to read the dominance of the descriptive school in the Italian states as an unmediated sign of their socio-political "backwardness."[61] But, needless to say, this would

[60] "Introduction to Volume 1," in Krüger *et al.*, eds., *The Probabilistic Revolution*, vol. 1, p. 3.

[61] On the heritage of descriptive statistics in the German states see T. M. Porter, "Lawless Society: Social Science and the Reinterpretation of Statistics in Germany, 1850–1880," in Ibid., pp. 351–375.

imply a simplistic reading of the link between ideas and the context of their production. The task of this study will not be that of explaining an absence, but of accounting for a presence.

STRUCTURE OF THE STUDY

It must be said right away that this is a story without female agents or voices, as the world of nineteenth-century statists and statisticians was almost exclusively male – more so than was the case in other fields. Women, to be sure, began to participate in empirical social investigation in the second half of the century, but I am not aware of any woman who was able to make a mark in the world of state statisticians, with the exception of Florence Nightingale. As part of the populations observed and analyzed by statisticians, women were present as what today's quantitative social scientists call the "variable" sex. But although there are no female voices, we face, needless to say, a very gendered discourse, as the "nation" is usually identified with conventionally feminine attributes, especially when confronted with that very male entity, the "state." I will have occasion to return to this later.

In Chapter 2 I examine the theories of statistics elaborated in the Italian states in the first half of the century, and their relation to neighboring discourses such as geography and political economy. I look at the political and intellectual location and backgrounds of the promoters of statistics, and account for the reception of the work of two intellectuals operating in Milan in the early part of the century, Melchiorre Gioia and Gian Domenico Romagnosi, who came to be seen as the main representatives of a particular inflection, if not of an "Italian school," of statistical science. The theoretical debates in which these figures were involved are familiar to scholars, but have been analyzed primarily in the context of the development of political economy. My contention is that they point to the relevant presence of statistics in the culture of the Risorgimento educated elites.

In Chapter 3 I move to a discussion of what I call the methodological concerns that attracted the attention of Italian statisticians working within the theoretical frameworks analyzed in Chapter 2. These concerns centered around the question of the principles that ought to organize statistical descriptions and especially the issue of the order of exposition. A closer examination of these principles or logics of description shows the close connection between epistemological and ideological stances. I emphasize the prescriptive nature and the pedagogical function of the descriptive works produced during the Risorgimento. Their taxonomies entailed a re-ordering of the "social body" along "natural" lines which suited a liberal and bourgeois outlook.

Chapter 4 focuses on the institutions that produced a great part of the figures that were published and circulated in the statistical literature of the first half of the century. I describe in particular the circumstances that brought about the creation of statistical offices and committees in the Italian states, and their investigative activities. Although, for polemical purposes, private promoters of statistics complained about the paucity and secrecy of the states' engagement in these activities, the newly created statistical offices also witnessed the diffusion of an interest in statistics among the ruling elites of the administrative monarchies. In characteristic fashion, however, the figures they released were used by patriotic statisticians to promote their own agenda.

In Chapter 5 I investigate the descriptive strategies of a genre of statistics that went beyond local and state boundaries to embrace an entity that did not yet exist politically, namely the Italian "nation." A call for the making of statistics of the nation was voiced by several intellectuals already in the 1830s and more consistently in the two following decades. A few authors actually undertook this enterprise privately, with their own resources and with a remarkable awareness of the constructive and constitutive nature of their work of representation. The chapter discusses the different modalities and outcomes of various attempts to achieve a description of the Italian peninsula in its entirety. It traces the new priorities of patriotic statistics *vis-à-vis* earlier local descriptions and discusses the distinctively abstract nature of the national space constructed by numerical statistics.

Chapter 6 deals with a favorite topic of nineteenth-century quantitative investigations, namely so-called "moral statistics." Focusing on a work published in Venice in 1865–1866 just before this city and its region were annexed by the Italian state, the chapter shows how the debate about moral statistics was centered around the meaning of civilization and eventually of national identity. As the field of moral statistics was, moreover, closely linked with the new statistics of Adolphe Quételet, the chapter also examines the impact and the rhetorical power of the introduction of a new methodology in the practice of statistical analysis.

The politics of the representation of the country in national statistics during the first decade of existence of the Kingdom of Italy is the subject of Chapter 7. After unification and in an intellectual climate dominated by the launching of the debate on positivism, the most prominent representative of patriotic statistics, Pietro Maestri, presided over the statistical service of the new nation. Maestri and his collaborators consistently attempted to use statistics to favor the process of national integration through the scientific study of the country. In fact, far from being able to spur this process, national statistics contributed to stabilizing and giving definite contours to the image of a country partitioned into regions.

Chapter 8 focuses on another aspect of the new state elites' attempt to use statistics as a means to stimulate processes of political modernization and national integration at the local level. I analyze the debates that took place at a particular event, the Sixth International Congress of Statistics, held in Florence in 1867. As the difficult relation between the national "whole" and its parts had by then become apparent, Maestri and his collaborator Cesare Correnti assigned what they called the "anatomy" of nations new tasks in order to confront the problem of persisting local identities and to control the government performance of local elites.

Finally in the epilogue I raise some considerations of the role of statistics in the construction of the image of the "Two Italies" which has strongly conditioned the reading of the Italian national past. Members of the Lombroso school of criminal anthropology, who used the statistics accumulated by the state statisticians to pursue their own scientific and political agenda, made a great contribution to the creation of this image. The Lombrosians looked at the world with categories that were not part of the cultural baggage of the state statisticians *à la* Maestri. Yet by then the language of numbers had become a dominant means for the illustration, if not definition, of sameness and difference and was fostering an oppositional way of thinking. That national statistics, far from promoting a complete unification of Italy – as its original makers had intended – ended up highlighting a profound internal division is an irony that the early statisticians were probably not able to appreciate.

CHAPTER 2

A SCIENCE FOR THE "CIVILIZING" OF NATIONS

In 1826 in the pages of the *Annali universali di statistica*, the most important journal of the liberal and rationalist-oriented Lombard intelligentsia, the editor, Francesco Lampato, related the diffusion of an interest in statistics to the new "spirit of the time": "Men, once bellicose, now industrious, address their curiosity to the progress of industry, where earlier they addressed it to the successes of war; and this is demonstrated by the general passion for statistics."[1] Having featured articles on the theory of statistics and statistical publications from its start in 1824, along with articles on political economy, geography, voyages, history, and so on (the complete title of the journal was *Annali universali di statistica, economia pubblica, storia, viaggi e commercio*),[2] this journal also began in 1827 providing its readers with figures that could be of more direct interest to their commercial activities, such as those on the sale of silk in London, and on the condition of the silk market in other Italian states. From 1831 onwards a special section in the journal assembled quantitative information concerning various European as well as Italian states.

The "general passion for statistics" underscored by the editor of the *Annali* was not limited to Lombardy–Venetia, the most commercially

1 [F. Lampato], "Su i progressi dell'industria in Inghilterra," *Annali universali di statistica* (henceforth *AUS*), 7 (1826), pp. 234–235.

2 This is the title it acquired with vol. 6 and retained until 1871, the last year of publication. The original title of the journal was *Annali universali di viaggi, geografia, storia, economia pubblica e statistica*; it became with vol. 2 *Annali universali di statistica, economia pubblica, storia e viaggi*. On this journal see S. La Salvia, *Giornalismo lombardo: gli "Annali universali di statistica" (1824–1844)* (Rome, 1977), and K. R. Greenfield, *Economics and Liberalism in the Risorgimento. A Study of Nationalism in Lombardy 1814–1848*, 2nd edn. (Baltimore, 1965), pp. 160–171. Lampato provided his own statistics on the subjects of the articles published in the journal between 1824 and 1829: statistical sciences, 253; public economy, 136; geographical sciences, 62; travels, 145; history, 118; biography, 19; industry and commerce, 128; miscellaneous, 180; total, 1,041 (from *AUS* 22 [1829]).

developed area in Italy. Around the same period, analogous periodicals and publications in the other Italian states – such as the *Antologia* in Florence, *Il progresso delle scienze, delle lettere e delle arti* in Naples, and the *Giornale di statistica* in Palermo – began to feature articles on statistics, on its theory and practice, and to offer numerical information on all sorts of subjects. Along with presentations and discussions of numerical data of a serial kind on population, commerce, crime, education, foundlings, and so on, they offered comprehensive statistical descriptions of villages, towns, provinces, and states. These descriptions of states and their subdivisions appeared not only in publications for the educated elites but also in the pages of popular almanacs. As books, they were often sold by installment, and received careful and sympathetic reviews from the compilers of the periodicals we just mentioned.[3]

These works constituted the kind of literature that most distinctively illustrates what the promoters of statistics in the Italian states meant when they used the word "statistics": a mainly descriptive endeavor aimed at constructing a comprehensive picture of a country, or of any of its subdivisions, with numbers, in order to evaluate its overall condition, and, most importantly, its degree of "civilization." I must stress that I use the term "descriptive" mostly out of convenience, as a shorthand to indicate a specific tradition of writing. One must be aware, however, that the expression was not employed by the practitioners themselves, who thought of what they were doing as statistics *tout court.* It was employed, on the contrary, by their critics, both contemporary scholars such as the French economist Jean-Baptiste Say and later statisticians, to refer to what they considered a practice devoid of a truly scientific status. In fact, there was more than pure description to descriptive statistics.

What there was can be understood by looking first at how this science was conceptualized, at the theories of statistics which were available to and were developed by Italian practitioners in the first half of the century. Contemporary writers and commentators liked to refer to a specifically Italian contribution to statistics, if not of a specifically Italian "school" of statistics; they listed among its chief representatives Melchiorre Gioia, Gian Domenico Romagnosi, and Luca de Samuele Cagnazzi, and described its distinctive approaches in glowing terms. They tended to stress the role that Italian authors played in the "regen-

3 A fairly good (although not complete) catalogue of these works is in an old official publication of the Direzione Generale della Statistica (the General Directorate of Statistics of the Kingdom of Italy): *Annali di statistica. Saggio di bibliografia statistica italiana*, ed. L. Bodio, 2nd edn. (Rome, 1885), pp. 18–54.

eration" of the "science" of statistics, to use their own terminology. In his 1828 review of Gioia's *Filosofia della statistica*, a collaborator of the *Annali universali di statistica*, Giuseppe Sacchi, pointed out the "sad" state of this science in other nations, and contrasted it with the solid principles on which it was based in Italy thanks to the treatise of Gioia and the contributions of Romagnosi.[4] In 1838, in the pages of the *Giornale di statistica* of Palermo, Raffaele Busacca claimed the superiority of the Italians with respect to the theoretical discussion on statistics: "the theory of statistics has not been so profoundly discussed by French or English authors as it has been *presso noi* by Gioia and Romagnosi."[5] In the late 1850s the Milanese Cesare Correnti contrasted the works of the Italians with the "narrow-mindedness and prolixity of what others call political arithmetic,"[6] and the Sicilian Gaetano Vanneschi maintained that Romagnosi had "the glory of having founded a school of statistical science" which was "quintessentially *Italian* [original emphasis]."[7] The Italian school of statistics was to be added, as yet another reason for national pride, to the Italian school of political economy and to what was regarded as the "Italian doctrine" of *incivilimento* (civilization).[8]

There are of course good reasons to be skeptical about claims of this kind especially when they are combined with concurrent claims put forward by some authors about the primacy of their own states or cities in the creation of statistics.[9] As the various Italian states possessed to a

4 *AUS* 15 (1828), pp. 309–323. See also La Salvia, *Giornalismo lombardo*, ch. 3.

5 Review of *Archives statistiques du Ministère des Travaux publics de l'agriculture et du commerce, publiées par le ministre secrétaire d'état de ce département*. Paris, Imprimerie royale [sic], 1837, in *Giornale di statistica* 3 (1838), p. 147. One should note that this kind of chauvinism did not prevent the author being critical of the Italian theorists as well.

6 *Annuario statistico italiano. Anno I 1857–58* (Turin and Milan, [1858]), p. 14.

7 *Elementi di statistica* (Palermo, 1859).

8 On the creation of the Italian school of political economy see R. Romani, *L'economia politica del Risorgimento italiano* (Turin, 1994), pp. 43–45.

9 See for example the work of Antonio Quadri, functionary and publicist from Vicenza, who claimed that statistical investigations were first undertaken by Venetian rulers in the thirteenth century and were then imitated by other nations (*Storia della statistica dalle sue origini sino alla fine del secolo XVIII per servire d'introduzione ad un prospetto statistico delle provincie venete* [Venice, 1824], chs. 2–3); and the Tuscan enlightened reformer Aldobrando Paolini who claimed that descriptive statistics for governmental use was born in Tuscany in the Middle Ages ("Studio e progressi della statistica in Toscana," *AUS* 36 [1833], pp. 209–230) and that present-day Tuscany continued to be a leader in the production of statistics – claims that Romagnosi (in a note added at the end of Paolini's article) declared unfounded as the author did not provide "positive" evidence and referred to something that could not properly be considered statistics.

greater or lesser extent their own cultural and scientific traditions, these traditions were likely to have made their impression upon the development of statistics. A certain diversity of practice and conceptualization can be expected also in relation to a diversity of foreign interlocutors; possibly several Italian "schools" could be reconstructed, or perhaps none at all considering that in some cases it is hard to find a unified stance even in the work of any one individual author. My purpose here, however, is not that of trying to establish whether an Italian school of statistics really existed or whether it was pure invention. Instead I will take the claim for what it is, namely an expression of the cultural nationalism of those who made it, and focus on what Italian scholars and intellectuals did when they debated statistics, what the promoters of statistics were after, what they had in common from both an intellectual and a socio-political standpoint. It has recently been argued that political economy constituted the dominant presence in the culture of Risorgimento patriots.[10] In fact statistics too contributed in a significant way to the shaping of their outlook. Indeed, through the examination of the theory and practice of statistics, we can also identify aspects of this outlook which a focus on other disciplines and discourses would not allow us to see.

THEORIES: THE TEACHINGS OF MELCHIORRE GIOIA AND GIAN DOMENICO ROMAGNOSI

It was a clergyman from Apulia, Luca de Samuele Cagnazzi (Altamura 1764 – Naples 1852), a self-taught mathematician and political economist, who authored the first treatise on statistics written in Italian. This work, *Elementi dell'arte statistica*, was published in Naples in 1808–1809 when Cagnazzi held a chair of political economy at the University and had also been appointed adviser on questions of economics and statistics – first in the government of Napoleon's brother, Joseph Bonaparte, and then under Murat.[11] His chair of political economy was trans-

[10] See Romani, *L'economia politica del Risorgimento*, p. 43.

[11] Cagnazzi, the son of an army captain, studied at the University of Altamura and at Naples; his patrons made him follow an ecclesiastical career. Victim to Bourbon repression against the revolution of 1799, he fled the country and lived for a while in Florence where he met with the most prominent Tuscan intellectuals of the time. With the return of the French he went back to Naples. He told the story of his life in a memoir (*La mia vita*, ed. A. Cutolo [Milan, 1944]). On Cagnazzi see *Dizionario biografico degli italiani*, s. v. by C. P. Scavizzi; B. Salvemini, *Economia politica e arretratezza meridionale nell'età del Risorgimento. Luca de Samuele Cagnazzi e la diffusione dello smithianesimo nel Regno di Napoli* (Lecce, 1981); see also E. Lombardo, "Il primo trattato italiano di statistica di Luca de Samuele Cagnazzi ed i suoi interessi demo-

formed into a chair of "statistics and economics" from 1811 to 1816.[12] Cagnazzi dedicated his work to the "peoples of the Kingdom of Naples" inviting them to "revive the ancient opulence of Southern Italy."[13] The first step towards this goal was that of "studying ourselves and what we have," creating in other words what he called a "national statistics"; the government, he announced, was about to initiate this "glorious enterprise."[14] He then went on to provide a brief history of the origins of statistics, and detailed all the objects to be included in a statistics, from the physical and climatic characteristics of a country to its population, agriculture, industry and commerce, and the state of its "public spirit."

Cagnazzi retained his post as statistical adviser on the return of the Bourbon monarchy, until 1821 when he was hit by the repression of the constitutional movement. The works he published in the 1820s and 1830s on questions of political economy and the demographic and economic conditions of the Kingdom of the Two Sicilies were known and reviewed in journals published across the peninsula.[15] Some writers hailed him as the "Nestor" of Italian statisticians or placed him among the pioneers.[16] His statistical treatise, however, was never reprinted and was not familiar to the Italian writers of the first half of the century. Even the economist Francesco Ferrara, who exercised his *vis polemica* so strongly against other authors, never found an occasion to mention the political economist and statistician from Apulia.[17]

In contrast, two authors operating in Milan, and with a background very similar to that of Cagnazzi, came to command the attention of Risorgimento writers. The teachings of these two authors, Melchiorre Gioia (Piacenza 1767 – Milan 1829) and Gian Domenico Romagnosi

grafici," in C. Corsini, ed., *Da osservazione sperimentale a spiegazione razionale. Per una storia della statistica in Italia* (Pisa, 1989), pp. 33–48.

12 M. G. Ottaviani, "Note per una storia dell'insegnamento della statistica in Italia: la statistica nell'ordinamento didattico dell'istruzione superiore dalle origini al 1938," *Statistica* 47 (1987), pp. 619–647.

13 *Elementi dell'arte statistica*, p. xii.

14 Ibid., p. xix.

15 His most important work of this period is *Saggio sulla popolazione del Regno di Puglia ne' passati tempi e nel presente*. The first volume (*Parte prima che contiene lo stato de' tempi passati*) was published in Naples in 1820; the second (*Parte seconda che contiene lo stato presente*) appeared in Naples almost twenty years later (1839). On the theory of statistics he wrote only a short essay in *Il progresso delle scienze, delle lettere e delle arti* 1 (1832), pp. 44–57.

16 See A. Padovani, *Introduzione alla scienza della statistica* (Pavia, 1819), and P. De Luca, *Principii elementari di statistica* (Naples, 1857). "Nestor" is a term appearing in a note by G. Sacchi in *AUS* 30 (1831), p. 12; in a review in *AUS* 36 (1833), p. 119, and in C. Cantù's review of *Saggio sulla popolazione del Regno di Puglia ne' passati tempi e nel presente* by L. Cagnazzi, *AUS* 67 (1841), p. 161.

17 According to Salvemini, *Economia politica e arretratezza meridionale*, p. 157.

(Salsomaggiore 1761 – Milan 1835), shaped the development of important Milanese intellectuals such as Carlo Cattaneo, and publicists such as Pietro Maestri and Cesare Correnti, who would play a significant role in the making of Italy and of Italian statistics. Their works reached as far as Sicily, where they became the object of intense debate among the liberal intellectuals who collaborated on the *Giornale di statistica.*[18] While both authors were the product of the same intellectual milieu and conceived of the disciplinary position of statistics in similar terms, their diverse training and experiences conferred a different character upon their contributions. Gioia had first-hand experience as an actual collector of quantitative information and was interested in providing guidelines for research; Romagnosi approached the discipline from an almost purely theoretical or philosophical standpoint.

An enthusiastic follower of Bentham and author of diverse works – from books on etiquette to treatises of political economy, from dissertations on merits and rewards to manuals of logic for the young[19] – Melchiorre Gioia is notorious for his extreme passion for quantifying, his unquenchable appetite for constructing numerical indicators of the condition of the "body social" and the efficiency of the administration. The son of an artisan from a provincial town, he first became known as the winner of a literary competition organized by the French-dominated government of Lombardy in 1796. The theme of the competition was "Quale dei governi liberi meglio convenga alla felicità d'Italia" ("Which form of free government is best adapted to the happiness of Italy"). Gioia responded with an essay in which he supported the idea of an Italian republic with a constitution modeled after the French constitution of 1795.[20] Influenced by the views of the *idéologues*, Gioia

[18] See E. Di Carlo, *L'influsso del pensiero di Romagnosi in Sicilia* (Palermo, 1959), and R. Salvo, "Melchiorre Gioia nel dibattito politico-economico in Sicilia (1824–1831)," in *Melchiorre Gioia (1767–1829). Politica, società, economia tra riforme e Rivoluzione. Atti del Convegno di studi, Piacenza 5–7 Aprile 1990*, special issue of the *Bollettino storico piacentino* 85 (1990), pp. 343–375. On the *Giornale di statistica* and Sicilian statisticians see also ch. 4 below.

[19] The book on etiquette was the *Nuovo Galateo* (Milan, 1802); the work on logic appeared a few years later and was, significantly, entitled *Logica statistica abbassata da M. G. alla capacità de' giovani agricoltori, artisti, commercianti, novizi in ogni altra professione privata o pubblica* (Milan, 1808). His major work on political economy is the *Nuovo prospetto delle scienze economiche* (Milan, 1815–1817). His most Benthamite work is *Del merito e delle ricompense* (Milan, 1818–1819). For Bentham's influence on Gioia see P. Barucci, *Il pensiero economico di Melchiorre Gioia* (Milan, 1965), and F. Sofia, "Melchiorre Gioia e la statistica," in *Melchiorre Gioia (1767–1829)*, pp. 249–268.

[20] The text of this essay can be found in A. Saitta, *Alle origini del Risorgimento: I testi di un "celebre" concorso (1796)* (Rome, 1974). There are no recent biographical studies on Gioia. Some information can be obtained from Sacchi's obituary, "Melchiorre

participated fully in the elaboration of the conception of statistics as an administrative science, and particularly a science functional to an *état dirigiste* such as the Napoleonic one.[21] His work, as Roberto Romani has recently emphasized, was inscribed in a post-revolutionary climate and rooted in an intellectual milieu of bourgeois state functionaries opposed to the political "excesses" of revolution, but solidly tied to its ideology and innovations.[22]

In a work first published in 1808 as a plan for the making of a statistics of the Kingdom of Italy and reprinted in 1834 and 1854, the *Tavole statistiche ossia norme per descrivere, calcolare, classificare tutti gli oggetti d'amministrazione privata e pubblica*,[23] Gioia invited the prospective statistician to "describe, calculate, and classify" (as the title says) all possible objects of "private and public administration." These were the most diverse, ranging from the "degree of the slope of hills" to the "age at which males and females start to have an inclination for marriage," from "the quantity of manure scattered on a hectare" to the number of "bonnet makers," from the number of "mothers who argue with their daughters-in-law" to that of "foreigners who have no acquaintances and support." Appearing at first as an emblematic embodiment of an obsessive classificatory passion, of a mad desire to reduce everything to its quantitative and tabular dimension, Gioia's *Tavole* are in fact a perfect translation of a Benthamite productivist and rationalizing utopia.[24]

Indeed Gioia's tabular statistics (which, according to a commentator, were "admirably conceived and assembled" and better than those of the German *Tabellenstatistiker* [tabular statisticians])[25] aimed at an identification of the causes of variations in social phenomena, of the ele-

Gioia," *AUS* 19 (1829), pp. iii–xxviii, reprinted as "Notizie storiche intorno alla vita e alle opere di Melchiorre Gioia," in the second edition of Gioia's *Filosofia della statistica* (Milan, 1829), pp. iv–xxx; from Barucci, *Il pensiero economico di Melchiorre Gioia*, and from some of the essays collected in *Melchiorre Gioia (1767–1829)*.

21 For an account of theories and debates on statistics in the Napoleonic period see F. Sofia, *Una scienza per l'amministrazione. Statistica e pubblici apparati tra età rivoluzionaria e restaurazione* (Rome, 1988).

22 Romani, *L'economia politica del Risorgimento*, ch. 2.

23 The first edition was published in Milan; the edition of 1834 was published in Lugano as part of the complete works; the edition of 1854 was published in Chieri.

24 On the rationalizing task of the state in Gioia see M. Pasini, "La filosofia della statistica di Melchiorre Gioia," in G. Tarello, ed., *Materiali per una storia della cultura giuridica*, vol. v (Bologna, 1975), pp. 473–532. For a very influential interpretation of Bentham's "utopia" see M. Foucault, *Discipline and Punish. The Birth of the Prison* (New York, 1977), esp. pp. 200–210.

25 See Sacchi, "Notizie storiche intorno alla vita e alle opere di Melchiorre Gioia," p. xxiv.

ments which exercised constant or variable influence on the body social, in order to locate the sites where intervention was needed and reforms could succeed. The goal was that of rationalizing knowledge about society and debates on policy by subjecting them to the discipline of systematic empirical investigation and analysis:

> If writers on civil and criminal law instead of collecting in a scattered way a few historical facts had exposed their reasoning in regular tables, and put in the vertical column countries in similar circumstances, and in the horizontal column the annual results ... we would not have so many useless volumes, crude reasonings, or stupid admirers [of authors he mentions later] ... Instead of oppressing me with authority, show me the tables of the crimes taking place before the application of your favorite principle, and of those taking place while this was being applied, and from the comparison of monthly and annual figures, I will be able to evaluate its efficiency.[26]

To a contemporary reader Gioia's *Tavole statistiche* cannot but evoke a sort of statistical panopticon: the desire to know everything in order to control everything, and the desire to control everything in order to make the body social more productive and useful. The idea of total surveillance exercised through statistical investigation appeared in the very words of its theorists. Listen to one of the collaborators of the *Annali universali di statistica*, Giuseppe Sacchi, a student and follower of Gioia and Romagnosi:

> To municipal authorities and public functionaries who have to provide for local needs, statistical tables must offer all those minute details that concern that part of the country which is in their care. As they are elected to enter the hovel, in order to administer the office of public tutelage, *nothing may escape their gaze*, and as particular sources of knowledge they have to answer the most detailed questions in a judicious way. In contrast, as individual powers coalesce in the state, statistical tables have to get rid of small details, and present only large masses. As in the making of pictures, detail may be suppressed to a greater or lesser degree, so also descriptions of the modes of being and social production gradually gather together in compact unity as they enter the study of the statesman [emphasis added].[27]

The reference in the quote to the elimination of details in the passage from the description of localities to that of the whole state introduces us to the idea of the *colpo d'occhio*, namely of seeing something at a glance. According to most theorists, a good statistics was one that made it possible for the reader to grasp the situation of a country or a town at

26 *Tavole statistiche*, p. xii (I am quoting from the 1854 edition).
27 See G. Sacchi, review of *Filosofia della Statistica* by M. Gioia, in *AUS* 15 (1828), p. 320.

a glance. The use of *tavole sinottiche* (synoptic tables) aimed at exactly this instant knowledge. At times the attempt to reach this goal, especially when applied in literal fashion to a large mass of data, generated fantastically absurd outcomes, as in Giovanni Salari's *Statistica generale della regia città e provincia di Milano* (1839). This work – by an employee of the central accounting office of Lombardy–Venetia – took the form of a gigantic numerical table in the shape of an edifice containing an enormous quantity of information on all the subjects imaginable for a statistically inclined mind of the time; yet the size of the table (5 ft x 5 ft 2 in; 1.52 m x 1. 57 m) was such that the *colpo d'occhio* was virtually impossible (see Figure 1).[28] Excesses aside, the idea of the *colpo d'occhio* fully expresses the virtues of statistics as a new technology simultaneously of knowledge and power.[29]

The above quotation actually comes from a review of one of Gioia's later works, the *Filosofia della statistica* (1826), two ponderous volumes (a total of 700 pages) that he significantly dedicated to Bacon. The *Filosofia della statistica* constituted an exuberant call for the collection of a great variety of data concerning topography, population, natural and agricultural products, crafts and trades, government institutions, and habits (intellectual, economic and moral) of the population. All these data had to be arranged into an orderly and comprehensive description, one progressing from "the most simple to the most complex" and supposed to show in this way the causal relations among various aspects of society. Statistics was an "art of describing," a "descriptive logic"[30] that administrators needed to learn in order to do their job effectively. The application of this logic gave rise to a statistics which Gioia defined as an "*economic description of nations* at a given time [original emphasis]" adding that he used the word "economic" to underline what he considered the primary object of any statistics, namely the analysis of the sources of wealth.[31]

28 There is an enthusiastic description of this work in *AUS* 61 (1839), pp. 184–191. The idea of the *colpo d'occhio* appears also in official instructions on the compilation of statistical reports: see "Istruzione della Regia Segreteria di Finanze, ai signori Intendenti generali . . . per la compilazione della Relazione statistica" (4 marzo 1820), in *Raccolta degli Atti del Governo di S. M. il Re di Sardegna dall'anno 1814 a tutto il 1832*, vol. x (Turin, 1845). The idea that statistics should be inquisitive about every aspect of society and try "to penetrate in every hiding-place" did not cease to be popular even later: see A. Zuccagni Orlandini, *Elementi di statistica* (Florence, 1869), pp. 225–226.

29 For a detailed analysis of the matrices of this idea see also M. Quaini, "Appunti per una archeologia del 'colpo d'occhio.' Medici, soldati e pittori alle origini dell'osservazione sul terreno in Liguria," in L. Coveri and D. Moreno, eds., *Studi di etnografia e dialettologia ligure in memoria di Hugo Plomteux* (Genoa, 1983), pp. 107–125.

30 *Filosofia*, vol. I, p. iii.

31 Ibid.

In the framework of Gioia's productivist outlook, the *Filosofia della statistica* provided a grand "symptomatology" of the body social and politic, a manual for the identification and the correct interpretation of all the imaginable "symptoms" indicating the state of a given society. Although not always of a numerical kind, these symptoms were, whenever possible, expressed by figures. The ratio of a profession to the total number of a population, for example, hinted at the "monstrosity or the normality, the weakness or the strength of the body social";[32] the proportion of pre-marital conceptions was a sign of the more or less "caring attitude of the parents";[33] "the ratio of the number of crimes to the number of inhabitants" constituted a "measure of immorality."[34] This construction of indicators, while developing procedures existing in the tradition of political arithmetic, bears remarkable similarities to the traits of statistical thinking that we associate with that typically nineteenth-century invention, "moral statistics," and with the work of Adolphe Quételet, who made extensive use of moral statistics. Moral statisticians did not limit themselves to counting, they attempted to measure the abstract qualities of a population – such as, indeed, its "morality" – through numbers. These numbers were means, ratios and rates, which statisticians endowed with the power of representing (in the sense of "standing for") the collectivity. Because of his interest in the construction of quantitative indicators, Gioia has been considered a "precursor" of a more mathematically oriented statistics which emerged in Italy only in the second half of the century.[35] Whether this is true or not, it is not very relevant here since we are more interested in those features of his work that his contemporary readers understood and actually used than in what was more or less "advanced" in his work.

A remarkable part of Gioia's *Filosofia* (almost 200 pages out of 700) was devoted to a detailed treatment of topography whose importance he was careful to stress in order to reject the criticisms of statistics expressed by the French political economist Jean-Baptiste Say. Intent on

32 Ibid., p. 207.

33 *Filosofia*, vol. II, p. 377.

34 *Filosofia*, vol. I, p. 207.

35 On the evaluation of Gioia's work by late nineteenth-century statisticians see Pazzagli, "Statistica 'investigatricc' e scienze 'positive' nell'Italia dei primi decenni unitari," *Quaderni storici* 15 (1980), p. 793. Because of Gioia's interest in quantification, Barucci maintains that he did not belong to the "descriptive" school of statistics (*Il pensiero economico*, p. 189). The issue, however, as I see it, should not be one of assigning Gioia to a category, but of considering his work from the standpoint of its reception among his contemporaries. On Quételet and the French moral statisticians see Porter, *The Rise of Statistical Thinking 1820–1900* (Princeton, 1986), pp. 24–30, 40–70.

constructing a new identity for political economy that would contrast with the strong empirical orientation prevailing among economic writers of Jacobin and Napoleonic origins, Say maintained that it was an essentially deductive discipline modeled after the "exact sciences."[36] While statistics dealt only with "particular facts" that often could not be true for more than an "instant," political economy dealt with "general facts" and was able to identify the immutable "laws" of economics. In an article published in the *Annali universali di statistica* in 1826, Gioia set out to refute especially Say's condemnation of statistics as a description of fleeting realities. In this article and in the *Filosofia*, Gioia insisted that statistics actually comprehended some quite "stable [*invariabili*] elements," namely all the information about the topography of a place. As land and climate exercised an influence on economic, intellectual, and moral phenomena, the description of these elements could not and should not be avoided.[37]

Say's attack on the empiricism of statistics and Gioia's response provided also the immediate context for Romagnosi's contribution. It must be said that in contrast to Gioia the actual collector of numbers, the philosopher and jurisconsult Gian Domenico Romagnosi wrote on statistics almost exclusively from a theoretical standpoint, and then only late in his life.[38] The son of a notary and state functionary from a provincial town in the Duchy of Parma and Piacenza, Romagnosi made an illustrious career as law professor in the Napoleonic Kingdom of Italy. Like Gioia an extremely prolific (his published work amounted to about 12,000 pages) and sometimes fairly cryptic writer, his publications were mostly on the subject of public and private law – his *Genesi del diritto penale*, first published in 1791, was very well known. However, like any Enlightenment intellectual, he addressed

[36] J.-B. Say had stated his views on the relations between statistics and political economy as early as in the first edition of his *Traité d'économie politique ou simple exposition de la manière dont se forment, se distribuent et se consomment les richesses* (Paris, 1802). This work had several editions; the fifth appeared in 1826. See also "De l'objet et de l'utilité des statistiques," *Revue encyclopédique* 9 (1827), pp. 529–553.

[37] *Filosofia*, vol. I, p. xi; the essay to which I refer is entitled "Esame di un'opinione intorno all'indole, estensione e vantaggi delle statistiche," *AUS* 8 (1826), pp. 3–57.

[38] G. D. Romagnosi's "Questioni sull'ordinamento delle statistiche" appeared in the *AUS* between 1827 and 1830 (14 [1827], pp. 281–298; 15 [1828], pp. 113–131; 16 [1828], pp. 170–191; 17 [1828], pp. 3–15; 25 [1830], pp. 131–202), and were subsequently published as a separate volume in Milan in 1830, and as an appendix to the second edition of Gioia's *Filosofia* in the same year. Romagnosi's only "exercise" in numerical analysis can be found in a series of articles on the silk market in London published in the *AUS* between 1827 and 1830. There are no recent biographies of this author, but one can find information in A. Levi, *Romagnosi* (Rome, 1935), and R. Ghiringhelli and F. Invernici, eds., *Per conoscere Romagnosi* (Milan, 1982).

also a vast range of issues ranging from philosophical to economic policy questions and was director of the *Annali universali di statistica* during the last years of his life from 1827 to 1835. His reputation among nineteenth-century Italian liberals was primarily linked to his philosophy of *incivilimento* which maintained the inevitability of progress as long as there was free trade and political power was in the hands of the wise.

In Romagnosi's view, expressed at length in his "Questioni sull'ordinamento delle statistiche," the specific task of statistics was that of describing "the modes of being and the interesting productions of things and men among a given people" ("i modi di essere e le produzioni interessanti delle cose e degli uomini presso di un dato popolo").[39] By this he meant the description of a society embodied within a concrete state – and without a state, without a government, Romagnosi insisted, there was no society.[40] Since "barbarians," the enlightened philosopher thought, did not have states, statistics was the distinctive knowledge of people who had formed "civilized" communities, namely who had settled stably on a territory, were engaged in agriculture and commerce, and had a government that guaranteed the security of property.

A statistics could appear at first sight as an inventory, but, as Romagnosi stressed – in a passage that was to be reiterated quite often by statistical writers and was to become his mark of "originality" – a mere inventory of things was not sufficient to make a statistics: "Listing names and numbers of things, men, and natural and man-made productions will never form a civil statistics for the use of political authorities [*magistrale*], but [will only be] a storekeeper's material inventory."[41] A true *statistica civile* was a rationally ordered and comprehensive catalogue of the components of a "civilized" state, a description of its forces, namely its territory, population and government, which was supposed to teach a statesman or an administrator to understand the relation between the whole and its parts and "to act with confidence in all sectors of public administration."[42]

While the final purpose of statistics coincided with the very purpose of a state, namely that of insuring the satisfaction of the material and moral needs of individuals living in society, the evaluation of the "level

39 "Questioni," *AUS* 14 (1827), p. 283.

40 On Romagnosi's conception of the relationships between state and society see E. A. Albertoni, *La vita degli stati e l'incivilimento dei popoli nel pensiero politico di Gian Domenico Romagnosi* (Milan, 1979), pp. 46 ff., and L. Mannori, *Uno stato per Romagnosi*, vol. 1, *Il progetto costituzionale* (Milan, 1984).

41 Romagnosi, "Questioni," *AUS* 16 (1828), p. 178.

42 Romagnosi, "Questioni," *AUS* 14 (1827), p. 291.

of *incivilimento*" attained by a society constituted, in Romagnosi's view, the specific task of a civil statistics.[43] We can translate *incivilimento* as civilization, bearing in mind, however, that the Italian word puts greater stress than its English equivalent on the "process" as opposed to the "condition." The definition given by Romagnosi himself shows it clearly: "*Incivilimento* can be said to be a continuous *progress towards the best form of social life* [*avvicinamento alla migliore sociale convivenza*; original emphasis]."[44] In Romagnosi's philosophical vocabulary the term *incivilimento* carried also an intrinsic polemic meaning, as it opposed an idea of progress understood in merely economic terms. For our philosopher, industrial development alone did not lead to *incivilimento* and even England, although very advanced in economic terms because of its commercial activity and manufacturing industry, lay behind other societies because of the presence of blatant injustice and privilege.[45] True *incivilimento* comprehended the intertwined economic, political and moral advancement of human societies. This view of *incivilimento* constituted the basis on which Romagnosi constructed his "ideal model" of state, the necessary norm or criterion allowing statisticians to evaluate the degree of civilization of a society. Romagnosi's ideal state possessed the characteristics of an open agricultural and commercial society, based on competition and the division of labor, which was also supposed to insure the liberty and security of all its inhabitants.[46]

Given this ideal of society, it should not be surprising that one of the immediate targets of Romagnosi's polemical stance in the "Questioni" were the supporters of unqualified industrialism such as the French economist and statistician Charles Dupin who, in his *Les Forces produc-*

[43] Romagnosi, "Questioni," *AUS* 14 (1827), p. 298. Romagnosi's essays on *incivilimento* were collected in *Dell'indole e dei fattori dell'incivilimento con esempio del suo Risorgimento in Italia* (Milano, 1832); ed. A. De Giorgi (Palermo, 1859). On Romagnosi's notion of *incivilimento* see S. Moravia's introduction to G. D. Romagnosi, *Scritti filosofici*, vol. 1, *Gnoseologia, psicologia, morale* (Milan, 1974), pp. 45–47, and Albertoni, *La vita degli stati*, pp. 29–39.

[44] I am quoting from Romagnosi, *Dell'indole e dei fattori dell'incivilimento*, ed. De Giorgi, p. 127; also the Sicilian economist Francesco Ferrara used *incivilimento* in the sense of "progress." In the Italian dictionary edited by N. Tommaseo and G. Bellini in 1865 "incivilimento" is defined as "condition of people who take the paths of civilization [*civiltà*] ... Civilization rather refers to the condition [of being civilized]." On the debate on progress in early nineteenth-century Italy see F. Rigotti, *L'umana perfezione. Saggio sulla circolazione e diffusione dell'idea di progresso nell'Italia del primo Ottocento* (Naples, 1980).

[45] This kind of criticism was common among continental writers in the first half of the century and not rare among British ones either: as R. Romani has observed, there were many Romagnosi in Europe at the time (*L'economia politica del Risorgimento*, p. 16, n. 13).

[46] "Questioni," *AUS* 14 (1827), p. 292.

tives et commerciales de la France (1827), claimed that the productive forces of a nation constituted the true indicator of its power and that consequently the primary task of statistics was that of counting and measuring them. Rejecting this claim, Romagnosi argued that one needed the knowledge of other elements besides output in order to judge the condition of a state and its people. Dupin's approach ended up measuring only the final product of a process, unable to take into account its equally important preconditions, which Romagnosi listed as "real and personal security, freedom of contracts, force of opinions, of honor, of religion, of national pride, and so on," all of which were difficult to translate into a numerical expression.[47] Elsewhere Romagnosi even drew attention to the need to include in statistical works reference to the local glories of a place, namely its illustrious antecedents:

> We do not know the unfortunate circumstances under which some conceived merely of territorial and confusedly statistical descriptions, thus erasing the past glories, the clue by which we could distinguish a more advanced people from a barbarian or less civilized one. Not without sorrow we observe in these bare tables a sad uniformity between a civilized country and one merely embarking on the road to civilization.[48]

Concerns of this kind certainly could not give rise to the type of mostly numerical statistics that were developing at the time in other European countries, and were not present in Gioia's work. They expressed a very cautious attitude, a reluctance to give up, as Aurelio Macchioro has observed, an older and more "aristocratic" *criterium comparationis*.[49] As we will see in more detail in Chapter 6, this caution surfaced again and again in the writings of Italian statisticians: while extolling numbers as the impartial witnesses of governments' deeds, they also constantly warned against their mistaken and improvident use and application.

This appeal to caution, however, was not a peculiarity of the Italians; in fact a similar sensitivity was to be found elsewhere. Romagnosi indeed expressed a critique resembling that of early nineteenth-century German nationalist and romantic intellectuals who had opposed the so-called *Tabellenknechte* (serfs of tables), the allegedly Francophile representatives of a highly numerical mode of descriptive statistics.[50] Yet

47 Romagnosi, "Questioni," *AUS* 15 (1828), p. 117.

48 G. D. Romagnosi, review of *Atlante geografico, fisico e storico del Granducato di Toscana* by A. Zuccagni Orlandini, *AUS* 37 (1833), p. 242.

49 "L'economia politica di Melchiorre Gioia," in Macchioro, *Studi di storia del pensiero economico e altri saggi* (Milan, 1970), pp. 245–246.

50 And yet it was from an author who had attempted to reconcile the two modes (the qualitative and the numerical), A. L. von Schlözer, a scholar from Göttingen and a

in his reflections on *incivilimento*, Romagnosi never came to reject the very notion of civilization in the way the German intellectuals described by Norbert Elias did at the turn of the eighteenth century, preferring what they considered to be the loftier notion of *Kultur*.[51] In fact he remained solidly in the Enlightenment tradition.

This affiliation was evident also in his epistemology. As he remarked in a passage of the "Questioni," statistics presented a "hieroglyph" which needed to be deciphered with the help of a comprehensive theory. This theory, needless to say, already existed in the shape of Romagnosi's own *civile filosofia*. As Sergio Moravia has observed, the ambition of Romagnosi's civil philosophy was that of "combining all kinds of knowledge concerning man, organizing them in the practical perspective of a better societal existence."[52] It was a normative knowledge of which political economy was a component, along with history, law, morals and statistics. This was the reason why political economists such as Say were "mistaken" in claiming that their discipline was the only foundation of statistics; in fact, statistics had in its turn its own lessons for political economy. By defending statistics with the kind of argument we have so far illustrated, Romagnosi was defending that belief in a unified "social science" which was the legacy of the eighteenth-century *philosophes*.[53] He had no part to play in the incipient process that made the idea of a comprehensive science of man crumble in the face of the organization of an array of new disciplines, of which the version of political economy heralded by Jean-Baptiste Say was but an example. As Romagnosi's conceptions were imbued with the program of a former generation of Enlightenment intellectuals, he could appear to be fighting a rearguard battle; in fact, if his idea of a unified social science was eventually to encounter defeat,

critic of despotism, that Romagnosi probably borrowed the idea of statistics as a rationally ordered inventory of the things of the state. Never keen to mention his sources (a habit not uncommon in his time), Romagnosi made no references to this author in his work. The German edition of Schlözer's *Theorie der Statistik* appeared in Göttingen in 1804 and was translated into French the following year. On the debate on the *Tabellenknechte* see A. Gabaglio, *Teoria generale della statistica*, 1st edn. (Milan, 1880), pp. 116–118; L. Marino, *I maestri della Germania. Göttingen 1770–1820* (Turin, 1975), pp. 346–358; J. Hoock, "D'Aristote à Adam Smith: quelques étapes de la statistique allemande entre le XVIIe et le XIXe siècle," in *Pour une histoire de la statistique* (Paris, 1977), pp. 477–491.

51 On the meaning of the word "civilization" in the French and German contexts see N. Elias, *The Civilizing Process*, vol. 1, *The History of Manners* (New York, 1978), pp. 1–50.

52 Moravia, Introduction to Romagnosi, *Scritti filosofici*, vol. 1, p. 37.

53 On the eighteenth-century origins of this idea see K. M. Baker, "The Early History of the Term 'Social Science,'" *Annals of Science* 20 (1964), pp. 21–226.

his views were put forward again in the work of several Italian writers of statistics and economics during the Risorgimento.[54]

Given Gioia's stress on quantification, on the one hand, and Romagnosi's very critique of tabular statistics, on the other, the pair could appear as unlikely candidates for the making of a distinctively Italian approach to statistics. Contemporary writers actually saw differences between them although, understandably, not with respect to the issues I have emphasized so far; they constructed a hierarchy between the two by assigning the latter the merits of a higher theoretical approach which the former, in his "excessive" tendency to classify and quantify, allegedly lacked. Romagnosi himself pointed out the limitations of Gioia's work maintaining that he lacked theoretical unity.[55] His critical evaluations informed also the interpretive line subsequently adopted by Romagnosi's follower, Giuseppe Sacchi, who became a major compiler of the *Annali universali di statistica* from the 1830s onward, and its director from 1852 to the last year of its publication, in 1871.[56] However, as we will see later in more detail, Gioia's views on the actual writing of statistics had a greater hold on average practitioners than those of Romagnosi as they reflected widespread epistemological conceptions and were formulated in a more comprehensible language

54 On his importance for the pre-1848 generation see G. Berti, *I democratici e l'iniziativa meridionale nel Risorgimento* (Milan, 1962). E. Sestan holds a different view on this subject: see his introduction to *Opere di Giandomenico Romagnosi, Carlo Cattaneo, Giuseppe Ferrari* (Milan and Naples, 1957). More recently R. Romani has made a strong case for Romagnosi's influence on Messedaglia (*L'economia politica del Risorgimento*, ch. 3).

55 [Anonymous: but G. D. Romagnosi] "Necrologia. Melchiorre Gioia," *Biblioteca italiana* 52 (1828), pp. 392–407. It was reprinted as "Elogio storico di Melchiorre Gioia," in *Florilegio di eloquenza italiana*, vol. II, (Pistoia, 1839), pp. 55–86. Although Romagnosi never mentioned Gioia in his *Questioni*, that work undoubtedly contained an implicit critique of the latter – evident for example in the references to the excessive importance given by some authors to topography (see "Questioni," *AUS* 25 [1830], pp. 186ff.).

56 See Sacchi's obituary of Gioia (details in note 20 above). G. Sacchi (Milan 1804–1891) studied law at the University of Pavia and had a career as inspector of elementary schools in Lombardy–Venetia first, and subsequently in the Kingdom of Italy; he was also involved with several philanthropic institutions, campaigned for literacy, and headed the actuaries' society in Milan; politically he was close to the programs of the moderates. See La Salvia, *Giornalismo lombardo*, pp. 289–295; and the obituary in *Bulletin de l'Institut International de Statistique* [Rome, 1890 (*recte* 1891)], pp. 327–328). Barucci has analyzed the reception of Gioia – but mainly as an economist – in nineteenth-century Italy pointing out the greater favor enjoyed by Romagnosi; he maintains that Gioia's reputation fell with the rise of a free-trade orthodoxy among Italian economists from the early 1830s (*Il pensiero economico di Melchiorre Gioia*, pp. 169–189). Gioia's defense of protectionism would have then become a complete heresy.

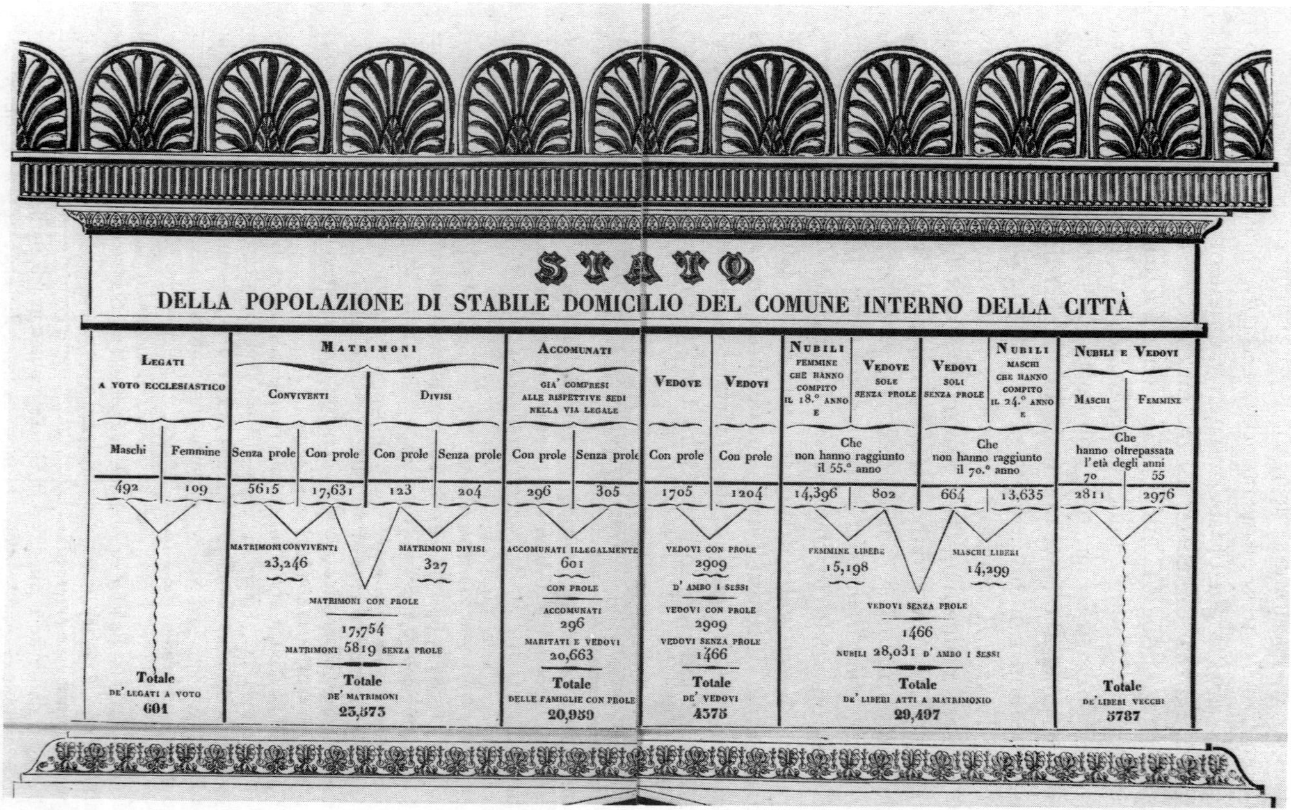

Figure 1 Detail from Giovanni Salari, *Statistica generale della città e provincia di Milano* (Milan, 1839)

– an acknowledgment in practice, we could say, which compensated for criticism at the level of theory.

But beyond the differences, these two authors shared a similar faith in the value of the inductive method and an identical conception of the autonomy of statistics in its relation to other moral and political sciences. These features, however, would probably not have been enough to make the pair part of the same "Italian" tradition had they not been helped by the cultural nationalism of Italian intellectuals. Indeed, the popularity that Gioia and Romagnosi enjoyed among a patriotic readership was certainly due to their being construed and perceived as "opposition intellectuals" symbolizing a cultural tradition penalized by the order established by the Congress of Vienna. What these two authors provided through their writings were not only ideas, but the very possibility of constructing a discourse endowed with specifically Italian points of reference. Writers of statistical monographs rarely failed to quote or refer to these authors' works (which appeared in several editions) often with overtly patriotic aims: Gioia's *Filosofia della statistica* ran through at least seven editions in six cities between 1826 and 1852, while Romagnosi's "Questioni sull'ordinamento delle statistiche," which first appeared on the pages of the *Annali universali di statistica* in 1827–1830, was reprinted as an appendix to the 1829–1830 and 1839 editions of Gioia's *Filosofia*, as a separate work in 1830, and in the several different editions of Romagnosi's collected works published in Milan, Florence, Palermo, and Naples between the 1830s and 1860s.[57] Although we do not have exact figures on the circulation of these works, they definitely became part of the consumption of a growing audience of readers.

AUTHORS AND PUBLISHING

It is ironic that, while uttering his lofty conception of statistics as an administrative science (the "eyes of politics," he called it once),[58] Romagnosi was very far removed from any administrative or political responsibilities. In fact he was speaking as a private individual, a private teacher and publicist, for he had been deprived of the public positions

[57] Editions of Gioia, *Filosofia della statistica* were as follows: Milan, 1826 and 1829; Naples, 1833; Lugano, 1837; Mendrisio, 1839; Livorno, 1839; Turin, 1852. Romagnosi's collected works (*Le opere complete*) were first published by Piatti in eight volumes in Florence in 1832–1840 and had several other editions; another collection edited by A. De Giorgi was published in Milan (*Opere edite ed inedite*) in the 1840s and reprinted in Naples and Palermo in 1859–1867 and 1869–1877.

[58] Romagnosi, "Questioni," *AUS* 15 (1828), p. 117.

he had held in the Kingdom of Italy (when he had taught jurisprudence at the University of Pavia and then had held a special chair of higher legislation in Milan) by the restored Austrian rulers. The same can be said of Melchiorre Gioia, former official historiographer and then head of the statistical bureau in the Kingdom of Italy, who in the 1820s could not rely on any government post and made his living by collaborating on various journals. This experience was not at all unusual in the milieu of the former functionaries of the Kingdom of Italy. Another one of these, the already mentioned Francesco Lampato, former war commissioner of the Kingdom, became publisher and managing editor of the *Annali universali di statistica*, as well as of several other periodicals propagating what were known as "cognizioni utile," namely practical scientific knowledge.[59]

In this milieu, although statistics was formulated primarily as an administrative science, publications of and on statistics addressed a "lay" public before addressing public officials; writers of statistics looked for their audience in that civil society which was emerging and organizing itself in spite of the limitations imposed by absolutist regimes. Melchiorre Gioia actually maintained in his *Filosofia della statistica* that statistics ought to serve the needs not only of statesmen and administrators, but also of members of society in general, and particularly of those engaged in professional and commercial activities: "Statistics thus includes all those knowledges concerning a country that in the course of everyday business can be useful to everyone or to the majority of its members, or to the government which is its agent, its procurator, or its representative."[60] And elsewhere he mentioned that statistical knowledge was useful for "the farmer, the artist, the merchant, the landlord, the physician, the engineer, the government, the inhabitants of a country [*nazionali*] and foreigners"[61] – a list which could not be more explicit about the need to include the distinctive figures of the bourgeoisie.

All publications were of course subject to censorship, but as long as publicists avoided debating directly political issues and openly criticizing the authorities, they could discuss a wide variety of subjects. In fact they did discuss politics and criticize governments in an indirect

59 On Lampato (Milan 1774–1852) see S. La Salvia, *Giornalismo lombardo*, pp. 18ff. On the characteristics and diffusion of a literature of "cognizioni utili" in the Kingdom of Lombardy–Venetia see Greenfield, *Economics and Liberalism*, especially part 2, and A. Galante Garrone and F. Della Peruta, *La stampa italiana del Risorgimento* (Bari, 1979), pp. 134–139. In 1830 there were about fifteen periodicals in Milan devoted to the applied sciences and "cognizioni utili."

60 *Filosofia*, vol. I, p. iii.

61 "Esame di un'opinione," p. 57.

fashion (besides doing it in clandestine pamphlets or abroad, that is); writing about statistics often allowed just that, as one of these patriotic authors, the Milanese Cesare Correnti, recalled years later:

> Forty years ago ... our revered master Gian Domenico Romagnosi had begun to make us realize how this statistical weapon was less worn out and blunt than the lamentations of historians and the denunciations of poets ... For this reason I resigned myself as an impatient conscript to the discussion of averages, tables and numbers, which allowed us to speak in jargon and in code, and to avoid being mutilated by censorship, which through long practice was capable of sniffing only at phrases and epithets. The thing was done. Numbers told their secret only to those who knew how to read their code; they were the true language of mutes.[62]

The "mutes" to which Correnti refers were actually quite vocal. In most Italian states partisans of liberal ideas and supporters of moderate programs of reform diffused their beliefs in growing numbers of books and periodicals. The growth of publishing, however, was not distributed in an even fashion across the Italian states. In the period between 1815 and 1848, the most dynamic center of cultural production in Italy was Milan. The more intense commercialization of the Lombard economy along with the milder working of Austrian censorship – at least until the revolution of 1848 – furthered the development of relatively larger and more modern publishing enterprises.[63] In the same period, the combination of an increasing number of law graduates and a reduction of jobs in the Austrian bureaucracy produced a fairly large stratum of individuals in search of alternative occupations. Journalism was one of these, although it often offered only meager monetary rewards. Frustrated career prospects, moreover, fostered political discontent which in turn fed an interest in journals and publications expressing a liberal orientation, however moderate and disguised.[64]

By the late 1830s a third of the periodicals published in the Italian states came from Lombardy. The Milanese periodicals could reach an audience larger than could those of the other states, and some of them

62 "Cesare Correnti a Luigi Bodio," in *Scritti scelti di Cesare Correnti in parte inediti e rari*, ed. T. Massarani, vol. I (Rome, 1891), pp. 213–215 (originally published in *Archivio di statistica* I [1876], pp. v–xvii).

63 For a detailed reconstruction of the publishing milieu in Restoration Milan see M. Berengo, *Intellettuali e librai nella Milano della Restaurazione* (Turin, 1980).

64 M. Meriggi, *Il Regno Lombardo-Veneto* (Turin, 1987), pp. 151–169, and "Funzionari e carriere nella Lombardia della Restaurazione (1816–48)," *Società e storia* 7 (1980), pp. 61–96. Meriggi points out that a source of frustration for those who were able to enter or re-enter (as several former officials of the Napoleonic administration did in the 1820s) the ranks of the Austrian administration was its increased bureaucratization which limited the prestige of the highest positions.

had very long life spans. The *Annali universali di statistica* were published without interruption from 1824 to 1871. We should not, however, exaggerate the level of this journalistic activity. The *Annali universali*, after a first run of 600 copies, fell to 500 with the second issue, went through a crisis in the early 1830s, when only 300 copies of the journal were printed, and then stabilized around 400 copies in the 1840s. Another prestigious Milanese periodical, the *Politecnico*, reached 700 copies. But these figures were fairly good by Italian standards; most journals did not reach that modest total, and even the *Antologia*, the most important vehicle of liberal ideas in the Tuscany of the 1820s and early 1830s, did not go beyond 530 copies at the peak of its popularity.[65]

We have less reliable data on book production, but they also show the primacy of Lombardy–Venetia in this sector. In 1836 book production of this state accounted for about half the overall production of the whole peninsula; the same was true ten years later.[66] The Kingdom of the Two Sicilies came after Lombardy with a share of about 17 percent in 1836. In this state, the 1830s saw the activism of a modernizing intelligentsia committed to the diffusion of the principles of political economy and statistics. Its organ was *Il progresso delle scienze, delle lettere e delle arti* which at its appearance in 1832 seemed to promise to become a leading cultural institution of liberal public opinion in Italy. Intellectuals such as Luca de Samuele Cagnazzi collaborated on this journal, which hosted also articles by *literati* and scientists from other Italian states. In the early 1830s, in fact, Naples was home to more journals than Milan (twenty-four versus twenty-two).[67] The 1830s were also lively in Palermo, the other major city of the Kingdom of the Two Sicilies, where a group of young liberals, convinced supporters of free trade, were the moving force behind the *Giornale di statistica*, a periodical which besides publishing official statistics made room on its pages for an intense debate on issues of economic development and state policies.

The political and socio-economic environment of the Kingdom of the Two Sicilies, however, did not allow for the development of large and long-lasting publishing undertakings such as those of Lombardy. *Il*

[65] La Salvia, *Giornalismo lombardo*, pp. 71–72.

[66] The data are from G. Ragone, "La letteratura e il consumo: un profilo dei generi e dei modelli nell'editoria italiana (1845–1925)," in *Letteratura italiana*, vol. II, *Produzione e consumo* (Turin, 1983), p. 691.

[67] See Galante Garrone and Della Peruta, *La stampa italiana del Risorgimento*, p. 188. See also G. Oldrini, *La cultura filosofica napoletana dell'Ottocento* (Bari, 1973), esp. pp. 115ff.

progresso became a casualty of censorship: its liberal director Giuseppe Ricciardi was arrested in 1834 and the journal was placed under strict control after less than three years of existence. It closed in 1846.[68] For its part the *Giornale di statistica* was a publishing enterprise of very limited dimensions: in the early 1850s no more than one hundred copies of the journal were printed; of these only about twenty to twenty-five were sold, the others being sent to various state functionaries and to the island's municipalities.[69] Moreover, the circulation of the journal outside the island was fairly difficult, as appears from the numerous letters which have been preserved from people complaining that they had not yet received the promised issues of the journal.[70] In the period between 1836 and 1845 book production in the Kingdom of the Two Sicilies decreased from a share of about 17 percent to 5 percent. The figure is likely to underestimate the reality of publishing (at least as we can estimate it from the figures on statistical publications reported in the appendix), but is nevertheless an indication of the overall trend with respect to other states.[71]

States which saw growth in the publishing sector were the Kingdom of Sardinia–Piedmont which went from a share of about 13 percent in 1836 to 21 in 1846, and the Grand Duchy of Tuscany which went from about 5 to about 10 percent. In the relatively tolerant political climate of the 1820s, Florence rose to intellectual prominence in the peninsula thanks to the *Antologia.* Although this journal was suppressed for political reasons in the early 1830s, the activity of liberal publicists did not decrease since the cultural entrepreneur who founded the *Antologia*, Gian Pietro Vieusseux, and his circle of friends launched other periodicals, such as the *Giornale agrario toscano* and the *Archivio storico italiano.*[72] In contrast to the relatively tolerant climate of Tuscany, extreme conservatism and lack of cultural vitality characterized the Kingdom of Sardinia–Piedmont in the 1820s. The intellectual scene began to change slightly in the 1830s and 1840s when a moderate lib-

68 Galante Garrone and Della Peruta, *La stampa italiana del Risorgimento*, pp. 188–194. See also the introduction by U. Dotti to *Il progresso delle scienze, delle lettere e delle arti (1832–34)* (Rome, 1970), pp. 12–57.

69 Admittedly, in that decade the journal lost part of the readership it had acquired in earlier times within liberal circles; but its publication had been irregular from the beginning.

70 Archivio di Stato, Palermo, Direzione Centrale di Statistica, no. 144. It is today almost impossible to find copies of this journal in libraries outside Palermo.

71 Any underestimation is due to the fact that a large proportion of book production in the Kingdom of the Two Sicilies consisted of counterfeited editions (Ragone, "La letteratura e il consumo," p. 690, footnote 2).

72 For an overview of Tuscan periodicals see Galante Garrone and Della Peruta, *La stampa italiana del Risorgimento*.

eralization of government took place under the reign of the new king, Carlo Alberto. Publishing activity intensified especially thanks to the initiative of figures such as Giuseppe Pomba and Lorenzo Valerio; the journals that appeared in those years, however, had usually a very brief life span. It was only in the 1850s, in fact, after the concession of a moderately liberal constitution in 1848 and the arrival of a multitude of political exiles from all parts of Italy, that publishing activity in the Piedmontese capital rose to national prominence. In that decade the number of periodicals published in Piedmont was much higher than in any other state of the peninsula and some of them, such as the *Rivista contemporanea*, reached the considerable circulation of about a thousand copies.[73]

By means of censorship and tariffs the governments of the Italian states put up barriers to the free circulation of periodicals and books across their boundaries. Published material of various kinds, however, always found readers outside the state of origin. This was especially true for the works produced in northern and central Italy which were easily reprinted in non-authorized editions in the Kingdom of the Two Sicilies where there was no legislation protecting copyright – while the other Italian states and Austria reached an agreement for the protection of literary property in 1840.[74]

Although by no means the only explanatory factor, the publishing environment that we have briefly sketched out helps account for the greater circulation and popularity of works produced by intellectuals operating in Milan before 1848, and in Turin in the 1850s; and it is thus not surprising that from what has been called the "cultural capital" of Italy in the first half of the nineteenth century came most of the authors of the texts which will be the object of our analysis in the following chapters.

Beside the layer of professional intellectuals who operated in the world of publishing, and even before the states began publishing official statistics, other individuals started to engage on their own in the collection and publication of statistical information on all sorts of subjects. In some cases they were zealous public officials who, eager to speed up the process of social "improvement," preceded the states' initiative. More frequently they were members of the liberal profes-

[73] I. De Luca, ed., *La vita letteraria in Piemonte e in Lombardia 1850–1859. Carteggio inedito Tenca-Camerini* (Milan and Naples, 1973); M. Berengo, "Intellettuali e centri di cultura nell'Ottocento italiano," *Rivista storica italiana* 87 (1975), pp. 132–166; Galante Garrone and Della Peruta, *La stampa italiana del Risorgimento*, esp. pp. 467ff.

[74] A. Scirocco, *L'Italia del Risorgimento 1800–1871* (Bologna, 1990), p. 212.

sions, primarily lawyers and medical doctors. With respect to his occupational location, the "typical" Italian statistician – if we accept for a moment such an abstraction – had many resemblances with the French practitioners of the science of numbers; during the July Monarchy these too were mostly members of the liberal professions, or administrators and local notables.[75]

The high visibility of lawyers and physicians in the ranks of the statistical practitioners partly depended on the obvious fact that these were the favored careers open to the sons of the middle and upper-middle ranks of society.[76] But there were other reasons too. In the universities of Lombardy–Venetia students of law were exposed to a discipline called "General Statistics of Europe and of the Austrian Empire." The Habsburg rulers introduced it at the end of 1816 as part of a wider reorganization of the universities of the Italian provinces. This reorganization was aimed at making their curricula homogeneous with those of the other universities of the Empire, where *Statistik* had been taught since the late eighteenth century.[77] The chair of statistics was part of the juridico-political faculty, and professors were usually jurists who often also taught courses in law or political sciences. The first professors of statistics in Restoration Pavia and Padua – respectively Adeodato Ressi and Antonio Marsand – had actually taught public economy in the same institutions during the Napoleonic Kingdom of Italy, and had not hidden their sympathy for it.[78] The presence of this

75 See the essays by B. Lécuyer and M. Perrot in *Pour une histoire de la statistique* (Paris, 1977). In England, reflecting a different composition of the middle class, there was a larger presence of industrialists: see M. J. Cullen, *The Statistical Movement in Early Victorian Britain. The Foundations of Empirical Social Research* (New York, 1975).

76 On the liberal professions in nineteenth-century Italy see P. Macry, "Notables, professions libérales, employés: la difficile identité des bourgeoisies italiennes dans la deuxième moitié du XIXe siècle," *Mélanges de l'Ecole Française de Rome* 97 (1985), pp. 341–359; P. Frascani, "Les Professions bourgeoises en Italie à l'époque libérale (1860–1920)," ibid., pp. 325–340; on physicians in particular see A. Forti Messina, "I medici e la professione del medico nell'Ottocento," *Società e storia* 8 (1984), pp. 101–162.

77 M. C. Ghetti, "Struttura e organizzazione dell'Università di Padova dal 1798 al 1817," *Quaderni per la storia dell'Università di Padova* 17 (1984), pp. 135–185. Ghetti provides information on other universities in addition to Padua.

78 Antonio Marsand, former professor of public economy and commercial law, taught statistics in Padua between 1816 and 1825; Adeodato Ressi, former professor of public economy, taught statistics in Pavia in 1816–1817 and was replaced the following year by Antonio Padovani, former professor of commercial law in a secondary high school. In 1822 the latter was appointed to the chair of civil and notarial procedure. Information on the teaching careers of Padovani and Ressi can be found in Archivio di Stato, Milan, Studi. Parte Moderna, nos. 415, 966, 969. On Marsand's teaching see *Orario delle Prelezioni dell'Imperial Regia Università di Padova per l'anno scolastico 1815–1816* (Padua, 1815), and the volumes for the following years.

discipline in the university curriculum (not to be found in other Italian states at this time)[79] must have helped foster the widespread interest in statistics among Lombard publicists: Cesare Correnti and Giuseppe Sacchi had both received law degrees at the University of Pavia.[80]

In the case of doctors, according to Giuseppe Sacchi, "the study entirely based on observation to which they [medical doctors] devote their life makes them more inclined than students of letters to investigations concerning the being and doing of men and their productions."[81] If we consider that statistics was a study of society based on naturalistic premises, the author of this remark is certainly right in stressing that it was the method of observation, which medical doctors shared with the practitioners of the natural sciences, that explained their statistical propensity. But more to the point is the fact that in the medical community at large, both in the Italian states and in other European countries, doctors were increasingly engaged in a debate over the value of the "numerical method" for establishing the validity of therapies and the causes of illnesses.[82]

79 In the late 1850s Placido De Luca, a professor of jurisprudence in Naples, appears to have been asked by the government to add "the principles of statistics" to his teaching of political economics (see preface to his *Principii elementari*). To my knowledge, however, this was an isolated episode.

80 We need to add that the Habsburg Monarchy itself, as a first-class producer of statistics – which it did not start to publish, however, until 1846 – may also have contributed to foster this interest. Before 1848 some Lombard publicists and intellectuals had first-hand experience of the Austrian administration and in certain cases developed fruitful exchanges with its more enlightened functionaries. For example Carlo Cattaneo established a friendly relationship with the presidential secretary of the Lombard government, Baron Carl von Czoernig, who became the director of the office of statistics in Vienna in 1841 and subsequently launched the publication of the official statistics of the Austrian Monarchy. On this figure and his relation to Cattaneo see L. Faccini, "Carl Czoernig e la statistica agraria in Lombardia," *Società e storia* 21 (1980), pp. 931–950, and Meriggi, *Il Regno Lombardo-Veneto*, pp. 107ff. Cesare Correnti was employed in the Austrian administration – first as clerk in the provincial deputation of Bergamo, and then as secretary of a committee on the public debt of Lombardy–Venetia – before fleeing to liberal Piedmont after the defeat of the revolution in Milan in 1848. On Correnti's career see N. Raponi, *Politica e amministrazione in Lombardia agli esordi dell'Unità* (Milan, 1967), p. 168. For more on Austrian statistics in Lombardy–Venetia see ch. 4 below.

81 See review of *Commentari dell'Ateneo di Brescia per l'anno accademico 1829*, *AUS* 30 (1831), p. 131.

82 Some references to this debate in Italy can be found in G. Cosmacini, *Storia della medicina e della sanità in Italia. Dalla peste europea alla guerra mondiale, 1348–1918* (Bari, 1987), pp. 328ff. G. C. Marino has underlined the vanguard role of medical doctors in the introduction of new "positive" methodologies in Italy (*La formazione dello spirito borghese in Italia* [Florence, 1974], pp. 115–116). See also G. Panseri, "Il medico: note su un intellettuale scientifico italiano nell'Ottocento," in *Storia*

Moreover, wherever doctors had been assigned public functions as municipal and provincial medical officials, they were bound to be concerned with issues of public health and trained to observe and collect data, if only in order to make periodical reports to their superiors. Again G. Sacchi offers us a glimpse of this active milieu in Milan:

> For some time now our most meritorious doctors have been publishing statistical works of the greatest importance. Dr. Giuseppe Ferrario was the first in his *Statistica medica di Milano* to collect a very rich inventory of news and facts which directly concern public health. Our collaborator Dr. Capsoni has also published precious statistical writings about diverse topics concerning the sanitary condition of the rural population and the state of the great lunatic asylum in Milan. Councilor Gianelli made public important statistical studies of the condition of the working classes and of the charity institutions of the whole of Lombardy. The provincial physician, Dr. Buffini, after having published his statistical memoirs on the orphanages of Milan and Brescia, has also published the statistical reports of the great Hospital of Milan for the years 1849, 1850, 1851...[83]

In the Kingdom of the Two Sicilies too medical doctors engaged in analogous activities. We should mention Salvatore De Renzi, physician at various hospitals of the city of Naples and secretary of the Istituto Centrale Vaccinico of the Kingdom of the Two Sicilies, who wrote a well-known *Topografia e statistica medica della città di Napoli* where observations on the effects of climate were mixed with statistical data on mortality, foundlings, and suicides.[84] Not only as custodians of public health, but also because of the very nature of their occupation, which put them in contact with all strata of the population, physicians comprised that part of the educated elite best situated to observe and report on its condition.

In any event, whether exposed to quantification for professional or for other reasons, authors of statistical monographs were essentially amateurs. Administrators, notables and members of the liberal professions appeared as "naturally" endowed with the qualities needed to be a good statistician, which, according to one of the compilers of the

d'Italia. Annali IV. Intellettuali e potere (Turin, 1981), pp. 1135–1155. On the debate in France see W. Coleman, *Death is a Social Disease. Public Health and Political Economy in Early Industrial France* (Madison, 1982).

83 Review of *Rendiconto sulla beneficenza dell'Ospitale Maggiore e del Pio Istituto di Santa Corona in Milano* by A. Verga and of *Sulla beneficenza del pubblico manicomio La Senavra* by C. Castiglioni, *AUS*, 3rd s., 2 (1854), pp. 3–4. The large presence of physicians among liberal writers in Lombardy–Venetia was related more in general to the discontent of young doctors who had difficulty finding decent jobs (see Meriggi, *Il Regno Lombardo-Veneto*, pp. 169–173).

84 The complete title is *Topografia e statistica medica della città di Napoli con alcune considerazioni sul Regno intero, ossia guida medica per la città e pel Regno* (Naples, 1845).

Annali universali di statistica, amounted to "conscientious spirit, healthy judgment, secure critical attitude, and love of order."[85] These were very generic requisites, which could have been asked of any scholar or investigator. There was no such thing as a "professional" statistician, if we exclude the professors of statistics established in the universities of Lombardy–Venetia, who, however, were themselves humanists trained in law, history, philosophy, and so on. In fact, when I use the term "statistician" or "practitioner of statistics" to refer to one who wrote on or compiled statistics I could just as well use the term "statist," to which Italian writers often resorted. "Statist" and "statistician" were terms used interchangeably also in other European countries in the first half of the century, a use pointing to the dominant conception of statistics as a governmental science.[86]

This science was thus very far from being a very technical knowledge in the hands of a group of specialists. Authors and readers of statistics came from the same social groups, espoused Enlightened ideas of economic, social and political improvement, and believed in the value of "useful knowledge." Statistics, however, did not have a monopoly of useful knowledge: other disciplines scored highly in this respect, and it is to statistics' relations and exchanges with these other contiguous "sciences" that we must now turn in order to better understand what statistics had to offer to its public.

TWO CLOSE RELATIVES

A question that was only marginally touched on by the theorists introduced earlier, but which was a matter of some debate in the first half of the nineteenth century, was that of the relation between statistics and geography. Since the mid-eighteenth century geographical descriptions had begun incorporating numerical information. A well-known example of this trend, actually a pioneer in this respect, is the work of the Prussian Anton Friedrich Büsching (1724–1793), author of a general geography of the earth in several volumes, the *Neue Erdbeschreibung* (1754–1792), which had several editions in many European countries, Italy included.[87] This tendency increased in the following

85 See review of *Statistica generale della Francia* by Schnitzler, *AUS*, 2nd s., 13 (1847), p. 126.

86 See V. Hilts, "*Aliis exterendum*, or, the Origins of the Statistical Society of London," *Isis* 69 (1978), pp. 24–25.

87 On the innovation brought about by Büsching see M. Bowen, *Empiricism and Geographical Thought. From Francis Bacon to Alexander von Humboldt* (Cambridge, 1981), p. 155; G. Lutz, "Geographie und Statistik im 18. Jahrhundert. Zu Neugliederung und Inhalten von 'Fächern' im Bereich der historischen Wissenschaften," in

century when publications on geography became very numerous. The Venetian Adriano Balbi (1782–1848), who for several years worked in Paris and Vienna, was one of the main representatives of this type of geo-statistical description in the first half of the nineteenth century; he was also the author of very popular geography textbooks, translated into several languages.[88] Indeed, when outcomes are considered, the boundaries between statistics and geography were rather blurred, the main distinction being one concerning the depth or scope of the descriptive coverage. Authors of treatises of geography and of statistics made sure to include a chapter or at least a few paragraphs on what each discipline was supposed to cover in its descriptions. For Balbi, for example, a geographical work did not need to include the mass of informative details that a statistics must provide, and should limit itself to the "description of the earth in general and of its political subdivisions in particular."[89] According to another geographer–statistician, the Swede Jakob Gråberg till Hemsö (1776–1847) – who, as author of the *Annali di geografia e statistica* (1802), claimed to be the first to introduce the word "statistics" into Italy[90] – geography had to limit itself to a "pure description of the earth;" it should not deal with the forces of a state, its income, expenditure, industry and so on. Statistics, on the contrary, focussed exactly on those objects and relations that "could either increase or decrease the well-being of civil society."[91]

Although not examining the question of the relations between geography and statistics directly, but only through the mediation of the debate on the status of statistics *vis-à-vis* that of political economy, Gioia and Romagnosi too did not see the matter in conflictual terms. We will recall that against Say who denied any role to topography in a statistical description, Gioia reaffirmed vigorously the importance of

M. Rassem, ed., *Statistik und Staatsbeschreibung in der Neuzeit* (Paderborn, 1980), pp. 249–263, and Hoock, "D'Aristote à Adam Smith," p. 481.

88 One of Balbi's most popular textbooks was the *Abrégé de géographie* (Paris, 1833) which had several editions. His works had titles such as *Atlas ethnographique du globe* (Paris, 1826), and *Balance politique du globe* (Paris, 1828).

89 *Compendio di geografia compilato su di un nuovo disegno conforme agli ultimi trattati di pace e alle più recenti scoperte*, 2nd edn., vol. II (Turin, 1840), pp. viii–x.

90 The claim was repeated in several places: *Della statistica e dei suoi progressi in Italia* (Tangier, 1818), p. 11; "Dell'attuale condizione della scienza statistica, e di alcune opere statistiche novellamente pubblicate," *Il progresso delle scienze, delle lettere e delle arti* 3, no. 8 (1834), p. 238. Among his works are *Lezioni elementari di cosmografia, di geografia e di statistica* (Genoa, 1819), *Specchio geografico e statistico dell'Impero di Marocco* (Genoa, 1834); he published essays in various Italian journals, especially in the *Antologia* between 1828 and 1831.

91 *Della statistica e dei suoi progressi in Italia*, p. 15. On this author see G. Natali, "Jacopo Graberg di Hemso," *Rivista d'Italia* 20 (1917), pp. 175–189.

knowing the physical setting of a society in order to understand its characteristics;[92] Romagnosi's juridical background made him less sensitive to the conditioning role of physical space, but he too rejected the arguments of those who wanted to exclude topography altogether from statistics.

In spite of the efforts on the part of some theorists to define and distinguish the tasks and scopes of the two disciplines, or perhaps we should say thanks to these efforts, uncertainties dominated in practice, and overlappings were inevitable; authors resorted extensively to hyphenation in the titles of their works. It was only with Alexander von Humboldt and Carl Ritter, the two greatest German geographers of the first half of the nineteenth century, that clearer boundaries between the two disciplines began to be defined, with the appearance of a geography that studied the earth in terms of dynamic relationships and not in order to draw up a catalogue of places.[93] This was not true in the Italian states, however, where what was called geography continued very much to resemble descriptive statistics, and vice versa.[94] There was indeed very little conflict between the two disciplines, which continued to overlap and to coexist peacefully.

Things were different with political economy, since this field was undergoing, as I have already mentioned, a process of more intense definition and boundary construction. In 1834 Jakob Gråberg complained in the pages of *Il progresso delle scienze, delle lettere e delle arti* that the "true creator of the modern statistical science," the German Gottfried Achenwall, had a tiny following in Italy. In the states of the peninsula almost anyone thought he could be a statistician. The low popularity of German statisticians among Italians was the result of the influence there of "French writers," who confused statistics with political economy.[95] The works of Melchiorre Gioia and Luca de Samuele

[92] Some historians consider him a major representative of a "geo-anthropological approach" to the study of society: see A. Macchioro, "La 'philosophia naturalis' gioiana dell'economia," in *Melchiorre Gioia (1767–1829)*, p. 275.

[93] For this interpretation see H. Beck, "Geographie und Statistik. Die Lösung einer Polarität," in M. Rassem, ed., *Statistik und Staatsbeschreibung in der Neuzeit* (Paderborn, 1980), pp. 269–276; on the characteristics of the new geography of Humboldt and Ritter see also Bowen, *Empiricism and Geographical Thought*, pp. 210–259.

[94] On the history of geography in Italy in the first half of the nineteenth century there is very little: see L. Gambi, "Uno schizzo di storia della geografia in Italia," in *Una geografia per la storia* (Turin, 1973), pp. 3–37; one can also consult an old work by G. Jaja, *Scritti di geografia e di storia della geografia concernenti l'Italia in onore di Giuseppe dalla Vedova* (Florence, 1908).

[95] Although he did not mention them, he was probably referring to the French theorists of statistics of the first decade of the century such as J. Peuchet and D. F. Donnant: on their conceptions see F. Sofia, *Una scienza per l'amministrazione*, ch. 1.

Cagnazzi exemplified this confusion. Gråberg's opposition to the intrusion of economists in the field of statistical observation was based on the view that their theoretical bent would bias the collection of facts:

> Thus the economist, that is, whoever cultivates political economy as a separate science, will always be a speculative man, author of systems, of theories, of projects and conjectures, who will collect and will let you see and examine only those facts and results that support his theories and conjectures, while he will not consider the rest; on the contrary the statistician, an essentially practical man who is learned and skilled by means of expertise and experience, will collect all [the facts], will arrange them in an orderly way for sensitive eyes, and will utilize them by presenting them to the economist and statesman.[96]

A political concern seemed to motivate this statement since in the same essay our author observed also that Italian statisticians, by making this confusion between the two sciences, had alienated governments:

> And if the statisticians of some Italian provinces had remained within the limits established by the same science [statistics], and if, above all, the Italian economists had not wanted by sheer force to be at the same time statisticians, and consequently [had not] mixed the mechanism of this plain and modest science with their sublime and ingenious concepts, *some governments that now seem to be nervous at the mere sound of the word statistics* ... would have been eager to protect and promote [this science] in their territories [emphasis added].[97]

Gråberg was writing only two years after the liberal Florentine journal *Antologia*, of which he was a collaborator, had been obliged to cease publication because of the tightening of censorship in the Grand Duchy. Political economy too had assumed the connotation of a politically suspect discipline in the eyes of the rulers of the peninsula. After the liberal unrest of 1820–1821, the teaching of political economy had been banned from all universities, with the exception of the Kingdom of the Two Sicilies.[98] It is thus likely that this scholar's denunciation of the "confusion" of Italian statisticians was actually meant as a tactical disavowal of the highly politicized character of the social science that had diffused in the Italian states. As a matter of fact the works that he later praised in this review were not mere compilations of facts, but presented, if not theories and projects, those critical views the authors claimed they had formed on the basis of the statistical information they had collected. In other words, they were works that did not stop at de-

96 "Dell'attuale condizione della scienza statistica," p. 239. 97 Ibid.

98 On these vicissitudes see M. M. Augello, M. Bianchini, G. Gioli, and P. Roggi, eds., *Le cattedre di economia politica in Italia. La diffusione di una disciplina "sospetta" (1750–1900)*, (Milan, 1988), p. 243.

scription, but analyzed causes and sometimes proposed remedies, thus fashioning a figure of an intellectual who was an active interlocutor of political power.

It is also possible, however, that Gråberg was constructing the classic straw man in order to be able to give more emphasis to his own standpoint. The fact that he never missed an occasion to boast of his primacy in having introduced the term "statistics" into Italy indicates that he had a high opinion of himself and intended to have others recognize his allegedly fundamental contribution. There was a lot of exaggeration in his claims. A quite prolific writer, known to practitioners of statistics and geography across the Italian states, Gråberg was not the only one to import the term "statistics" into Italy in the first years of the century; writers who collaborated with the French could claim to have done the same at exactly the same time: in 1802, for example, Vincenzo Cuoco proposed to the then vice-president of the Napoleonic Republic of Italy, Francesco Melzi d'Eril, the project for a statistics of the Republic.[99]

Whatever Gråberg's motivations were, the politically subaltern and conciliatory position he proposed to the Italian practitioners of statistics was destined to have little success at a time when Italian intellectuals were increasingly looking at statistics, along with political economy, as a means to further a discourse of opposition to the political order established by the Congress of Vienna, to promote and direct the formation of a liberal public opinion, and to lay claim to an increased participation in government. The main reason why I have dwelled at length on Gråberg's views is that they point to the continuing strength of the ideas he actually contrasted, namely the close complementarity between statistics and political economy maintained by Luca de Samuele Cagnazzi, Melchiorre Gioia, and Gian Domenico Romagnosi. For these authors political economy or, to use a term more common at the time, *economia pubblica* (public economy), was, exactly like statistics, a science of government, based on inductive procedures.[100] In other words, it continued to be, as it had been in the formulation of

[99] On these events see the introduction by V. Gatto to V. Cuoco, *Statistica della Repubblica italiana. Scritti inediti* (Rome, 1991), pp. v–xvi.

[100] Aurelio Macchioro has recently called this approach "statisticismo" (see his "La Raccolta Custodi di 'Scrittori Classici di Economia' fra la statistica e l'economia politica," in D. Rota, ed., *Pietro Custodi tra Rivoluzione e Restaurazione. Atti del Primo Convegno Nazionale* (Lecco, 1989). This author had already emphasized the tradition of the Italian economist–statisticians from his early studies collected in *Studi di storia del pensiero economico* (1970). On the characteristics of economic thought in Italy see also D. Parisi Acquaviva, *Il pensiero economico classico in Italia (1750–1860)* (Milan, 1984).

Adam Smith himself, a normative science which had the legislator, or in this case the aspiring legislator, as its principal target.[101] It indicated what the legislator had to do, or not do, in order to insure the wealth and well-being of the nation. Statistics shared the same overall purpose, but had different tasks – although, as we have seen, never of a purely empirical and subordinate nature. Cagnazzi put it particularly well: while statistics was the "art of analyzing the natural and political state of nations," political economy was "the science that teaches how to make [nations] happy through wealth and social order."[102] Gioia was more confused, as Gråberg would say, by making statistics into a source of remedies and not only of analyses.

In his study of Cagnazzi, Biagio Salvemini has linked this scholar's conception of the role and status of the two disciplines to his lack of trust in the "propulsive and harmonizing function of the self-interest of the entrepreneur" and his awareness that in the peripheral reality of the Kingdom of the Two Sicilies the market was not yet an autonomous, self-sustained mechanism, but needed to be encouraged and helped to grow.[103] Luigi Pucci has observed a similar awareness in Gioia's advocacy of extensive state intervention to protect a manufacturing industry which was still in its "infancy" and to promote the overall productivity of the body social. While the state ought not to interfere with the freedom of private entrepreneurship, it ought to create all the conditions favoring its activity.[104] Although a convinced free trader, Romagnosi too always insisted on the educational role of government and on its function of providing a stimulus for the advancement of "civilization."[105] From the attribution of an important function to government derived the need to equip state functionaries with an adequate and modern body ok knowledge of which statistics and political economy constituted the core.

Indeed the two sciences served to stimulate the emergence of a

[101] Recent revisionist scholarship on Adam Smith has especially emphasized Smith's description of political economy as "a branch of the science of a statesman or legislator": see for example D. Winch, "Adam Smith's 'Enduring Particular Result': A Political and Cosmopolitan Perspective," in I. Hont and M. Ignatieff, eds., *Wealth and Virtue. The Shaping of Political Economy in the Scottish Enlightenment* (Cambridge, 1983), pp. 253–259.

[102] Cagnazzi, *Elementi dell'economia politica* (Naples, 1813), p. 7–8 quoted by Salvemini, *Economia politica e arretratezza meridionale.*

[103] Salvemini, *Economia politica e arretratezza meridionale*, pp. 174–175.

[104] L. Pucci, "Un assertore della 'forza industriale': Melchiorre Gioia tra Chaptal e List," in *Melchiorre Gioia*, p. 339.

[105] See R. Ghiringhelli, "Gian Domenico Romagnosi e gli 'Annali universali di statistica' (1827–1835)," in Ghiringhelli and Invernici, eds., *Per conoscere Romagnosi*, pp. 208–209.

more "civilized" society in a situation of relative backwardness, to use a term that, although not present in the vocabulary of these intellectuals, expresses the perception they had of the place of the Italian states in relation to the centers of economic and political power in Europe.[106] In his *Storia della statistica* (1824), Antonio Quadri, while introducing his own empirical study, the *Prospetto statistico delle provincie venete*, proclaimed his desire to "enrich in this way also Venetia with those *modern works* [emphasis added] that are so numerous in other regions as they are so few in this one, although when Europe left the darkness of barbarism the first to cultivate this kind of study were, as we have shown, the Venetians."[107] In the preface to the first issue of the *Giornale di statistica* Gaetano Vanneschi announced to his Sicilian audience that the journal was intended to make them aware of "[their] good things and [their] evils" so that they could make a comparison with the condition of more civilized people and learn how far they still had to go "in order to be second to no one."[108] The reviewer of a descriptive work on the economic condition of a province of the Papal States complained in the *Annali universali di statistica* that

> the author had done only a part of his work, depriving the reader of the most vital and most fruitful part, [namely] the exposition of improvements that could be introduced today in the practice of Ferrarese agriculture, the demonstration of the needs of his fatherland and consequently of the *measures that should be demanded* [emphasis added], the comparison with the agrarian practices of the richer and more progressive countries.[109]

And the view of the eminently practical purpose of statistical knowledge persisted in later years. Still in the *Annali universali*, the reviewer of the *Giornale di statistica* of Palermo observed in 1853 that "By now statistics has ceased to be a simple arithmetical computation; instead it uses all civil and moral elements to make its investigations effective, and to be a powerful auxiliary tool of economics and of the science of government."[110]

The relation of close complementarity between statistics and political economy was rarely questioned. The only strong attack on the Milanese thinkers came in 1835 from a young employee of the newly

106 On the consciousness of this "backwardness" which in almost all Italian political economists translated into a rejection of *laissez-faire* (but not of free trade) see Romani, *L'economia politica del Risorgimento*, p. 47.

107 p. 240.

108 "Proemio," *Giornale di statistica* 1 (1836), p. xviii.

109 *AUS*, 2nd s., 5 (1845), p. 330. The work reviewed is A. Casazza, *Stato agrario economico del Ferrarese* (Ferrara, 1845). The reviewer is a certain Dott. Bonomi.

110 *AUS*, 2nd s., 35 (1853), p. 230.

established central statistical office of Sicily, Francesco Ferrara (1810–1890), who published in Palermo a booklet entitled *Dubbi sulla statistica* (*Doubts on Statistics*).[111] There Gioia was presented as an example of an unbridled and futile passion for numbers, of a useless empiricism that exercised a negative influence on the many people who followed him. In contrast Romagnosi was depicted as a proponent of a more sober and seriously motivated quest, although not exempt from major flaws. Ferrara praised Romagnosi's insistence on a theory to guide the collection of numbers, only to conclude that Romagnosi's approach too, however, was intrinsically destined to failure – a failure brought about by the impossibility of constructing the concrete term of comparison (the model) on the basis of which one could assess statistically the conditions of a country or its degree of "civilization."

To be sure, Ferrara also recognized statistics as the knowledge appropriate to make transparent the action of governments:

> As long as a people is deceived, it will not be told about statistics; and if [those in power] will pretend that they are interested in statistics, what will come out of it will not be a statistics in the correct sense of the word ... For, when the conscience of a public administrator is not right, the language of facts is dangerous for his reputation; one figure can reveal what had been hidden with much care.[112]

It was the Sicilian economist's strong opposition to the protectionist policies and the centralizing efforts of the Bourbon government that primarily motivated his harsh criticism of Gioia. In the 1820s Gioia had become in effect the hero of political economists of protectionist sentiment both in Sicily and in other Italian states.[113] A supporter of Sicilian autonomy and *laissez-faire* policies, Ferrara almost came to deny legitimacy altogether to a body of knowledge whose purpose seemed to be that of providing a secure base for state intervention in society. Ferrara's skepticism towards Romagnosi's conceptions was predicated on a different set of ideological issues. It expressed a rejection of any

111 Now in F. Ferrara, *Opere complete edite e inedite*, vol. 1, *Scritti di statistica*, ed. B. Rossi Ragazzi (Rome, 1955), pp. 3–34. The introduction to this volume provides useful information on this early period of Ferrara's intellectual activity; see also A. Signorelli, *La statistica preunitaria tra "assunto civile" e funzione burocratica* (Catania, 1983).

112 "Sulla teoria della statistica secondo Romagnosi," in *Opere complete edite e inedite*, vol. 1, *Scritti*, p. 48 (orig. publ. in *Giornale di statistica* 1 [1836]).

113 See A. Macchioro, "Francesco Ferrara," in Macchioro, *Studi di storia del pensiero economico*, pp. 342–376; P. L. Tedeschi, "Francesco Ferrara e la politica della statistica (1835–1846)," *Archivio storico per la Sicilia orientale* 78 (1982), pp. 94–151. For a thorough reconstruction of the reception of Gioia in Sicily see R. Salvo, "Melchiorre Gioia nel dibattito politico-economico in Sicilia," pp. 343–375.

idea of a "perfect state" as an unacceptable concession to utopian thought and therefore also to abhorred socialism.[114] In his polemic against what he called "dead numbers," Ferrara admitted only of a statistics of "effects" as opposed to a statistics of "causes": a study of indicators of the well-being of individuals such as average life expectancy, the proportion of the poor in a population, and the levels of real wages. One could then compare the results against a "*de facto* type," which was preferable to the vagueness that accompanied the idea of a "perfect state."[115] Yet even this approach did not guarantee reliable results, which by definition statistics could not provide. Ferrara's conclusion was that since figures and facts had been used to justify the most absurd opinions, only the rigorous deductive reasoning of political economy could assign causes to social phenomena.

Some of the contributors to the *Giornale di statistica* of Palermo joined Ferrara in the condemnation of Gioia's alleged empiricism, but at the same time most of them upheld Romagnosi's views. The reference to the futility of the statistical enterprise when not directed by solid principles became a commonplace in the journal. While praising the virtues of the Italian school of statistics, guided by Romagnosi, Emerico Amari (who with Ferrara shared the task of directing the journal) derided the motto of the London Statistical Society, *aliis exterendum*, which accompanied the graphic symbol of the association, a wheatsheaf; how to be sure, he asked rhetorically, that without the guiding light of theory one did not gather bales of hay instead of sheaves of wheat?[116] Amari also tried to respond to Ferrara's objections to Romagnosi by arguing that the idea of "a perfect civil state" which, according to Romagnosi, had to guide statistics, could be conceived dynamically as that "of a society in which no obstacle remains to a state of greater, indefinite, unlimited and ever growing perfection."[117]

Overall, Ferrara's rather devastating criticisms did not have much impact on the practical level at a time when the Sicilian economist was

114 See Romani, *L'economia politica del Risorgimento*, p. 145.

115 "Dell'unico modo in cui forse si potrebbe oggidì avviare utilmente la scienza della statistica," in *Opere complete*, vol. 1, pp. 357–379 (orig. publ. in *Biblioteca di scienze morali, legislative ed economiche* [1844], pp. 287–309; reprinted in *Giornale di Statistica* [1846]).

116 See E. Amari, "Società statistica di Londra, e i suoi lavori," *Giornale di statistica* 3 (1838), pp. 41–60; see also similar remarks by V. D'Ondes Reggio in his review of *Notices statistiques sur les colonies françaises, Giornale di statistica* 3 (1838), pp. 260–269. On the London Statistical Society see V. Hilts, "*Aliis exterendum*, or, the Origins of the Statistical Society of London." See also ch. 4in this study.

117 "Difetti e riforme delle statistiche de' delitti e delle pene," *Giornale di statistica* 3 (1838), pp. 270–334 (the quote is from p. 279).

still fairly unknown outside the local milieu in which he operated. While the label of empiricist thinker assigned to Gioia was reiterated in rather formulaic terms by other writers, in actual practice statistical writings were never restricted to information on those few objects admitted by Ferrara's dogmatic *laissez-faire* views. But even later when the Sicilian economist became more well known as professor of political economy at the University of Turin from 1848 to 1858 and as editor of the *Biblioteca dell'economista*, his theories never went unchallenged. Since his views on political economy contrasted with the teaching of Romagnosi, the editors of the *Annali universali di statistica* never embraced them wholeheartedly.[118] Then in the 1870s and 1880s, the vitality of the so-called "Lombard–Venetian school" stood as an important witness to the legacy of that earlier generation of Lombard thinkers. The ideas of the group of economists and statisticians which included Luigi Luzzatti, Emilio Morpurgo, Fedele Lampertico, and others, rested on the tradition of thought and the "culture of observation" which can be traced back to Romagnosi through the mediation of the economist and statistician Angelo Messedaglia.[119]

Although sometimes subjected to controversies which received meaning primarily from a local context, as in the Sicilian case, the debates on statistics that we have examined in this chapter clearly show that most publicists and intellectuals across the Italian states participated in a common dialogue and had common points of reference. If they did not form a "school", they certainly formed a community of discourse. They all understood statistics to be an indispensable science for measuring "civilization" and insuring its very advancement, an eminently practical science based on observation which aimed to increase the productivity of government and society. They agreed that statistics closely complemented political economy; in fact, it was sometimes difficult to distinguish precisely their respective tasks. But in order to better understand the specificity of statistics *vis-à-vis* its neighboring discourses, one must move beyond the analysis of theoretical debates, and look at the specific methodological issues that concerned Italian

118 See for example the comment by G. Sacchi to Ferrara's introduction to his course of political economy at the university of Turin in *AUS* 22 (1849), pp. 3–7. Historians have usually emphasized the dominance of Ferrara's views among Italian political economists in the 1850s and 1860s. For an interpretation which greatly limits his influence see Romani, *L'economia politica del Risorgimento*, ch. 3.

119 On this author see chs. 3 and 6. On the relation between Romagnosi, Messedaglia and the *scuola lombardo-veneta* see Romani, "Il Messedaglia dimezzato," *Venetica. Rivista di storia delle Venezie* 1 (1984), pp. 109–130, and Romani, *L'economia politica del Risorgimento*, ch. 3.

statisticians, at what they actually did in their works, at what they actually represented and counted. This will allow us to have a fuller understanding of the cognitive and ideological assumptions of descriptive statistics and of its implications as a practice of representation of society and the state. Statistics themselves were not descriptions hopelessly subjected to a directionless empiricism, as some claimed, but conveyed indeed a specific image of society and carried an important prescriptive content.

CHAPTER 3

THE LOGICS OF STATISTICAL DESCRIPTION

We saw in the previous chapter that for the Italian theorists and practitioners of statistics of the early nineteenth century a work of statistics amounted essentially to a description of a state, of a society enclosed in a given territory. As society, according to the predominant organicist metaphor,[1] was a body – the "body social" – statistics was its "anatomy" or "physiology" or even, some said, its "phrenology."[2] For Gioia it was a "sort of anatomy which dissects the social body ... it shows its vividness or its pallor ... its health or illness."[3] Romagnosi pushed the body analogy even further, claiming that in order to build a civil statistics one had to imagine "body, soul, life, functions, age, and thus good health or illness [of a nation] similarly to what one does with respect to an animal."[4] In his turn Jakob Gråberg remarked that statistics was "the anatomy and the inventory of states, the accurate and detailed examination of the parts constituting the body social and of its particular needs."[5] The statistical observer himself was construed as a kind of doctor of society, whose task was to investigate the causes of its good health and illnesses and to propose the remedies needed for its

1 Use of the metaphor does not imply the presence of an organicist conception of society, which was not at all strong in nineteenth-century Italian thought. For an analysis of the metaphor see J. E. Schlanger, *Les Métaphores de l'organisme* (Paris, 1971), and on Italy in particular see S. Moravia's introduction to G. D. Romagnosi, *Scritti filosofici*, vol. I (Milan, 1974), p. 48.

2 C. Correnti referred to statistics as a kind of "phrenological checking of a nation" in his "Teoria della statistica. Memoria," *Annali universali di statistica* (henceforth *AUS*) 71 (1842), p. 24.

3 "Indole, estensione e vantaggi della statistica," in Gioia, *Opere minori*, vol. VII (Lugano, 1834), p. 87. The essay appeared originally in Milan in 1809, and must be distinguished from a later one published in 1826 with a very similar title.

4 *Questioni sull'ordinamento delle statistiche* (Milan, 1830), p. 97.

5 "Dell'attuale condizione della scienza statistica, e di alcune opere statistiche novellamente pubblicate," *Il progresso delle scienze, delle lettere e delle arti* 3, no. 8 (1834), p. 242.

cure, and the interventions necessary to prevent any "diseases." At the same time, as conveyed by another fairly common analogy, the statistician's work resembled that of the painter engaged in an accurate representation of a country, statistics being nothing other than its mirror-image.[6]

Given these premises, statisticians facing the task of description needed more than common sense to accomplish it; they needed directions on what to describe and how to describe it, on where to start and where to finish. Indeed nothing would be more wrong than to think that the work of description is a simple endeavor. In fact it is a complicated matter even when it is not the principal component of a text;[7] it is even more so in the case of the statistics examined here which are primarily descriptions, descriptions obsessed with classifying and presenting things in their correct order. In light of this obsession, this kind of statistics appears to share some of the principles which, according to Michel Foucault, organized several scientific domains in the seventeenth and eighteenth centuries. In Foucault's view, *mathesis* (the ordering of simple natures) and *taxonomia* (the ordering of complex natures) informed at that time the common aspiration to a general "science of order"; they constituted a knowledge based on identity and difference and centered around the *table*, a major technique for the orderly display of accumulated learning.[8] Without going into the merits of this reading, it would seem to provide a suggestive clue to the taxonomic enterprise of descriptive statistics – the making of comprehensive and orderly inventories of the things of the state.

As long as statistics amounted to the making of these inventories, issues of classification and the order of exposition constituted the specific methodological arenas of this particular version of the science of statistics. Far from being idle or pedantic questions they were the terrain on which, during the Restoration, different visions of what the state and society were about confronted each other. The political order of the Restoration states, which allocated power and value according to the hierarchy of the *ancien régime*, confronted a civil society expressing values and ideologies more attuned to a bourgeois

6 These images are in M. Gioia, *Filosofia della statistica* (Milan, 1826), p. iv, and G. D. Romagnosi, review of *Grundzüge einer Allgemeinen Statistik* by W. E. Schlieben, *AUS* 44 (1835), p. 5.

7 On the literary theory of description see a special issue of *Yale French Studies* 61 (1981), especially the essays by M. Sternberg, "Ordering the Unordered: Time, Space, and Descriptive Coherence," pp. 60–88, and M. Riffaterre, "Descriptive Imagery," pp. 107–125.

8 M. Foucault, *The Order of Things. An Archeology of the Human Sciences* (New York, 1970), ch. 3.

world view. The tension could not but be reflected in and shape all practices of representation – and especially one that was so intrinsically political as the one that we are examining.

By the late 1820s Italian statisticians could rely on three more or less consolidated traditions which taught how to compile a statistical description, namely, the academic statistics of the German-speaking world, especially in its Austrian variety as taught in the universities of Lombardy–Venetia; the French statistics of the Napoleonic years; and an emerging, self-fashioning Italian "school" whose theories we described in the previous chapter. In the large synoptic tables at the end of the second volume of his *Filosofia della statistica* (1826), Gioia placed side by side the contents of a statistics firstly according to the Austrian Johann Zizius, as a representative of the "German nation," secondly according to the Baron de Ferrussac, as a representative of the "French nation," and finally according to himself as representative of the "Italian nation."[9] The comparison served to highlight the superiority and primacy of Gioia's own version of the statistical "order of things."

In his textbook of 1844 Giuseppe Zuradelli, a professor of statistics at the University of Pavia, in Lombardy–Venetia, also placed side by side a German (Gottfried Achenwall), a Frenchman (Jacques Peuchet), and an Italian statistician (M. Gioia) as representatives of different modes of classification.[10] Actually, the professor from Pavia should also have mentioned another emerging tradition – whose rationale in the early 1840s was expounded by the Frenchman P.-A. Dufau – that called into question what all the previous three had in common, and would eventually replace them. Interestingly, Zuradelli also observed that these statisticians had ordered their data in various ways according to the "concept that one has of the state, and of the task [*ufficio*] of statistics";[11] however, he did not elaborate on the nature of the different conceptions of the state and of the tasks of statistics which, in his view, lay behind these diverse modes of selecting and presenting statistical objects.

This is, in contrast, the task that I assume in this chapter in order to elucidate the underpinnings and implications of these diverse descrip-

9 The works cited by Gioia are the following: J. Zizius, *Teorie preliminari e introduttive alla statistica* (Pavia, 1822; transl. from the orig. German edn. *Theoretische Vorbereitung und Einleitung zur Statistik*, Vienna and Trieste [1810]); A.-E. Ferrussac, *De la nécéssité de fixer et d'adapter un corps de doctrine pour la géographie et la statistique* (Paris, 1819); M. Gioia, *Tavole statistiche* (Milan, 1808).

10 G. Zuradelli, *Preliminari alle teorie statistiche*, 2nd edn. (Pavia, 1844), pp. 119ff. The works from which he quoted had been written respectively in 1767 (Achenwall), 1805 (Peuchet) and 1826 (Gioia).

11 Zuradelli, *Preliminari*, p. 119.

tive logics and to understand more fully what was at stake in the work of statistical representation. I emphasize in particular the pedagogical and didactic character of this activity. To be sure, the pedagogical orientation of the social sciences in the early-nineteenth-century Italian states has not escaped the notice of historians who have observed how political economy and statistics, especially in their Gioian version, were supposed to teach public officials and a lay public how to make society – and the individuals in it – more productive, without infringing on the liberty of entrepreneurs.[12] Yet not enough attention has been paid to the way in which descriptive statistics were *intrinsically* pedagogical. By this I mean that their very organization was part of their content; the order of exposition was assigned, and actually played, a fundamental role for both epistemological and ideological reasons. Accompanied by the powerful rhetoric of the "natural," they visualized for the reader the necessary relations linking different parts of reality to each other and thus made immediately apparent a hierarchy of things and a scheme of causality.

TWO "FOREIGN" MODELS

In an imaginative analysis of the works of statistics produced by Napoleon's prefects, Marie-Noëlle Bourguet has shown the criteria and conceptions on which they were based, and the specific image of society they proposed. Shaped by the encyclopedic passion of the Enlightenment and by the widespread environmental determinism of the time,[13] the prefects' statistics maintained that the configuration of the territory and the climate exercised direct influence not only on production, but also on people, on their physical appearance and condition first of all, and then on their attitudes and moral inclinations.[14] Inspired in addition by physiocratic views of the economy and a desire on the part of the Napoleonic bureaucracy to instruct the administrators and the

[12] See R. Romani, *L'economia politica del Risorgimento italiano* (Turin, 1994), pp. 55–58. See also A. Macchioro, "La 'philosophia naturalis' gioiana dell'economia," in *Melchiorre Gioia (1767–1829). Politica, società, economia tra riforme e Restaurazione. Atti del Convegno di studi, Piacenza 5–7 Aprile 1990, Bollettino storico piacentino* 85 (1990), pp. 277–278.

[13] On the environmental determinism of the late eighteenth and early nineteenth centuries see L. J. Jordanova, "Earth Science and Environmental Medicine: the Synthesis of the late Enlightenment," in L. J. Jordanova and R. S. Porter, eds., *Images of the Earth. Essays in the History of the Environmental Sciences* (Aberdeen, Scotland, 1979), pp. 119–146.

[14] See M.-N. Bourguet, *Déchiffrer la France. La statistique départementale à l'époque napoléonienne* (Paris, 1988), pp. 82-87; she calls these conceptions a "Hippocratic code" (p. 238–244).

elites of the new nation, these works were informed by a liberal conceptualization of society within the shell of a *dirigiste* view of the role of the state. These epistemological and ideological presuppositions operated as organizing principles of the descriptions. For example a manual prepared for instructing local authorities in the making of a district statistics in Piedmont around 1801 listed the following subjects to be treated in this order: general topography; meteorology; natural history; population, professions, arts and crafts; agriculture; rural economy; domestic economy; mores and social habits; public health; financial administration; administration of waterways and forests; administration of roads and bridges; public administration; public police; military administration; judicial system; prisons; hospitals and hospices; public relief institutions; public schools; public worship; fine arts and antiquities; arts and crafts; commerce; fairs and markets; history.[15] Although a certain logic drives this list, works of this kind were often not far from resembling those "storekeeper inventories" condemned by Romagnosi. An effort was made by some French theorists, notably Jacques Peuchet, to put some order into these inventories. The "well-thought-out order" of Peuchet eliminated the historical and natural-historical sections and placed the headings which dealt with state power at the end of the description.[16]

The makers of French departmental statistics were familiar with an older tradition of descriptive statistics which had emerged in the universities of the German territorial states and the Habsburg Empire in the eighteenth century. This was a discipline born and developed in absolute states whose task was that of describing the organism of the state in its constitutive parts and resources. Along with cameralism and police science, which became subjects for study in Prussian universities in the early eighteenth century and in the Habsburg Empire in the mid-eighteenth century,[17] the academic discipline of *Statistik* was one of the state sciences supposed to provide graduates who would enter the administration with the kind of knowledge necessary to manage effectively the finances of the ruler and increase the wealth of a country.

15 R. Laboulinière, *Plan d'une statistique générale d'arrondissement communal publié par ordre du général Jourdan conseiller d'Etat, administrateur général de la 27e division militaire* (Turin, Year IX [1802]).

16 See J.-C. Perrot, "The Golden Age of Regional Statistics," in J.-C. Perrot and S. J. Woolf, *State and Statistics in France 1789–1815* (London, 1984), p. 47.

17 On cameralism, besides the old study by A. Small, *The Cameralists. The Pioneers of German Social Polity* (New York, 1909), see also P. Schiera, *Dall'arte di governo alle scienze dello stato. Il cameralismo e l'assolutismo tedesco* (Milan, 1968), and especially K. Tribe, *Governing Economy. The Reformation of German Economic Discourse 1750–1840* (Cambridge, 1988).

It was introduced in the universities of Vienna and Prague in the late eighteenth century, and in 1816–1817 it also entered, as we mentioned earlier, the universities of the Italian provinces of the Habsburg Empire, namely the Kingdom of Lombardy–Venetia. The teaching of statistics in these universities consisted of methodological notions about the objects to be included in a statistics and the mode of presentation of the material, and of the exposition of the actual data – both numerical and non-numerical – on Europe and the Austrian Empire. The professors of this discipline in Pavia and Padua used as textbooks not only their own publications, but also translations of texts compiled by Austrian writers and used in other universities of the Empire, Vienna and Prague. These statistical manuals were very homogeneous in their strongly state-oriented character. They devoted a lot of space – sometimes between one third and one half of the text – to the description of the administrative and constitutional arrangements of the Monarchy.[18]

Consider for example one of the Austrian manuals translated for the use of Italian students at Padua, *Teorie preliminari e introduttive alla statistica* by Johann Zizius, Gioia's representative of the "German nation."[19] The objects of description were organized around the idea of the power of the state. Zizius's principles of description followed the simple mnemonic formula – devised by one of Achenwall's students, Ludwig Schlözer – of *Vires unitae agunt* (forces united act). *Vires* referred to territory, population, and products, *unitae* to the constitution of the state, and *agunt* to the administrative function of the state.[20] This was a modified version of the old Aristotelian scheme of the four causes (material, final, formal, efficient) which ordered the descriptive model of earlier German academic statistics.[21] Territory and population

[18] See for example J. M. Lichtenstern, *Saggio di una statistica dell'Impero d'Austria considerato nelle sue attuali circostanze*, transl. from the 2nd German edn. (Milan, 1819); J. Zizius, *Teorie preliminari e introduttive alla statistica*; J. N. Schnabel, *Statistica generale degli Stati Europei*, transl. from the orig. German edn. (Pavia, 1835); J. Springer, *Statistica dell'Impero d'Austria*, transl. from the orig. German edn. publ. in Vienna in 1840 (Pavia, 1840).

[19] Zizius's work was adopted as a textbook by Carlo Augusto Bazzini, professor of statistics at Padua between 1829 and 1849: see *Prospetto degli studi dell'Imperial R. Università degli studi di Padova per l'anno scolastico* MDCCCXXIX--XXX (Padua, 1830) and following years.

[20] Zizius, *Teorie*, p. 129.

[21] See J. Hoock, "D'Aristote à Adam Smith: quelques étapes de la statistique allemande entre le XVIIe et le XIXe siècle," in *Pour une histoire de la statistique* (Paris, 1977), p. 482. On German academic statistics see also R. Horvath, "Statistische Description und nominalistische Philosophie," in M. Rassem, ed., *Statistik und Staatsbeschreibung in der Neuzeit* (Paderborn, 1980), pp. 37–52.

(the material forces of the state) opened the description, followed by information on the constitution of the state (final cause), government forms and laws (formal cause), and, finally, administration, production, and state finances (efficient cause). Following – although not uncritically – the German descriptive tradition, the Viennese academic presented economic matters as a branch of the state administration: the rubric "state of the national economy" is found in the final section entitled "state of the physical forces of individuals, their employment and attitude of the government towards them." Even this author, however, while proposing a very orthodox scheme, noticed how it was becoming a bit outdated (the "material cause" in particular included too many things and in a disorderly fashion) and added that since the "immortal Adam Smith with his ingenious inquiries into the elements of national wealth gave us a new reason to reflect on some branches of the state sciences," the principal task of statistics had become that of making known the progress of national wealth.[22]

Although clear-cut oppositions should be avoided, it appears, as Bourguet has observed, that the logics at work in the descriptive practices of the two traditions – French and German-Austrian – were quite distinct.[23] While German and Austrian statisticians subordinated what we would call "civil society" both *conceptually* and practically to the needs and goals of the administration (and, one should not forget, as academics they were state functionaries), the makers of Napoleonic statistics conceived of society as an entity autonomous from the state, although they also firmly believed that the knowledgeable gaze of the state facilitated the working of the social body. It is to this conception that Italian publicists such as Gioia were closer in their work, while managing to introduce also an imprint of their own into the matter.

"FROM THE MOST SIMPLE TO THE MOST COMPLEX": THE STATISTICAL ORDER OF THINGS

At the very beginning of his *Filosofia della Statistica*, Gioia confronted the question of the objects to be described in a statistics and their order of presentation. In his view, the "imperfection of our intellect" requires "that the most simple things precede the most complex [*composte*]; that the presentation of causes precede that of effects; that analogous entities be combined in distinct masses."[24] On the basis of these general cognitive rules – the explicit purpose of which was to

22 Zizius, *Teorie*, p. 131.
23 *Déchiffrer la France*, p. 119.
24 (Milan, 1826), p. vi.

make knowledge more accessible given the limitations of the human mind – Gioia constructed his descriptive scheme. The scheme was such that it contained also a model of causality. In the statistical description of a country, topography, namely the description of the territory's physical components and climatic characteristics, came first not only because it was the most "simple" element but also because it exercised the most general, basic influence on all the other components:

> As a matter of fact geographical position, the type of soil, the abundance or scarcity of water, the degree of heat or cold, humidity or draught, the more or less violent movements of the atmosphere, etc., influence to a higher or to a lower degree production, distribution and consumption of wealth, and all man's activities, and explain [*sono cause*] why a branch of agriculture is productive in one place, why an art is declining in another, why elsewhere a branch of commerce begins [*s'apre*], etc.[25]

As Gioia further explained, topography was to be followed by a description of the population. Population had a passive as well as an active side, as it was at the same time a "resource" of the state and the main "agent" of production, an element which participated in the physical nature of the environment but also marked a discontinuity with it because of its ability to make the former productive. Here one had to describe "the laws of births and deaths [*leggi delle nascite e delle morti*], the pattern [*andamento*] of marriages, the vicissitudes of health and sickness, emigration and immigration, etc."[26] Excluded from this section, on the other hand, were "mores, intellectual, economic and moral habits" because being also the result of legislation and opinions they could not be explained only through the action of topographical elements.[27] The demographic section in turn was to be followed by the description of the population's labor (agriculture), of its transformative activity on labor's products (arts and crafts) and of the exchanges to which the products of these activities were subjected (commerce). Next was a description of the force that kept human beings from doing harm to each other and defended a people from their enemies, namely the government. Three major aspects of a public authority's action (repression, relief, and direction) were considered here, as well as its finances. Finally, the last section was to be devoted to the character and habits of the population, the outcome of the influence of all the previous forces. The knowledge of the people from this standpoint constituted also the ultimate end of a statistics since it gave administrators the information necessary to deal with them effectively through legislative means.

[25] Ibid. [26] Ibid., p. vii. [27] Ibid.

In its emphasis on the ordering from the least to the most complex, Gioia's scheme reflected a fundamental epistemological procedure of the Enlightenment, namely the "structuring [of] the world of phenomena by ordering a system of signs from simple to complex."[28] Gioia did not seem to be using his ordering principle in an ontological sense, but some of his readers did. As a matter of fact the ordering of things on a ladder of increasing complexity bore similarities to the centuries-old idea of the "great chain of being," namely the cosmological conception that sees all things in the universe ordered according to the principles of continuity and gradation. This idea reached its maximum popularity in the eighteenth century; then ordering things from the most simple to the most complex became an exclusive criterion of serial progression in the speculation of European scientists, especially in the field of natural history.[29] Although the idea of a *scala naturae* as a taxonomic principle in the animal world was beginning to be questioned by Cuvier at the turn of the century,[30] it continued to be present and clearly available as a criterion for the description of the social world. This should not be surprising since many statisticians were, as we have seen, doctors or naturalists and understood society in naturalistic terms. But it was of course not necessary to be a doctor to share this conception. In his *Elementi dell'arte statistica*, Luca de Samuele Cagnazzi had maintained the need to follow the "nature of things" in adopting a system aiming to make a statistics.[31] And a professor of statistics at the University of Pavia, Antonio Padovani, expressed disdain in 1819 for those authors who "in the ordering of the materials did not follow that order that came from the nature of things, from their relations and from the purpose of the science."[32] Needless to say, Padovani's scheme had little in common with Austrian descriptive statistics and very closely resembled that of Gioia, whom he described as the "restorer [*ristauratore*] of economic sciences."[33] In 1829 Francesco Campagna, a state functionary in charge of the statistics of the province of Girgenti in the Kingdom of the Two Sicilies, spoke of the material

28 K. M. Baker, *Condorcet. From Natural Philosophy to Social Mathematics* (Chicago and London, 1975), p. 110. On the progression from the simple to the complex see also M. Foucault, *The Order of Things*, pp. 53–55.

29 See B. Balan, *L'Ordre et le temps. L'anatomie comparée et l'histoire des vivants au* XIXe siècle (Paris, 1979), pp. 53–54. On the idea itself the classic study is A. O. Lovejoy, *The Great Chain of Being. A Study of the History of an Idea* (Cambridge, Mass., 1936).

30 See E. Mayr, *The Growth of Biological Thought. Diversity, Evolution, and Inheritance* (Cambridge, Mass., and London, 1982), p. 202.

31 (Naples, 1808–1809), p. 31.

32 A. Padovani, *Introduzione alla scienza della statistica* (Pavia, 1819), p. 70.

33 Ibid., p. vi.

he collected and ordered in various tables as forming "a kind of *mystical chain* [emphasis added]" of objects all linked to each other in a gradual progression.[34] Appealing to Gioia, a Tuscan statistician, compiler of the *Almanacco aretino*, claimed in 1837 that his order of exposition followed the "natural *successione* [sequence] of things."[35] The same motifs appear some time later also in the work of another author, the lawyer Filippo Maria Deliliers, who, invoking the authority of the "two greatest writers of this science, Gian Domenico Romagnosi and Melchiorre Gioia," pointed out in the pages of the *Annali universali di statistica* in 1848: "If you want to compile a well-ordered statistics ... it is indispensable to make general ideas precede the particular ones, the causes precede the effects" following an order "in conformity to that of nature and man in their work."[36]

Other practitioners of statistical description praised Gioia's model for its "logical" features and commonsensical appeal. As a physician from Milan and zealous collector of figures – Giuseppe Ferrario – observed, Gioia's *Filosofia della statistica*, although burdened by a proliferation of examples, had the advantage of indicating what objects had to be observed, what categories used, what tables drawn up:

> From the beginning of this century, Melchiorre Gioia has given us statistical works of such quality that we can honestly say that he has prepared for all the nations of the world a *Catechism of Statistics* with the fairest norms and the clearest examples. Any literate man of common sense understands him, and also learns very soon through the models that he provides how to make a statistics according to his genius and his purposes with the healthiest logic [original emphasis].[37]

The effectiveness of Gioia's scheme derived, according to another

[34] "Memoria che accompagna le sei tavole statistiche della Valle di Girgenti redatte da Francesco Campagna vice-capo del I ufficio": this text introduced a little volume entitled "Statistica della Valle di Girgenti" [1829] containing seven printed tables. It is conserved in the Archivio di Stato, Palermo, Direzione Centrale di Statistica, no. 154.

[35] *Almanacco aretino* (Arezzo, 1837), p. 86.

[36] See F. M. Deliliers, "Discorso sull'insegnamento statistico," *AUS*, 2nd s., 25 (1850), p. 170. In this article the author – described as the secretary of the Chamber of Commerce of France – while repeatedly extolling the authority of Gioia and Romagnosi, complains about the "disorder" of French statistics, that had placed finances before commerce and industry, and population after the latter.

[37] G. Ferrario, *Statistica medica di Milano dal secolo* XV fino ai nostri giorni (Milan, 1838), pp. 17–18. This author was also the founder in 1845 of an "Accademia Fisio-Medico-Statistica," which, amidst several interruptions due to the unstable political situation, brought together medical doctors, publicists and liberal members of the Milanese aristocracy to discuss diverse topics of a scientific and social nature.

Tuscan commentator, from its ascending from the "most simple to the most complex ideas, whereby the very arrangement [of objects] is a demonstration."[38]

To be sure, Gioia's scheme bore many similarities to statistical works produced in the Napoleonic period. However, it also exhibited concerns and emphases that were absent from his French models. These, in spite of the recommendations of the theorists, were characterized by a remarkable instability in the order of presentation of their objects and by a tendency to cover all sorts of domains, in other words to be encyclopedic.[39] In contrast, Italian theorists were more normative; and in their turn, although variations of course existed, practitioners tended to apply the same model more consistently. Most of the works I have been able to examine are organized along lines very much resembling Gioia's scheme.[40] Even though there is very little or no theorizing at all in these empirical works, the references to Gioia, Romagnosi, and occasionally also to other theorists (Cagnazzi, Gråberg) show that their authors did not proceed in a purely empirical manner. Consider for example the work of the *Dottor Fisico* (medical doctor) Giuseppe Raffaele Raso from Casalnuovo, in the Kingdom of the Two Sicilies. He was a corresponding member of the *società economica* of Reggio Calabria – one of the several local societies originally established by Murat in 1810 and revamped by the Bourbon government to gather information on the economic conditions of the provinces and to promote agrarian

38 [X. X.], review of *Filosofia della Statistica* by M. Gioia, in *Antologia* 22 (1826), p. 72.

39 Perrot, "The Golden Age of Regional Statistics," pp. 45–49.

40 See in particular [Bicchierai], "Statistica delle comunità di Fivizzano e di Casola," *Calendario lunese per l'anno 1834* (Fivizzano, [1834]); Bicchierai, "Statistica delle comunità di Albiano, Bagnone, Filattiera, Groppoli e Terrarossa," *Calendario lunese per l'anno 1835* (Fivizzano, [1835]); C. Cattaneo, *Notizie naturali e civili sulla Lombardia* (Milan, 1844); M. Cevasco, *Statistique de la ville de Gênes* (Genoa, 1838); C. Correnti, "Indicazioni storiche e statistiche sulla provincia di Bergamo," *AUS*, 81 (1844), pp. 48–69; 82 (1844), pp. 138–165 and 265–278; 2nd s., 3 (1845), pp. 46–56; G. L. De Bartolomeis, *Notizie topografiche e statistiche sugli stati sardi*, 5 vols. (Turin, 1840–1847); F. D'Emarese, *Cenni statistici sulla provincia di Mondovì* (Mondovì, 1842); G. Eandi, *Statistica della provincia di Saluzzo*, 2 vols. (Saluzzo, 1833–1835); F. Mandarini, *Statistica della Provincia di Basilicata* (Potenza, 1839); L. Molossi, "Cenni statistici intorno ai Ducati di Parma Piacenza e Guastalla," in *Vocabolario topografico dei Ducati di Parma Piacenza e Guastalla* (Parma, 1832-1834); E. Morozzo della Rocca, *Saggio statistico della Valsesia* (Varallo, 1856); *Notizie economico-statistiche della provincia di Casale raccolte e pubblicate dal suo comizio nella solenne occasione del Quinto Congresso Generale dell'Associazione Agraria* (Casale, 1847); R. Petroni, *Censimento ossia Statistica de' Reali Dominii di Qua dal Faro del Regno delle Due Sicilie* (Naples, 1826); T. Plebano, *Statistica del Mandamento di Baldichieri provincia d'Asti* (Turin, 1832); A. Quadri, *Prospetto statistico delle provincie venete* (Venice, 1826); G. R. Raso, *Quadro statistico dei distretti di Palmi e Gerace nella Prima Calabria Ultra* (Naples, 1843).

education.[41] He was assigned the task of drawing up the statistics of two districts in the province of Calabria Ultra, situated in the continental part of the Kingdom. In his *Quadro statistico dei distretti di Palmi e Gerace nella Prima Calabria Ultra* – a description of the territory, mines, rivers and roads, population, natural products, agriculture, textile products, industry, commerce, public morality, and health of the two districts – he invoked several times the "profound Philosopher of Statistics" to support his statements.[42]

In contrast also to their French colleagues of the Napoleonic period who were inspired, to use Bourguet's expression, by a "Hippocratic code" – namely a strong environmental determinism – most Italian statisticians espoused a more complex model of causality. In agreement again with one of Gioia's recommendations, they generally presented information on the habits and mores of the people in the final descriptive section since these constituted the outcome of a whole series of natural influences and human – especially political – interventions. This had also been Cagnazzi's scheme. His model statistics opened with the *stato naturale* (topography, climate, natural products) followed by the *stato politico* which included population (its state, increase or decrease, mortality), "public subsistence and preservation" (food, dwellings, public health), "public industry" (economic activities from agriculture to manufacturing industry and commerce), and finally "public spirit," namely a description of popular mores and the forces able to modify and direct them.[43]

I have concentrated so far on Gioia's work because of his greater influence on the descriptive practices of statistical writers. Romagnosi's influence operated primarily at the level of largely theoretical concerns (namely the status, tasks, and purposes of the discipline) while his methodological observations with regard to the issues that we are discussing commanded less interest. In his *Questioni sull'ordinamento delle statistiche*, after expressing skepticism about the importance attributed to topography by certain authors, he recommended that a historical outline of the events leading to the present configuration of the territorial unit being described should introduce every statistical monograph.[44] Yet especially in the last part of his *Questioni*, that devoted to the actual compilation of a statistics, Romagnosi was very cryptic, and

[41] On these societies see M. Petrusewicz, "Agromania: Innovatori agrari nelle periferie europee dell'Ottocento," in *Storia dell'agricoltura italiana in età contemporanea*, vol. III, *Mercati e istituzioni* (Venice, 1991), pp. 303–311.

[42] Pages 16, 41, 104. [43] *Elementi dell'arte statistica.*

[44] (Milan, 1830), p. 92.

this did not help the actual practitioners, as we learn from one of those who tried to apply his teachings:

> As for the description of the modes of compilation of this statistics, it seems to me that Romagnosi in many places is an obscure metaphysician; and because he never gives an actual example of statistics, I think that there will hardly be a scientist without a statistical background who will be able to understand the inmost thought of Romagnosi well enough to compile a perfect statistics according to his mode of seeing.[45]

Obscure theorizing and convoluted language made Romagnosi a difficult model to follow in the actual compilation of a statistics. In any event, he himself was not too preoccupied with addressing an audience of laymen, in contrast to Gioia who very self-consciously made a career out of writing educational treatises for a large public on very diverse topics. In this respect Gioia more fully exemplified one trait of the early social sciences as they developed in the Italian states, the aspiration to provide a "useful knowledge" which had to educate the people, and to help deliver a more "civilized" society.

A PRESCRIPTIVE ORDER

If epistemological principles, ideas about causality, and ontological assumptions were closely intermingled in Gioia's scheme of statistical description, elements of a more directly ideological nature were also profoundly inscribed in it and intertwined with the others. In order to properly identify them we need to take a closer look at the rationale linking some of the components of the statistical chain of things. As we have seen earlier, for Gioia economic activities constituted, after population, a further rung on the ladder of complexity, a product, according to his theory, of the combination of natural resources and the agency of the population. We should not, however, be misled by the seeming self-evidence of this idea. It is here especially that one can observe how very *prescriptive* was the descriptive scheme of statistics. Gioia made a point of observing, and reviewers were sensitive to pick up the observation, that it would have been a mistake to place the description of the government after the population section in the way "German statists did," since this would mean "breaking the course of economic ideas, which require the examination of productive actions; it would forcibly separate the objects that analogy connects; it would intrude very diverse entities of which the

[45] G. Ferrario, *Statistica medica di Milano*, p. 17.

reader does not yet feel the need, and some of which are useless for economic description."[46]

The "German statists" to which he refers were in fact Austrian statisticians, professors of statistics in the University of Vienna, whose work he knew well. It is their model that Gioia rejected in the name, he claimed, of clarity and sound logic. In contrast to the subordination of society to the state inscribed within the practice of German statistics, Gioia's statistics was a picture of the social body in a rationalized form, following a natural order, ready to receive the stimulus of a rationalizing administrative power. By now also Zuradelli's remark quoted at the beginning of this chapter about the link between different conceptions of the state and different ordering of the data in statistical works begins to make sense. The order of exposition of a statistics depended on specific conceptions of the nature of the state and of its relations to society, as well as on the task of statistics. Even more strongly than his French models, Gioia expressed a view of the primacy of the sphere of production *vis-à-vis* the state that contrasted with the models of the German-speaking world, still strongly shaped by the cognitive and ideological primacy of the state as *raison d'être* of the statistical enterprise.

To conceive of the economy as an autonomous sphere did not necessarily require espousing a view of the natural harmony of the market and having a free-trade agenda; Gioia himself was no dogmatic free-trader, but in fact a pragmatic supporter of industrial protectionism.[47] Nor did awareness of the conceptual autonomy of the economy ever imply for Gioia a lesser role for the state since the latter had a very important function in shaping the social fabric, as well as in furthering the expansion of the market economy. In fact it was a "very dense network of micro-interventions" that Gioia's statistics, as well as his "voluntaristic political economy," to use Roberto Romani's description, prefigured[48] – without however infringing on the liberty of the entrepreneurs. But if neither the economy nor the body social were for Gioia self-sufficient entities, state power in itself was not the end to which society was subordinated, but an instrument for its development, and especially for the growth of productivity. Through their "tutelage" and educational functions, public authorities could, and should, affect decisively the intellectual, moral, and economic habits of

46 See *Filosofia della statistica*, p. v, and review of the same in *Antologia* 22 (1826), p. 73 [signed X. X.].

47 See L. Pucci, "Un assertore della 'forza industriale': Melchiorre Gioia tra Chaptal e List," in *Melchiorre Gioia (1767–1829)*, pp. 331–342.

48 See Romani, *L'economia politica del Risorgimento*, ch. 3.

the population, and thus increase the overall wealth of a collectivity.[49] This was the program conveyed through the writing of a well-ordered statistics.

Gioia's critique of the Austrian models expressed of course also his cultural nationalism, as well as a more general unwillingness (which he shared in any event with many of his contemporaries) to acknowledge his sources. While criticizing the Austrians, he referred only cursorily to the French and also British writers to whom he was highly indebted,[50] and it is not accidental that as representative of the "French nation" he chose a contemporary (the Baron de Ferrussac) and not one of the earlier promoters of the science of statistics in Napoleonic France. He clearly intended to make himself into the creator of a new, Italian tradition of statistics, an operation destined of course to be widely successful in the climate of the Risorgimento when Italians were eager to stress their own contribution to "civilization" in all scientific fields. This is why, even though the contributions to the political and social sciences made by German-speaking and especially by Austrian scholars were well known, they were often actually neglected if not altogether rejected as models for the works produced by the Italians. This attitude appears clearly in the reviewing policies of journals such as the *Annali universali di statistica.* Between 1824 and 1860, 265 works (journals included) of and about statistics were reviewed. A large majority of these were Italian (141), 67 were French, 21 Belgian, 20 German, 10 Austrian, and the rest British and Swiss. The contributors to this journal, despite being part of the cultural universe of the Habsburg Empire and therefore exposed to the time-honored tradition of German academic statistics, addressed French writers as privileged interlocutors.

The dozens of amateur statisticians who set out in the 1820s–1850s to describe the villages, cities, provinces and states in which they lived or where they traveled applied the scheme for which Gioia had provided a theoretical underpinning. Even in those works in which this scheme was not respected, the description of economic activity occu-

49 *Filosofia della statistica,* pp. 387–408 ("Influenza delle opinioni, delle leggi, de' governi, ossia cause morali sulle abitudini"). On the role of the state as a means for rationalizing society and making it more productive see also M. Pasini, "La filosofia della statistica di Melchiorre Gioia," in G. Tarello, ed., *Materiali per una storia della cultura giuridica,* vol. v, Bologna, 1975, pp. 473–532.

50 We have already mentioned the French tradition. As for British influences, Gioia knew J. Sinclair's *Statistical Account of Scotland* (Edinburgh, 1792) which had a descriptive scheme in which the author very self-consciously placed society before the state. References to both French and British works of statistics are scattered throughout the *Filosofia della statistica.*

pied an important and autonomous space.[51] It was followed by a picture of the institutions of government and public establishments (such as law courts, police forces, prisons, schools, hospitals, and so on), and this in turn – with few exceptions[52] – preceded a description of the habits and the moral and intellectual state of the population. As the objects described in the two last sections, on government and the habits and morality of the population, could be construed more subjectively than others, these latter sections exhibited greater variation. They ranged from a collection of few items to a vast array of data on every possible aspect of social life which could satisfy the most voracious appetite for "facts": such was the case for example of the *Notizie topografiche e statistiche sugli stati sardi*, a work by a captain of the Piedmontese army and librarian of the Royal Military Academy, which recommended that an extremely long and detailed series of people's habits and attitudes be observed, from the temperament of the inhabitants, to their prejudices, to their degree of hospitality, to the more or less frequent use of cold and warm baths.[53] The information about the education and moral qualities of the population as well as the instruments of its moral and social improvement (such as the number of schools, hospitals, charity institutions, savings banks, and so on) placed at the end of a monograph served to sum up the purpose of a statistics by indicating the overall level of *incivilimento* attained by a given community.[54] In this case the statisticians often paid homage, with the usual patriotic pride, to the scholar who had contributed most to develop the notion of *incivilimento*, namely Gian Domenico Romagnosi.

STATE AND SOCIETY

While there clearly was a polemical element in Gioia's work directed at the constraining conceptions of the economy and the body social underlying the scheme of Austrian academic statistics, it is evident

[51] See La Marmora, *Voyage en Sardaigne, ou description statistique, physique et politique de cette île avec des recherches sur ses productions naturelles et ses antiquités* (Turin, 1839); A. Perini, *Statistica del Trentino* (Trento, 1850); C. Roncaglia, *Statistica generale degli Stati Estensi* (Modena, 1849–1850); A. Zuccagni Orlandini, "Topografia della comunità di Certaldo esposta in prospetti statistici," in *Ricerche statistiche sul granducato di Toscana*, vol. v (Florence, 1854).

[52] Among the works listed in note 40, see Cevasco, *Statistique de la ville de Gênes*, and Eandi, *Statistica della provincia di Saluzzo*.

[53] De Bartolomeis, *Notizie topografiche e statistiche*, vol. I, pp. xxxix *et seq.*

[54] Ibid., pp. xxxix–xlii; "Statistica delle Comunità di Fivizzano e di Casola," pp. 156ff.

that his criticism was not motivated by a liberal preoccupation to insure the separation between state and society and the rights of citizen. We must wait for the years immediately following 1848 to witness the clear expression of this kind of concern on the part of a theorist of statistics. This theorist was a young scholar from Verona, Angelo Messedaglia (1820–1901), a law graduate of the University of Pavia who had absorbed the principles of Romagnosi's civil philosophy.[55] He spelt out the issues in an essay published in 1851 where he proposed a reform of the juridico-political faculty and the creation of a special department for the teaching of a science of administration modeled after the German liberal theorist R. von Mohl.[56] Messedaglia's purpose was twofold: on the one hand, to stimulate the activity of a modern state administration in the organization of a bourgeois society and the management of societal modernization; on the other hand, to limit the power of the state *vis-à-vis* civil society.[57] In relation to the first goal, he envisaged an efficient restructuring of the training of prospective administrators and bureaucrats and the possible creation of two schools of public administration in Milan and Venice. In relation to the second, he proposed to reform public law and to provide it with a theory devoid of the "confusion between *State and Society* [original emphasis]" that dominated the work of most writers; this task implied the precise definition of the jurisdiction of public authority and the protection of the personal autonomy of the citizen in order to satisfy the needs of government and liberty at the same time. Especially after 1848, he significantly observed, the extent to which "social questions" differed from "questions of state," and were much more extensive than the latter, had become clear. When he emphasized that the state was but an aspect of society, he implicitly manifested his criticism of the concentration of political

[55] R. Romani has recently underlined the link between these two authors (*L'economia politica del Risorgimento*, ch. 3). Messedaglia was to become a very distinguished scholar and statesman in post-unification Italy. For biographical information see ch. 6.

[56] See A. Messedaglia, *Della necessità di un insegnamento speciale politico-amministrativo e del suo ordinamento scientifico* (Milan, 1851). I am however quoting from A. Messedaglia, *Opere scelte di economia e altri scritti*, ed. L. Messedaglia, vol. 1 (Verona, 1920), pp. 207–307. On this work in the context of the development of a "science of administration" in Italy see C. Mozzarelli and S. Nespor, *Giuristi e scienze sociali nell'Italia liberale. Il dibattito sulla scienza dell'amministrazione e l'organizzazione dello stato* (Venice, 1981), pp. 29–34.

[57] Mozzarelli and Nespor, *Giuristi e scienze sociali*, pp. 30–33. The authors also underline how Messedaglia's criticisms were expressed in a language and rhetoric that prevented their being perceived as "subversive" by the Austrian authorities.

power in the hand of the monarch and his bureaucracy in the Austrian Empire.[58]

In the context of this discussion, Messedaglia acutely denounced the limitations of the discipline of statistics as it was taught in the universities of Lombardy–Venetia and put forward a proposal for a dramatically new way of teaching it. He proceeded to redefine the task of the science of statistics as one of working out the movements and laws of society, and not only of collecting and exposing information on the conditions of a state. To this end he proposed an integration between the new statistics based on the theory of probability as practiced by Quételet and the Frenchman P.-A. Dufau, and the tradition of descriptive statistics. The former would constitute the "theoretical" part of the discipline (which he called the "Filosofia della statistica") and the latter its "applied" part.[59] This should replace the "pompous program," to use his words, which went under the name of "Science of Statistics – General Statistics of Europe – Special Statistics of the Austrian Empire." The new science moreover would deal only with "social facts and results in their calculable elements."[60]

It is significant that Messedaglia's defense of the prerogatives of civil society was associated with a kind of statistics which represented a very self-conscious attempt to be the expression of society as distinct from the state. P.-A. Dufau, in a work which gained a certain popularity – the *Traité de statistique ou théorie de l'étude des lois d'après lesquelles se développent les faits sociaux*[61] – had explicitly rejected the German conception of academic statistics as a description of the power of a state, maintaining that it was incomplete and partial. Statistics should no longer be a science concerned with the territorial–political entity of the state, but should concern itself primarily with "civil society," which he defined as the population considered from the standpoint of its physical and moral characteristics.[62] Civil society was an entity which could be con-

[58] *Della necessità di un insegnamento*, p. 268. [59] Ibid., pp. 285–289.

[60] Ibid., pp. 233–234.

[61] (Paris, 1840). Dufau's work was quickly and positively reviewed in the *AUS*: a short review signed "Dott. B." appeared in vol. 64 (1840), pp. 270–271; a longer one by a young collaborator of the periodical, Cesare Correnti, in vol. 68 (1841), pp. 53–57. Dufau's work also inspired Correnti to publish an essay the following year on the theory of statistics ("Teoria della statistica. Memoria") where he stressed in particular the French statistician's insistence on the new focus on society as being characteristic of nineteenth-century statistics, in contrast to the statistics of the eighteenth-century German school.

[62] Dufau, *Traité de statistique*, p. 93. Interestingly, this conception of civil society is distinct from the dominant meanings attributed to the term in the same period. These were associated with issues of the preservation of individual liberties against the encroachment of the state (on which see J. Keane, "Despotism and Democracy. The

ceived in isolation from the physical space in which it was contained; in fact, Dufau underlined, it was not the territory which made the population, but the population which made the territory. Similarly, Messedaglia drew attention to the need to get "society" directly into focus by radically reorienting a discipline which had been shaped by *raison d'état* and which had grown increasingly rigid and unable to portray any longer a changing and growing organism.

We need also to mention here the only other author who, besides Messedaglia, called for a renovation of the discipline along similar lines. This was a liberal writer from Basilicata in the Kingdom of the Two Sicilies, Giacomo Racioppi, who in a volume published in Naples in 1857 criticized those authors who conceived of statistics as an "encyclopedia of the state."[63] In contrast he maintained that it should be more properly understood as "the systematic picture of all those manifestations of social activity that can be expressed in numbers [*che capaci di quantità ponno esprimersi in numeri*]; in order to show in a rigorous way the state and movement of facts [*lo stato e il movimento dei fatti*], to measure social progress; and in order to draw from these facts the general laws of social vitality." He too criticized the inclusion of topography in statistical descriptions, claiming that topography should only be part of geographical works, and not of the social sciences.[64]

Messedaglia's sensitivity to the new and his movement away from a purely descriptive conception of statistics by proposing to adopt a different kind of quantification were virtually ignored at the time. His radical proposal did not change the way academic statistics was defined and taught. When he was appointed to teach political economy and statistics at the University of Padua in 1858, he certainly referred in his lectures to the methodological innovations and the "discoveries" of the new Quételetian statistics, but continued also to provide students with the conventional statistical description of the Austrian Empire – what the Austrian government required of teachers

Origins and Development of the Distinction between Civil Society and the State 1750–1850," in Keane, ed., *Civil Society and the State. New European Perspectives* [London and New York, 1988], pp. 35–71). With Dufau we see instead a major object of statistical research (and increasingly of state intervention) – the population – being made into the very substances of civil society.

63 *Del principio e dei limiti della statistica*, p. 44. This was the only work on statistics published by this author who wrote mostly on political history. After unification, he was director of the central statistical division of the new state for the brief period between the death of the first director, Pietro Maestri, in 1871 and the appointment of Luigi Bodio in 1872.

64 Ibid., pp. 44–57.

like him.[65] Messedaglia and Racioppi anticipated at the level of theoretical awareness a conception and practice of statistics that would become more widespread if not dominant in the 1860s – once a national society had finally been assembled through the creation of a new state inspired by liberal principles. But in so doing they made fully evident the political assumptions and implications that descriptive statistics in its academic version had had all along, and they confirmed the intrinsic ideological and political character of all "statistical epistemologies."[66]

LEARNING FROM STATISTICS

In the climate of the Restoration and in the hands of individuals excluded or marginalized from the exercise of political power, the writing of statistical monographs clearly embodied a "will to represent" which expressed implicitly their aspiration to a greater role in the conduct of public affairs. The making of these inventories conveyed the ability of their authors to know the realities of state and society. These realities were characterized in terms of resources to be known, managed, and increased, and of institutions to be brought under the eye of the "public" and rationalized. Statistical knowledge was to shape an enlightened public opinion, and instruct the makers of political and administrative decisions. It was to teach a correct method of observation and help detect weaknesses and strengths of the body social in order to favor its development through the adoption of policies based on "positive" knowledge.

Statistics were not simply collections of "positive" information, catalogues of "facts" on the condition of a state in order to measure its degree of "civilization," but conveyed a normative and prescriptive view of reality. Statisticians attempted to put order into the social world by assigning each object the position it ought to have in the "natural order of things." The invocation of nature as universal measure and model of rationality – a hallmark of Enlightenment thought of which statistics was a child[67] – prefigured a reordering of

65 See the notes he prepared for his students, now conserved at the Biblioteca Civica, Verona, Fondo Messedaglia, no. 1249: "Note di statistica dell'Impero Austriaco precedute da alcune nozioni teoretiche fondamentali per servire all'insegnamento del Prof. Angelo Messedaglia. Padova 1861 per cura di T. A."

66 I borrow an expression of J.-C. Perrot, "The Golden Age of Regional Statistics," p. 44.

67 For an analysis of the rhetoric of nature in scientific and literary writings of the late eighteenth and early nineteenth centuries see L. J. Jordanova, ed., *Languages of Nature. Critical Essays on Science and Literature* (London, 1986).

the social body along more "rational" lines. The establishment of a correct representation of things within statistical texts constituted an important epistemological step on the road towards the "improvement" of society. That the order of things in statistical representation was supposed to prefigure a new order of things in the real world appears also in the representation and ranking of social groups. While Austrian academic statistics still divided the population according to the scheme of a society of orders (nobles, city dwellers, and peasants), no such representation is to be found in the works published by Italian statisticians. In the latter, occupational groups, or classes devised on the basis of socio-economic status, or according to the criterion of utility, summoned up the image of a bourgeois society in which "castes" no longer existed.[68] To be sure, in most Italian states feudalism was a thing of the past; and yet social groups that were already dominant in society still did not exercise full control of political institutions. The battle they conducted to gain this full control was fought also through the symbolic reordering of society in its representations.

Statisticians had a clear awareness of the pedagogical potential of statistics. They thought that the reader of a well-ordered statistics would absorb not only information, but a correct idea about the relationships linking different kinds of phenomena and people – in other words, he or she would learn how to think about the world. In this respect statisticians seemed to be aware of what Elisabeth Eisenstein would call "the power of the format," namely that "the thoughts of readers are guided by the way the contents of a book are arranged and presented. Basic changes in format might well lead to changes in thought pattern."[69] A contributor to the *Annali universali di statistica* observed, while commenting on a work of French statistics, that the reader of that work would afterwards not only be well informed about the political and administrative institutions of his or her country, but also, "*without realizing it*, he [would] have learned to love the forms of government, where law reigns impartially, where openness has become a duty, and legal liberty a habit; he finally [would] have the basis for be-

[68] A particularly good example is D. Rabbeno, *Delle condizioni della statistica nell'Italia centrale* (Parma, 1861), pp. 25–27. For Austrian statistics see in particular Springer, *Statistica dell'Impero*, pp. 227ff. Even this author, however, acknowledged the existence of other schemes of classification beside the feudal one and divided the population into "productive" and "unproductive" categories.

[69] *The Printing Press as an Agent of Change* (Cambridge, 1979), pp. 88–89; I was directed to this text via a brilliant essay by C. J. Koepp, "The Alphabetical Order: Work in Diderot's *Encyclopédie*," in S. L. Kaplan and C. J. Koepp, eds., *Work in France. Representations, Meaning, Organization, and Practice* (Ithaca and London, 1986), p. 235.

coming a more enlightened man, and a better citizen [emphasis added]."[70]

It is partly because of the pedagogical value inscribed in the very fabric and organization of a description that statistics found a place even in relatively "popular" publications, such as the growing literature of almanacs.[71] Periodicals such as the *Calendario lunese* and the *Almanacco aretino* published in Tuscany in the 1830s, the *Almanacco etrusco, cronologico, statistico, mercantile* which appeared in Florence in the 1850s, *Il nipote del Vesta-Verde, Il pronipote del Vesta-Verde*, and *Il nuovo Burigozzo*, published in Milan in the late 1840s and 1850s, all contained statistical descriptions of villages and towns. For example, the "Statistics of the Communities of Fivizzano and Casola" took up most of the first issue of the *Calendario lunese* published in 1834 in the small town of Fivizzano near Lucca. Abundant information was given on the geographical location of the two communities, the present and past condition of their population (number, structure by age, sex, and domestic status, occupations, mortality trends and the seasonality of conceptions, state of public health and the factors influencing it, emigration), their agriculture, industry and commerce, and their level of *incivilimento*. The author, who signed himself "avv. [lawyer] Bicchierai," complained about the "incredible reluctance" to help him with his investigations rooted in the "prejudices of the multitude" and wished that the same problem would not arise again so that he "would not be discouraged."[72] And discouraged he was not. The 1835 issue of the same publication added even more statistical information on nearby communities along with the usual moral advice and agronomic observations. And he did the same the following year. Other *almanacchi* offered statistical information on whole provinces, or states, or even on all the Italian states, as did the 1857 issue of the *Almanacco etrusco*. The 1848 issue of the *Pronipote del Vesta-Verde*, a small volume of 246 pages sold in Milan at the price of 50 *centesimi*, contained extensive data on population, agriculture, and industry in the province of Milan, and also information on Tuscany.

"Everybody needed to know" certain "statistical cognitions": this was the explanation put forward by the Accademia Reale delle Scienze

70 Review (signed M. C.) of *Statistica generale della Francia* by Schnitzler, in *AUS* 2nd s., 13 (1847), p. 126.

71 For a general view of this expanding sector see M. Rak, "Il popolo del libro. L'editoria popolare italiana nella prima fase della società industriale (1790–1890)," in Rak, *La società letteraria. Scrittori e librai, stampatori e pubblico nell'Italia dell'industrialismo* (Venice, 1990), pp. 101–145.

72 Page 5.

of Turin in 1838 while offering a prize for a work that would give an "exposition of the most common elements of *vita civile* [civil life] abridged for the teaching of young people who are not destined to the higher professions." Such statistical notions included, among other things,

> the distinction among states, and trades, and conditions, with an emphasis on the utility of each of them, the reciprocal aid that they owe to each other, and the excellence of all when they are laudably exercised; then the distinction which is found among inhabited places, namely the different order in which they are ranked according to population, civilization, and *proprietà* [propriety], by observing the mores, the particular habits which can establish a well-founded difference among them. Then they will have to indicate the principal causes of that previously mentioned prosperity as they are linked to the nature of the soil, to the circumstances of the place and climate, to the character, education and morality of the inhabitants.[73]

Even though, as the previous quotations indicate, not all proponents of statistics shared the same vision of the well-ordered society, they agreed on the general utility of statistics as a means to educate the people, namely to teach them "useful notions" and how to make proper distinctions and see the correct relationship between things. Then there were authors who went even further claiming that the very compilation of statistics had therapeutic properties; a medical doctor writing in the *Annali universali* in 1848, Giovanni Capsoni, maintained that "to draw up synoptic tables, to apply the results of an exact distribution of figures and correct arithmetic calculations" had curative effects among the inmates of lunatic asylums.[74]

The diffusion of statistics in the literature for the popular classes published in the 1840s and 1850s and the projects concerning its teaching to a popular audience represented also a response to the growing dissemination of "dangerous" ideas that questioned the "natural order of things" even before its implementation. Authors of statistics intended to teach the laboring classes about their place in the world and to remind "dreamers" of all kinds about the need to study the "facts" in order to avoid proposing "inept" – as well as dangerous – schemes of social and political change. In his "Discourse on the Teaching of Statistics," significantly published in 1850, the previously mentioned Filippo Deliliers proclaimed that this teaching was most educational and was intended "to insinuate the spontaneous cooperation of all members of

[73] "Programma dell'Accademia Reale delle scienze in Torino," *AUS* 57 (1838), p. 237. In the end the prize was not awarded.

[74] *AUS* 2nd s., 15 (1848), p. 173.

society towards the discovery of their national physiology, of that light that must illuminate their minds so that they can better their existence."[75] For many moderate liberals nothing seemed more effective against the "fabricators of utopias" than the science of statistics, "the cold witness" – to use the words of one of these writers, Ignazio Cantù – "of progressive developments; impartial [source of] praise or blame, and of humble advice to governments."[76] As in other fields, a desire for stimulating progress never came alone; it was always, and especially after the Year of Revolutions, strictly linked to a preoccupation about order.

In fact these dreams of reforming the populace through the teaching of statistics were not translated into practice, at least not at the time. Even the almanacs, which could have runs of several thousand and even tens of thousand copies, were able to reach only a small segment of the still predominantly illiterate laboring classes, and those authors who had originally thought of addressing an audience of peasants often had to be satisfied with producing some "impressions in the mind and the heart of the most educated of [their] fellow-countrymen, so that they will decide to promote through any means [*con ogni studio*] the beneficial enterprise [*opera*] of progress in their country."[77] The laboring classes continued to be primarily an object of description – most often as part of the chapter on population or of that on popular habits – in the statistical inventories of the time.

We have examined so far the ideological and epistemological battles which individuals waged by means of statistics, individuals who were often, and not of their own choice, external to the state apparatuses. It is now time to look at the role of state administrations in the production of statistics, since from the 1830s onwards most Italian states established their own institutions and developed their own agenda for creating inventories of their resources and subjects.

75 "Discorso sull'insegnamento statistico," p. 169.

76 See C. Cantù, review of the *Saggio sulla popolazione del Regno di Puglia ne' passati tempi e nel presente* by L. de Samuele Cagnazzi, *AUS* 67 (1841), p. 167.

77 *Calendario lunese per l'anno 1834* (Fivizzano, 1834), p. 4.

CHAPTER 4

OFFICIAL NUMBERS

There is a question that we have not asked so far, but that needs to be asked: where did the numbers come from, the numbers published in the descriptive works whose logics and rationale we have tried to account for in the previous two chapters? They came to a large extent from the very administrations that were criticized implicitly or explicitly by liberal commentators for their alleged unwillingness or reluctance to promote statistics and make them fully public. In fact, in most pre-unification states the 1830s and 1840s marked the beginning of the official era of statistics. This was not the first time statistics entered the "public sphere":[1] as we have mentioned, already in the Napoleonic period official and non-official statistical publications had flourished with the aim of enlightening a diverse audience of administrators and private individuals. But now the enterprise took place in distinctive forms and modes, and, in spite of its ambiguities and hesitations, it showed that it belonged to a new, and, in the long run, irreversible trend common to all European countries.[2]

The first to start this trend among the Italian states was the Kingdom

[1] J. Habermas employs this term as a conceptual as well as sociological category indicating the sphere occupied by a "reasoning bourgeois public," which mediates between the state and civil society: see *The Structural Transformation of the Public Sphere. An Inquiry into a Category of Bourgeois Society* (Cambridge, Mass., 1989 [orig. German edn.: Neuwied, 1962]). In my use of the expression I do not intend to convey the idealized representation of the bourgeois public which critics have recently detected in Habermas's construct: see in particular G. Eley, "Nations, Publics, and Political Culture: Placing Habermas in the Nineteenth Century," and N. Fraser, "Rethinking the Public Sphere: A Contribution to the Critique of Actually Existing Democracy," in C. Calhoun, ed., *Habermas and the Public Sphere* (Cambridge, Mass., and London, 1992), pp. 289–339 and pp. 109–142 respectively.

[2] On entry into the "public sphere" as a distinctive characteristic of the development of statistics in the first half of the nineteenth century see G. Leclerc, *L'Observation de l'homme. Une histoire des enquêtes sociales* (Paris, 1982), p. 82.

of the Two Sicilies, which established a Direzione Centrale di Statistica in Palermo in 1832 with the purpose of coordinating the collection of statistical information on the island. In 1836, the Kingdom of Sardinia created a special committee which included high-ranking state functionaries and scholars, whose aim was to direct efforts towards the collection of statistical data on the physical, moral, economic, and civil conditions of the state. In the reformist conjuncture of the late 1840s, statistical committees and offices were established also in the Duchy of Modena (1846), the Grand Duchy of Tuscany (1847), and the Duchy of Parma (1847).[3] Even the Papal States, that quintessentially "reactionary" government in the eyes of liberal public opinion, saw the creation of an "Istituto Statistico, Agrario e di Incoraggiamento" in 1846 with the purpose of collecting information that could serve the aims of economic and social improvement.[4] Beginning in 1846 Austria-dominated Lombardy–Venetia too had its numbers made public in the large folios of the statistics of the Austrian Empire.

In fact the restored rulers never stopped collecting quantitative information; though at first they did so in secrecy since they ostensibly rejected everything that was associated with the Napoleonic era. Indeed, in order to maintain their power, most Restoration rulers realized that they could not simply go back to the structures of the old regime; therefore several institutional innovations introduced by the Napoleonic regimes continued to inform state institutions. Beginning in the 1830s, pressured by the example of other European states, rulers increasingly felt the need to show that they were sensitive to the requirements of "civilization" and proceeded to make some statistics public, although they did not always publish those numbers that could be a source of embarrassment or difficulty for their governments.

For the nineteenth-century Italian patriots who, with national pride, claimed Italy's statistical primacy, tracing it back to the medieval communes and republics,[5] the establishment of statistical committees and

3 An old work by a nineteenth-century patriot and statistician, Pietro Castiglioni ("Introduzione storica dei censimenti delle popolazioni italiane dai tempi antichi sino al 1860," in *Statistica del Regno d'Italia. Popolazione. Censimento degli antichi Stati Sardi (1 gennaio 1858) e censimenti di Lombardia, di Parma e di Modena (1857–58)*, vol. I [Turin, 1862]), provides the basic information on the establishment of these offices. See also ISTAT, *Decennale 1926 IV – 1936 XIV* (Rome, 1936), and ISTAT, *Dal censimento dell'unità ai censimenti del centenario. Un secolo di vita della statistica italiana*, text by R. Fracassi (Rome, n.d. [1961]). These are mostly celebratory works.

4 A. Bellettini, "Contenuto e tecnica degli ultimi censimenti dello Stato Pontificio," in Comitato Italiano per lo Studio della Demografia Storica, *Le fonti della demografia storica in Italia. Atti del seminario di demografia storica, 1971–72*, vol. I (Rome, 1974), pp. 463–478.

5 Castiglioni, "Introduzione storica," is among them. Other examples of the same

the official publication of data on society represented a first victory of the liberal stance, the end of the secrecy that surrounded the collection of information in absolutist regimes. According to the liberal view, under those regimes this practice aimed at a knowledge of society's resources exclusively for fiscal, military, and police reasons. In contrast to this use of statistics, liberals believed that statistical knowledge should be a means of enlightening public opinion and of making the working of power transparent. The publication of these numbers would provide tools for evaluating the effectiveness of the government; some even dreamt that thanks to an objective statistical knowledge, eventually statesmen would be able to forget their ideological and political differences: "in the conflict of opinion and amidst the doubts of reason, we will resort to the logic of figures."[6]

Of course things were more murky than allowed by the liberals' neat binary opposition. The interest in statistics on the part of the administrative monarchies (as the pre-unification states are called to indicate their attempt to combine absolutism with some of the administrative and political institutions of the Napoleonic era)[7] had various origins and impulses, some of which were closer to those of the liberals than they were themselves aware of. In the following pages, by focusing on four Italian cases – Sicily (not a state of its own, of course, but a distinct component of the Kingdom of the Two Sicilies), the Kingdom of Sardinia or Piedmont, the Grand Duchy of Tuscany, and the Kingdom of Lombardy–Venetia, a *Land* of the Austrian Empire – I describe what these origins and impulses were. These cases represent variants of a general phenomenon and there is no reason to believe that the states not included here exhibited very different patterns of development.

In each case I look at the interactions between local intellectual and administrative traditions and the reasons that drove governments and members of the liberal educated classes and social elites to engage in, or

rhetoric can be found in A. Balbi, *Miscellanea italiana. Ragionamenti di geografia e statistica patria. Raccolti e ordinati da Eugenio Balbi* (Milan, 1845), esp. pp. 40ff. Also J. Burckhardt in his *Civilisation of the Renaissance in Italy* praised the Italian communes of the Middle Ages – Venice and Florence in particular – for having produced "for the first time a true science of statistics" (orig. edn.: Basel, 1860; I quote from the English translation by S. G. C. Middlemore [London, 1921], p. 76).

6 E. Basevi, *Della necessità d'un uffizio di statistica in Toscana sue immediate applicazioni nelle riforme municipali e daziarie* (Florence, 1847), p. 7.

7 For the meaning of this term see R. Romeo, "Il Risorgimento: Realtà storica e tradizione 'morale,'" in *Dal Piemonte sabaudo all'Italia liberale* (Bari, 1974), p. 286; see also L. Riall, *The Italian Risorgimento. State, Society and National Unification* (London, 1994), ch. 2.

to resist, the production of official statistics. I emphasize the models (both Italian and European) these various undertakings relied upon, and stress their successes and failures. The elements that help explain the outcomes of these undertakings in the individual case are no different from those that explain more generally why some states were more successful than others in implementing policies of conservative reform and modernization. The Bourbons in Sicily had to face very strong opposition among the local landed elites who had no intention of releasing information which could be used by the state; the bureaucracy moreover did not offer much cooperation. In contrast, the collaborative effort between state administrations, social elites, and educated classes established in the 1830s did not encounter substantial obstacles in Piedmont. The experience of the Grand Duchy of Tuscany owed more to the strength of a local tradition of erudite scholarship and intellectual involvement on the part of the landed elites than to an autonomous impulse on the part of the Restoration rulers. In the case of Lombardy–Venetia, the establishment on the part of the Austrian administration of an apparatus of statistical monitoring of the state and the eventual publication of its numbers could not avoid becoming the very target of criticism by a growing public opinion of nationalist orientation.

STATE STATISTICS IN SICILY: THE EXPERIENCE OF THE DIREZIONE CENTRALE DI STATISTICA IN PALERMO

It was in the reforming climate of the early years of his reign that Ferdinand II, king of the Two Sicilies, instituted in 1832 the Direzione Centrale di Statistica (Central Directorate of Statistics) in Palermo, following a suggestion of his Minister of the Interior, Nicola Sant'Angelo.[8] The creation of this institution had already been proposed in 1822 by the head of the office for the census of the city of Naples, the economist Saverio Scrofani.[9] When Sicily was reunited with the Kingdom after the Congress of Vienna, the Neapolitan rulers, in an attempt to strengthen their control over the power of the local aristocracy, imposed on the island (which shared with Sardinia the peculiarity

8 Castiglioni, "Introduzione storica," p. 162. In contrast in the continental provinces of the Kingdom only in 1851 was a committee for the compilation of a general statistics created within the Ministry of the Interior (ibid., pp. 117ff.). Until then some information was collected by the various ministries on specific topics (mostly population and justice).

9 On this episode see R. Salvo, "Nota su un progetto di Saverio Scrofani per l'istituzione in Sicilia di una direzione di statistica e del censimento (1821–23)," *Nuovi quaderni del Meridione* 100 (1988), pp. 529–533.

of not having been subjected to French rule) an administrative organization inspired by the Napoleonic regime, although still partly distinct from that of the continental provinces.[10] In 1818 Sicily was divided for administrative purposes into seven valleys or provinces, and the state officials in control of each valley – who were called intendants – were assigned some statistical tasks and the help of a *redattore statistico* (statistical compiler). However, until the matter came to be regulated by a new central office in the island's capital, their statistical investigation was limited to the collection of figures on population and grain supplies, and was rather sparse.[11] In 1832, with the creation of the Direzione Centrale di Statistica in Palermo, an attempt to undertake systematic social counting became a reality: regulations, instructions, and printed forms started to flow from the center to the periphery of the administration, generating in their turn a counterflow of letters, complaints, and apologies, as well as, of course, figures.

As head of the new office was appointed the previously mentioned Saverio Scrofani (1756–1835). Scrofani united his knowledge of political economy with an expertise in the practice of statistics which he had acquired during the very last year of the Napoleonic regime in Naples, when he was appointed director of the census of the capital, an office which he maintained during the Restoration years.[12] He thus represented a real *trait d'union* with the eighteenth-century Enlightened tradition as well as with the experience of the Napoleonic period. Like Gioia and Romagnosi, who shared a similar experience and culture, he viewed statistics as a "science of facts," and an indispensable complement of political economy as a tool of the legislator. From the first half of the eighteenth century, political economy had developed in the

10 See D. Mack Smith, *A History of Sicily. Modern Sicily after 1713* (London, 1968), pp. 362ff., and V. D'Alessandro and G. Giarrizzo, *La Sicilia dal Vespro all'unità d'Italia* (Turin, 1989), pp. 668–675.

11 According to Castiglione, "Introduzione storica," p. 163, and to the members of the Direzione: see the account given by its director to the Intendant of Palermo in Archivio di Stato di Palermo (henceforth ASP), Intendenza di Palermo, no. 938, letter dated 10 September 1835, and the "Lettera ufficiale del Direttore della Statistica scritta individualmente alle pubbliche autorità, a' sindaci de' Comuni, ed a varii dotti di Sicilia (November 30, 1855)," *Giornale di statistica* (henceforth *GdS*), n. s., 1 (1858), p. 4.

12 According to Castiglioni, "Introduzione storica," p. 117, this office continued the work of the *burò* of statistics established within the Ministry of the Interior in 1810 and assigned to the direction of L. de Samuele Cagnazzi. On this office see A. Scirocco, "La *Statistica murattiana* del Regno di Napoli," in S. Martuscelli, ed., *La popolazione del Mezzogiorno nella statistica di Re Murat* (Naples, 1979), pp. vii–xx. On Scrofani see the introduction by G. Giarrizzo to S. Scrofani, *Memorie inedite* (Palermo, 1970), pp. 7–37.

Kingdom of the Two Sicilies with a reforming stance that later in the century won the brief support of the Bourbon monarchy. In the 1780s and the early 1790s Giuseppe Maria Galanti, a student of one of the major Neapolitan political economists, Antonio Genovesi, had compiled, with royal support, a wide-ranging description of the economic state of the kingdom which was regarded by his student Vincenzo Cuoco as a seminal work in the field of statistics.[13]

As envisaged by the decree that established the Direzione and by its regulations (1832), the collection of statistical data on the island was to be assigned to a staff recruited through a system of competitive examinations.[14] The first of these examinations saw twenty-six candidates tackle questions of political economy and agriculture formulated by a committee composed of the director of the statistical office and two professors of the University of Palermo, one of political economy and the other of agriculture. The questions tested the candidates on controversial issues such as the promotion of manufacturing industry in Sicily, which placed economists of free-trade conviction in opposition to the protectionist policies of the government.[15] Owing to the so-called "autonomist compromise" of the early 1830s whereby the government placated the Sicilian aristocracy by meeting some of their demands, the winners of this examination ended up being a mixed group of supporters of free trade and Sicilian autonomy (such as Francesco Ferrara, Emerico Amari, Raffaele Busacca) and supporters of government protectionist policies (such as Gaetano Vanneschi and Giuseppe Palmeri). As for the position of provincial *redattore statistico*, although the regulation did not include any provision on this matter, the Direzione put pressure on the intendants to test the "expertise" of the employees.[16]

13 See the introduction by D. Demarco to G. M. Galanti, *Della descrizione geografica e politica delle Sicilie* (Naples, 1969), vol. 1, esp. pp. lxiv–lxv, and R. De Lorenzo, "Strategie del territorio e indagini statistiche nel Mezzogiorno fra Settecento e Ottocento," in De Lorenzo, ed., *L'organizzazione dello stato al tramonto dell'Antico Regime* (Naples, 1990), pp. 129–185.

14 The texts of both the decree and the regulation were published by F. Ferrara as an appendix to his article "Cenno sulla miglior maniera di formar uffici statistici," *GdS* 3 (1838), pp. 190–200 (now in F. Ferrara, *Opere complete edite e inedite*, vol. 1, *Scritti di statistica*, ed. B. Rossi Ragazzi [Rome, 1955], pp. 247–260).

15 The minutes of the works of the committee and the texts of the questions are in ASP, Direzione Centrale di Statistica, no. 2. On the vicissitudes of this examination and the struggles and deals behind it see R. Salvo, "Alcune notizie sul concorso per i posti di 'commesso' alla Direzione centrale di statistica di Palermo (1832–33)," in P. F. Asso, P. Barucci and M. Ganci, eds., *Francesco Ferrara e il suo tempo. Atti del Congresso Palermo 27–30 ottobre 1988* (Rome, 1990), pp. 491–510.

16 As it appears in the account given in a letter of the Luogotenente Generale to the Director of the Central Statistical Office, December 29, 1842 (ASP, Direzione Centrale di Statistica, no. 14).

Thus, also at the provincial level, other competitive examinations were held in which questions of both a theoretical and practical nature were asked, such as whether statistics was a "science of comparative facts," what caused an increase in marriages, and how to draw up a table on crime statistics. Both at the central and at the provincial level, the "theoretical" questions were drawn from M. Gioia's well-known *Filosofia della statistica.*[17]

Among the newly appointed employees of the Direzione there was a nucleus of committed intellectuals eager to make known their views on statistical investigations, but also more generally on vital issues of economic policy. In 1836 two of these, Francesco Ferrara and Emerico Amari, took the initiative to start a periodical, the *Giornale di statistica,* which would publish the statistics that they were collecting, reviews of foreign statistical works, and articles on economic and social issues.[18] One of the members of the Direzione, Gaetano Vanneschi, stressed in the first issue of the journal that the periodical was the outcome of an autonomous initiative on the part of the employees. They aimed at overcoming the scarce and delayed circulation of foreign works and ideas in Sicily, and at promoting the development of a statistical culture on the island, filling the void of specialized publications in the field.[19] The journal became an arena for intense debate and criticism: there were statistics which the Direzione was bound to compile but which did not find the approval of all its members or of the periodical's collaborators. This was the case, for example, of trade statistics, the so-called "balance of commerce," attacked by the political economists of strict *laissez-faire* orientation as a product of a despicable institution which was bound to disappear when all artificial barriers to commerce were at last eliminated in favor of the sound dictates of free trade.[20] Ferrara, as we have seen in Chapter 2, was the most vocal of these critics, but his radical critique found only limited support among the other members and collabora-

[17] The texts of the essays and the minutes concerning the examinations are preserved in ASP, Direzione Centrale di Statistica, nos. 2 and 3.

[18] Six volumes were published between 1836 and 1846, and five more between 1852 and 1864. See the introduction by B. Rossi Ragazzi to F. Ferrara, *Opere complete,* vol. 1, pp. xxi–xxxvi.

[19] G. Vanneschi, "Proemio," *GdS* 1 (1836).

[20] See the review by R. Busacca of *Archives statistiques du Ministère des Travaux publics de l'agriculture et du commerce,* in *GdS* 3 (1838), pp. 146–169, and G. Bruno, "Difetti e riforme delle statistiche commerciali," *GdS,* 2nd s., 7 (1852), pp. 14–67. On the debate on statistics among Sicilian political economists see P. L. Tedeschi, "Francesco Ferrara e la politica della statistica (1835–1846)," *Archivio storico per la Sicilia orientale* 78 (1982), pp. 94–151.

tors of the journal, who professed to follow the teachings of Gioia and Romagnosi, and even used Gioia's works, as we have seen, as a source of subjects for the questions to be set in the competitive examinations.[21] As it appears from the voluminous archival records left by the Direzione, there were few imaginable topics that escaped the quantifying attempts of the Sicilian statisticians. The functionaries of the Direzione collected, or at least attempted to collect, figures on topography, population, agricultural production and consumption, wheat prices, stock-raising, industry, sulfur mines, the professional structure of the population, commerce, education, charitable institutions, foundlings, hospitals, the clergy, crime and the administration of justice, and landed property.[22] In its range, the program of the Direzione di Statistica was as all-embracing as the researches that would be undertaken in other Italian states in the following years, and this in spite of the ideological divergences among its employees concerning the extent and limits to which the enterprise was to be subjected.

Besides publishing critical articles and reviews, the *Giornale di statistica* was supposed to present the results of the quantifying activity of the Direzione. The figures published in the journal are, however, a poor indication of the range of topics on which the Direzione tried to collect data in order to document comprehensively the resources and conditions of the island. The numbers on population movements and the level of trade, being based on the registration activity of established offices (registry and customs), were gathered and appeared with a certain regularity. From time to time scattered figures on the production of the sulfur mines (one of the few industrial activities of the island), on the merchant navy, crimes, hospitals, the clergy, cholera-related mortality, smallpox, and landed property found their way into the pages of the journal. However, what was considered the most important statistical operation, that is the census, was never performed by the Direzione – in fact the only available figures for the whole island in the first half of the century were those of enumerations carried out by the intendants just before the creation of that institution.[23]

The scarcity and fragmentation of the statistics published in the

21 Another dogmatic free-trader, very close to Ferrara, was R. Busacca. Close to Romagnosi were F. Perez (author of "Idea del Perfetto Civile riguardata come norma della statistica," *GdS* 5 [1840], pp. 140–207) and E. Amari (see his "Difetti e riforme delle statistiche de' delitti e delle pene," *GdS* 3 [1838], pp. 278–279).

22 The materials concerning the compilation of these statistics are preserved in about 133 of the 168 bundles catalogued under the heading Direzione Centrale di Statistica at the ASP.

23 See L. Izzo, *Introduzione allo studio della popolazione nel Mezzogiorno d'Italia nel secolo XIX* (Naples, 1975), p. 88.

journal reflected the many difficulties in which the officials of the Direzione found themselves entangled from the very beginning of their work. They met much resistance at various levels. At a very early stage the director, noticing the "obstacles that hamper the course of things, among which the principal ones are the prejudices still widespread among certain lower authorities, as they are among the people,"[24] had to remind the intendants and other functionaries of the aims of the enterprise to be communicated to their subordinates:

> It is well known, and I do not need to mention it to persons of such great knowledge, how the most civilized European nations acknowledge the immense advantages to be reaped from the adoption of government measures based on the important information provided by statistics, which are far from being dangerous to the interest of the individuals. I could compare the state of a whole nation to that of a private family: if its head should ignore the number of people in it, its means of subsistence, and everything which is relevant for its internal state, could he ever be aware of the several needs which require intervention?[25]

The representative of the central government on the island and even the Ministry of the Interior in Naples periodically requested assessments of their progress from the officials of the Direzione. From the start, these were busy urging peripheral functionaries and local authorities to quicken the pace of their data collection in exchanges of letters which sometimes went on for years. The provincial *redattori statistici* responded by drawing discouraging pictures of the situation they had to face: to the indifference and lackadaisical attitude of the mayors[26] they added the difficulties of the sometimes surreal classifications invented by the Direzione, which tried the patience and sometimes the intelligence of the data collectors, as in the case of the first very detailed statistics on the professions.[27] In their turn the directors of the Direzione in Palermo protested their zeal to the government while complaining about the lack of collaboration on the part of many intendants.[28] To them the problem appeared to be one of authority: the Direzione had

24 ASP, Direzione Centrale di Statistica, no. 76, Circular to the "Intendants, Bishops, Archbishops, Justices, and other administrative and judicial officials of Sicily," May 18, 1835.

25 Ibid.

26 Ibid., the *redattore statistico* of Caltanisetta to the Direzione Centrale di Statistica, December 23, 1835.

27 Ibid., the intendant of Trapani to the Direzione Centrale di Statistica, May 8, 1835.

28 ASP, Direzione Centrale di Statistica, no. 1, Director Scrofani to the minister secretary of state of the Luogotenente Generale in Sicily, October 22, 1833; Director Palmieri to the government in Naples, October 10, 1838.

very little, if any, control over the intendants and the mayors, who were figures of fundamental importance for the successful collection of data.[29] It does not appear that this difficulty was ever resolved.

Although the figures collected fell short of constituting the general statistics that the Direzione was supposed to compile, this institution nevertheless functioned as a center of diffusion of a language and an ideology similar to the one that the educated elites were adopting all over Italy as well as abroad. Its members were particularly eager to establish exchanges with the statisticians who operated in other Italian states and elsewhere: they obtained a favorable response from Piedmontese functionaries and from the collaborators of the *Annali universali di statistica*; they sent copies of their journal to the London Statistical Society, and even to Harvard College.[30]

Following the defeat of the 1848 revolution, most members of the Direzione left Sicily and went into exile in Piedmont and Tuscany. Ferrara, who had fought for an autonomous Sicily in a new confederation of Italian states, went to Turin. However, the statistical program of the Palermo office was not abandoned in the 1850s. In fact, as the employees of the Direzione made new attempts to count the population and collect data on the industry in the island, they also tried to revamp the *Giornale di statistica*: the director, Federico Cacioppo, Baron of Antalbo, who had been in charge of the office since 1841, sought the collaboration of intellectuals who had shown "affection for the very important science of statistics," and, more importantly, had not compromised themselves with the revolution.[31] As usual, he promoted the diffusion of the journal among the municipalities of the island.[32] Yet, without the contribution of the original group of founders, the periodical lost its critical edge and probably also many of its readers, although it continued to be on the mandatory subscription list for the municipalities of Sicily.

The experience of the office of statistics of Palermo shows the possibilities and limits of the convergence between liberals and administrative monarchy in the Kingdom of the Two Sicilies. It confirms the relative

[29] See the complaints contained in an unsigned manuscript entitled "Organizzazione attuale della statistica in Sicilia" (n.d., but probably written in the early 1840s), ASP, Direzione Centrale di Statistica, no. 1.

[30] Ibid., no. 144.

[31] See the letter offering the position of collaborator to the *GdS* to the director of the Real Osservatorio Astronomico of Palermo, Domenico Ragusa, published in *GdS* 10 (1858).

[32] ASP, Direzione Centrale di Statistica, no. 144 contains several lists of people and institutions to which the periodical was sent.

isolation in which functionaries of the state bureaucracy had to operate due to lack of collaboration and straightforward opposition on the part of the social elites, especially after the end of the "autonomist compromise" of the early 1830s.[33] This helps explain why in spite of the vast ambitions and early activism of the Direzione, the actual results were meager, and most works remained incomplete. By the time the new Italian government imposed the closing down of the Direzione Centrale di Statistica in Palermo (as well as the offices of statistics that had been established in Naples, Florence, Parma, and Modena),[34] the work of this institution had already been undermined by the resistance that it faced in society at large, and even within the administration itself.

ELITES AND THE STATE IN THE KINGDOM OF SARDINIA: THE COMMISSIONE SUPERIORE DI STATISTICA AND BEYOND

From the very beginning of the reign of the moderately reform-oriented Carlo Alberto, in 1831, several functionaries of the Kingdom of Sardinia became busy publishing statistics of the provinces they administered. The best example of this semi-official production is Giovanni Eandi's *Statistica della provincia di Saluzzo*, a two-volume publication of more than a thousand pages dedicated to the description of the area over which the author exercised his authority as *vice-intendente*.[35] The Napoleonic *statistique départementale* was the immediate historical antecedent of Eandi's work; its actual model was a later expression of that enterprise, Chabrol de Volvic's two-volume description (1824) of a Piedmontese *département* he had administered in 1810.[36]

In fact, the provincial administrators' involvement in the description of their provinces had a fairly long history in the Subalpine state. In the mid-eighteenth century the central authority started to require peripheral functionaries to make periodic reports on the economic and de-

33 On the more general limitations on the intendants' power in Sicily see P. Pezzino, "Monarchia amministrativa ed *élites* locali: Naro nella prima metà dell'Ottocento," in Pezzino, *Il Paradiso abitato dai diavoli. Società, élites, istituzioni nel Mezzogiorno contemporaneo* (Milan, 1992), pp. 95–176.

34 See Castiglioni, "Introduzione storica," p. 13.

35 Eandi's *Statistica* was published in Saluzzo between 1833 and 1835. Other works of the same type were A. Piola, *Statistica della provincia d'Alessandria* (Alessandria, 1831); T. Plebano, *Statistica del mandamento di Baldichieri, provincia d'Asti* (Turin, 1832); C. Racca, *Notizie statistiche e descrittive della Valsesia* (Vigevano, 1833). On these works see U. Levra, "La 'statistica morale' del Regno di Sardegna tra la Restaurazione e gli anni Trenta: da Napoleone a Carlo Alberto," *Clio* 28 (1992), pp. 353–378.

36 *Statistique des provinces de Savone, d'Oneille, d'Acqui et de partie de la province de Mondovì formant l'ancien département de Montenotte* (Paris, 1824).

mographic conditions of their administrative units, reports which came to be known as *relazioni statistiche.*[37] Regular information on the economic conditions of the state was sought not only for immediate tax purposes but also in order to support economic policies aiming at the development of the overall resources of the state. It was, indeed, in the context of eighteenth-century absolutism and the flourishing of cameralistic literature that the idea of the compilation of a *statistica generale,* namely a general picture of the resources and conditions of the state, had originated. Between 1750 and 1755 the intendants were required to send a very detailed description of their provinces, mostly providing information of an economic nature but in a highly confidential manner. They were supposed to get the necessary information "without any publicity, and without obliging anybody to release any more information than that already prescribed by the Royal orders."[38] They were also asked to summarize in eleven tables all the information they collected. On this basis, the functionaries of the Ministry of Finance compiled a final global prospectus presenting the general conditions of the kingdom.

The general statistics of 1750–1755 remained, however, an isolated episode, and the statistical reports were produced very irregularly. The second half of the century even saw, in the absence of the state intervention, a "privatization" of demographic statistics by several economists and "political arithmeticians."[39] Later, the annexation to France brought to Piedmont the previously mentioned *statistique départementale,* or "prefects' statistics," in which the Napoleonic version of the intendant was assigned the task of the description of the territory under his administration.[40] Then statistical works entered a new

37 An enthusiastic account of the successes of Piedmontese administrators in the collection of statistical information is given by G. Prato, *La vita economica a mezzo il secolo XVIII* (Turin, 1908); see especially the introduction: "Le indagini statistiche nell'antico Piemonte," pp. 1–30. The first instructions to the intendants on the compilation of annual reports were issued in 1742.

38 Ibid., p. 13.

39 On this "privatization" see G. Levi, "Gli aritmetici politici e la demografia piemontese negli ultimi anni del Settecento," *Rivista storica italiana* 86 (1974), pp. 201ff.

40 Examples of this literature are Saussay, *Statistique du département du Mont-Blanc par le citoyen Saussay, préfet, publié par ordre du Ministre de l'Intérieur* (Paris, Year IX [1801]); Jacquet, *Mémoire sur la statistique de l'arrondissement de Suse addressé au général Jourdan conseiller d'Etat . . . par le citoyen Jacquet sous-préfet du même arrondissement* (Turin, Year X [1802]); Palluel, *Annuaire statistique du département du Mont-Blanc, rédigé par le secrétaire-général de la préfecture* (Chambery, 1803–1806); D. Destombes, *Annuaire statistique du département de la Stura pour l'an 1809, faisant suite à celui de 1806* (Cuneo, n.d. [but 1809]). For a comprehensive bibliography see F. Sirugo, ed., *L'economia degli*

public sphere formed by functionaries and members of the educated elites who supported the regime. Because of its association with the Napoleonic regime, statistics was banned again from public life during the Restoration. Yet this did not impede the government from collecting statistical information. The reorganization of the administrative apparatus undertaken in 1818 assigned to the re-established general intendants the task of compiling an annual statistical report.[41]

It should not come as a surprise that the appearance of the statistical works of the Piedmontese intendants in the early 1830s received great commendations in liberal journals all over Italy: the *Annali universali di statistica* praised the commitment of these authors to the development of a useful body of knowledge,[42] while the Sicilian *Giornale di statistica* underlined their role in the absence of a governmental statistical office.[43] The preface to Eandi's work opened with a celebration of Melchiorre Gioia who, in his *Filosofia della statistica*, had shown

> the nature and extent [of statistics], and its advantages; [he] required this kind of works to be based on a reasoned knowledge of many elements; that we describe a country with the precision used by a painter in making the portrait of a person; that we evaluate matters with the impartiality used by judges towards defendants; that we finally search for the causes of poverty or weakness, and that we propose remedies, as doctors look for the causes of sickness and the means to cure them.

As Eandi emphasized, only a first-hand knowledge of the province, together with the collaboration of local notables and technicians, had made his work possible, a work directed not only at administrators but also at a larger, "lay" public of notables and scholars.[44] To them the author made his apologies for his writing style, which could not be particularly elegant and ornate not only because of his own literary

stati italiani prima dell'unificazione. I. Stati sardi di Terraferma (1700–1860). Saggio bibliografico (Milan, 1962).

41 See "Regio Biglietto col quale S. M. approva le Istruzioni formatesi dalla Regia Segreteria di Finanze il 2 corrente mese per gl'Intendenti generali . . . 3 Dicembre, 1818," in *Raccolta degli Atti del Governo di S. M. il Re di Sardegna dall'anno 1814 a tutto il 1832*, vol. VIII (Turin, 1818), pp. 324ff.; further instructions followed in 1820: see "Istruzione della Regia Segreteria di Finanze ai Signori Intendenti generali, Intendenti e vice-Intendenti per la compilazione della Relazione Statistica, 4 Marzo 1820," ibid., vol. x, (Turin, 1820), pp. 121ff.

42 See review by G. Sacchi of both G. Eandi's *Statistica della provincia di Saluzzo* and T. Plebano's *Statistica del mandamento di Baldichieri, provincia d'Asti, Annali universali della statistica* (henceforth *AUS*) 37 (1833), pp. 185–186.

43 Review by Gaetano Vanneschi of *Statistica del mandamento di Baldichieri* and *Statistica del mandamento di Riva presso Chieri* by T. Plebano, in *GdS* 2 (1837), pp. 104–105.

44 See preface in Eandi, *Statistica della provincia di Saluzzo*.

limitations, but also because of the very "aridity of the subject and the language of calculations."[45]

The very appearance of these works indicated that the Piedmontese state was changing its attitude towards statistics. Although Eandi and the other Piedmontese intendants seem to have published these works mostly at their own expense, they also received support from the state: for example Antonio Piola, author of the *Statistica della provincia di Alessandria* (1831), was able to convince the Secretary of State to buy twenty-five copies of his work.[46] In 1836, the same year that the last installment of Eandi's *Statistica* was published, the Minister of the Interior, Count Pralormo, instituted the Regia Commissione Superiore di Statistica (Royal High Committee on Statistics).[47] Its members were mostly high-ranking magistrates, administrators and scientists, many of whom – in contrast to the very conservative Pralormo – were active in that "moderate party" which began to coalesce in Piedmont after 1830: suffice it to mention here the most famous among them, Camillo Cavour.[48] Alongside this central committee, which had a directing and coordinating function, were created thirty-seven *giunte provinciali*, provincial committees headed by the intendants and composed of local notables and members of the liberal professions (officials, priests, doctors, lawyers),[49] a choice highly praised, a few years later, by the editor of the French *Journal des économistes* who observed:

45 Ibid.

46 See Archivio di Stato, Turin (henceforth AST), Materie Economiche-Statistica, no. 5 (A. Piola to the Secretary of State).

47 Castiglioni, "Introduzione storica," pp. 247ff., claims that he could not find any decree of institution or regulatory orders concerning the Commissione. Although Count Pralormo belonged to an extremely conservative faction in the government, he was on very friendly terms with liberals such as the young Count Camillo Cavour (see R. Romeo, *Cavour e il suo tempo (1810–1842)* [Bari, 1969], pp. 691 and 784).

48 On this "party" see Romeo, *Cavour e il suo tempo*, pp. 790ff.

49 The original members of the Commissione were: the Minister of the Interior, Count Pralormo (chairman), Count Petitti di Roreto, state councilor (vice-chairman), Baron Giuseppe Manno, a high-ranking magistrate (vice-chairman), Abbot Genevois, Count Avogadro di Quaregna, a magistrate and physicist, Intendant Ghia, Secretary of State, Inspector of the Royal Mines Despine, Marquis Cesare Alfieri, statesman and student of social reform issues, Chevalier Bonafous, Director of the Royal Botanical Garden, and Intendant Eandi. See *Informazioni statistiche raccolte dalla Regia Commissione Superiore per gli Stati di S. M. in Terraferma. Censimento della Popolazione* (Turin, 1839), pp. vii–xxx (it has also a list of the members of the provincial committees). According to Castiglioni, "Introduzione storica," p. 248, other members were subsequently appointed to the Commissione: Alberto della Marmora, an army colonel, Carlo Boncompagni, a magistrate, and Count Camillo Cavour (all in December 1836); Felice Muletti, an army officer, and Gian Giacomo Bonino, a physician and inspector of the military hospitals

These *giunte provinciali* have been formed by enlightened men, who practice the so-called liberal professions, which presuppose education; by belonging to the bourgeoisie, which has the good idea, in a country where the roads to power are monopolized by the nobility, of devoting itself to strong and positive studies ... they accumulate knowledge and are much more apt to apply themselves conscientiously to official works.[50]

Both in Italy and abroad the institution of the Commissione and its works were greeted as a sign that Piedmont was advancing toward the condition of a modern, civilized nation;[51] only the previous year, we should recall, the French government had begun to publish its official figures, apparently in response to the publication, two years earlier, of British figures by the Board of Trade.[52] Baron Giuseppe Manno, one of the vice-chairmen of the Commissione, while emphasizing the continuity between the new enterprise and the statistical traditions of the Piedmontese state, located it within the larger European context in which the alliance of science and power had become reality:

Since we began to study the principles of good administration, we had come to recognize how necessary [statistical information] was to the course of a wise and enlightened government. Without going further back in time, it is with this intention that the instructions of 1742 and 1775 were given to the provincial Intendants by the general administration of finances ... Since then the progress of political economy has given a wider extent and a new shape to statistical research. It is thanks to this progress that we owe the works undertaken by the governments or by private individuals in England, in France, in Holland, in Belgium, in Germany, and in the Kingdom of Naples, researches that the governments, the learned, the welfare institutions have used in the interest of nations and for the improvements of the social state. H. M. Carlo

(1837); and finally the economist Carlo Ignazio Giulio, the Marquis di Santa Rosa, Baron Duport and Count Francesetti di Messenile (1840).

50 J. Garnier, "Statistique des Etats Sardes. Population," *Journal des économistes* 5 (1846), p. 382.

51 G. Sacchi had words of great praise for the "Instructions" to the provincial committees (*AUS* 58 [1838], pp. 129–142). See also the enthusiastic review of the first volume of the *Informazioni statistiche* by J. Gråberg in *Rivista Europea* (May 1840). In France, on the pages of the *Journal des économistes*, Joseph Garnier, reviewing the first two volumes published by the Commissione (the second volume concerned the analysis of vital statistics: *Informazioni statistiche raccolte dalla Regia Commissione Superiore per gli Stati di S. M. in Terraferma. Movimento della popolazione* [Turin, 1843]), praised the Piedmontese government for its initiative ("Statistique des Etats Sardes," p. 381). In the same issue of this journal see also the report on the *Movimento della popolazione* given by L.-R. Villermé at the Académie des sciences morales et politiques in Paris (pp. 344–366).

52 P. Marietti, *La Statistique générale en France*, (Paris, 1949), p. 24. On the Board of Trade see L. Brown, *The Board of Trade and the Free-Trade Movement 1830–42* (Oxford, 1958), esp. ch. 5.

Alberto has established that this course would be open to the nation he governs.[53]

As we read in the instructions especially written for the provincial committees which had to collaborate with the central institution, the statistical program of the Commissione was "meant to present in an orderly way the facts that reveal the physical, moral, and economic conditions of a nation, as far as these facts can be expressed and evaluated through precise quantities."[54] The emphasis on "precise quantities," on the need to limit the research to "those series of facts which can be the object of rigorous observation, and are thus reducible to sufficiently accurate figures,"[55] signaled a desire to translate reality into numbers that was absent from previous and even current statistical works of a more qualitative kind.

How exactly the Commissione was supposed to proceed in order to reach its goal appears to have been a matter of controversy. Because of the strength of the bureaucratic tradition, some could conceive of the new general statistics in terms of yet another task for the intendants. This was for example the position of one of the vice-chairmen of the Commissione, Count Carlo Ilarione Petitti di Roreto (1790–1850). A member of an aristocratic family of state functionaries, Petitti spent about fifteen years of his career as, first, vice-intendant, and, subsequently, general intendant, a position he occupied until he was nominated member of the Council of State in 1831.[56] As a provincial administrator he became familiar with the compilation of reports on the condition of the provinces under his control, a task that, as we have seen, had become mandatory in 1818.[57] Meanwhile he brought himself up to date with the statistical literature that was being published

53 AST, Materie Economiche-Statistica, no. 1, manuscript volume entitled "Instruction relative à la formation de la Statistique générale du royaume, no. 1 Recensement générale de la population" (subsequently published with some changes in *Informazioni statistiche . . . Censimento della popolazione*, vol. I, pp. xxxi–xlvii).

54 See *Informazioni statistiche . . . Censimento della popolazione*, p. l.

55 The quote is from a work-plan (*piano di lavori*) written for the Commissione by Camillo Cavour (now in *Tutti gli scritti di Cavour*, ed. C. Pischedda and G. Talamo, vol. II [Turin, 1976], p. 619).

56 A biographical note on Petitti is in C. I. Petitti di Roreto, *Opere scelte*, ed. G. M. Bravo, vol. II (Turin, 1969), pp. 1001–1025.

57 In 1826 he dedicated a particularly detailed *relazione statistica* to the heir to the throne, Carlo Alberto di Savoia-Carignano, who ten years later was to agree to the creation of the Commissione di Statistica. This manuscript (entitled "Memorie statistiche intorno alla provincia d'Asti compilate dall'intendente conte Petitti di Roreto. Omaggio rassegnato a S. A. S. il signor principe Carlo Alberto di Savoia Carignano volgendo l'anno 1826" [2 vols.]) is preserved at the Biblioteca Reale in Turin.

abroad. Reviewing the *Recherches statistiques sur la ville de Paris* by Chabrol de Volvic, he stressed the progress made by the "science of government" in those years. In England, he went on, statistics had already been usefully employed for several years; the same was true in France thanks to the government, which combined "all the knowledge and ability once ascribed only to the so-called liberals, with the good doctrines of legitimacy";[58] in Italy there was no lack of economists who had written on the subject in the eighteenth century, and a contemporary, Melchiorre Gioia, was an example of the progress of this new science.

A conservative reformer interested in the introduction of social legislation, Petitti saw statistics as an important tool in the hands of an enlightened and paternal administration. On the basis of his own experience as intendant, he conceived of the new general statistics of the Kingdom as a renovated version of the undertaking of the 1750s, one to which the gaze of the knowledgeable intendant was central. Informing chairman Pralormo of his doubts about the capacity of the Commissione to realize its program, he proposed using the skills of an "expert" in statistical descriptions such as intendant Eandi: he could collect the material for each province with the help of the provincial committees, and then assemble all the data necessary to produce the picture of the whole state.[59]

This proposal, however, was never implemented.[60] The views of another member of the Commissione, the young Count Camillo Cavour (1810–1861), were to be more influential. Cavour, who in the 1830s was involved in the management of his family estates and in business activities, was enjoying a growing reputation in the state for his knowledge of economic matters. This reputation and the favor of Pralormo probably brought him the honor of being appointed as member of the Commissione.[61] The report that he compiled at the

58 *Opere scelte*, vol. 1, p. 78 (orig. publ. as "Ricerche statistiche sulla città di Parigi e il dipartimento della Senna. Quadri sinottici compilati, riuniti e pubblicati volgendo l'anno 1823 d'ordine del prefetto del dipartimento", *L'amico d'Italia* 3 [1824], pp. 317–336).

59 Museo Nazionale del Risorgimento, Turin, Archivio Petitti, no. 341/31, minute of a letter presumably addressed to Pralormo dated October 30, 1837 ("Provvedimenti necessari per attivare i lavori della statistica gen.le dello stato").

60 In a letter addressed to Pralormo in January 1838 (Ibid., Archivio Petitti, no. 341/32), while complaining about his bad health and his heavy work load as state councilor, he expresses his regrets at not being able to continue the "beautiful enterprise" of the general statistics. Although not a member of the Commissione, in the following years he continued to be interested in the compilation of official statistics.

61 Romeo, *Cavour e il suo tempo (1810–1842)*, p. 691.

request of the Commissione on the existing statistical literature was not very flattering about the work of the intendants, who were accused of having collected scarcely reliable data.[62] Although Cavour did not deny the coordinating role of the bureaucracy, he demanded a role for civil society in the compilation of the new general statistics. The observations he presented to the Commissione in 1840 on the means of implementing the Kingdom's statistics emphasized the need for an extensive participation in the collection of the data: this should not have been the exclusive territory of members of the bureaucracy and state agencies, but should have included scientific institutions, agrarian and commercial associations, as well as individual landowners and manufacturers.[63] Statistical information ought not to be a state secret. Influenced by his admiration for the institutions of Britain and France and by his knowledge of the inquiries and research practices which had been developed in those countries,[64] Cavour was clearly concerned about the involvement of the elites, that is of people like himself, in the production of statistical knowledge.

How was the ambitious program of the Commissione translated into practice? The members of the Commissione appear to have overestimated the will of the peripheral administrators and of the notables to collaborate in the counting of the resources of the country. While a "statistical ideology" was certainly part of the intendants' outlook, other administrators were less enthusiastic about the prospect of assembling numerical information on all sorts of topics. A provincial town clerk, for example, addressing the president of the Commissione with some suggestions in 1848, referred ironically to the "very many and very diverse requests for statistical facts with the philosophical intent of collecting them in as many statistical tables, of making them the object of high speculations and then deriving from them severe pathological, economic, physiological, prophylactic, agronomic doctrines."[65] Neither did the provincial committees seem to be thrilled by their involvement in the tedious census operations: according to a high-ranking official of the Ministry of Public Works, Agriculture, and

62 The report is now in *Tutti gli scritti di Cavour*, vol. II, pp. 567ff.

63 Ibid., pp. 607ff. Three years later as one of the heads of the newly founded "Associazione agraria," he was among the promoters of a series of agricultural descriptions of the provinces of the Kingdom. On these researches see G. Prato, *Fatti e dottrine economiche alla vigilia del 1848. L'Associazione agraria subalpina e Camillo Cavour* (Turin, 1921), pp. 211ff.

64 On the young Cavour's studies and travels to France and England see Romeo, *Cavour e il suo tempo (1810–1842)*, chs. 5 and 6.

65 AST, Materie Economiche-Statistica, no. 7, text annexed to letter of the town clerk of Villafalletto.

Commerce, none of them had participated in the counting operations and "the intendants were obliged to get the employees in their own bureaus do the work by rewarding them with a bonus."[66] Indeed the provincial committees did not perform well. But by the late 1850s the memory of the poor performance of these committees had faded away since new promoters of statistics asked for their re-establishment in order to revamp the statistical service. More committed to social counting were the physicians, who collaborated with the civil servant and doctor Gian Giacomo Bonino in the quantitative description of the physical conditions of the Piedmontese population, enabling him to assemble two large volumes on medical statistics.

In spite of these limitations, the results of these efforts were considerable. Of the four sectors of inquiry (topography, population, economy, moral–intellectual state of the population) which, according to its original program, the Commissione was supposed to develop, two were realized to a remarkable extent: the one concerning the numbers and physical conditions of the population (censuses, population movements, medical statistics),[67] and the one concerning its intellectual and moral conditions (education, crime and the administration of justice).[68] Only the former, however, was pursued within the structure set up by the Commissione. The production of the latter took place instead within the ministries concerned with education and crime. The topographical part of the project was delegated from the very beginning to the military: as Cavour observed, the most qualified agency for this task was the general staff of the army, and it would have been a waste of labor trying to use members of the provincial committees. These could be usefully employed, however, in limited tasks such as the compilation of tables of meteorological observations.[69] However, the collection of topographical data and descriptions did not become a subject of a government publication.

The description of the economic condition of the state was the object of several efforts, but did not produce substantive results: in

[66] Ibid. (the letter is addressed to the Ministry of the Interior).

[67] The censuses were taken in 1838, 1848, and 1858 and the results published in 1839, 1852, and 1862 respectively; statistics on population movements were produced for the years 1828–1837; two volumes on the sanitary condition of the population were published between 1847 and 1852 (*Informazioni statistiche raccolte dalla Regia Commissione Superiore per gli Stati di S. M. in Terraferma. Statistica medica*): they concerned exclusively the institutionalized population (prisoners, foundlings, maniacs, etc.), and included also few moral statistics, such as the figures for suicides.

[68] This included several statistics on schools, and on civil and penal justice: a bibliography, although incomplete, is in ISTAT, *Dal censimento dell'Unità*, p. 41.

[69] See Cavour's work-plan (note 55 above), pp. 610–611.

practice it was reduced to the statistics of commercial movements constructed on the basis of customs data, and never extended much further.[70] Already at the time of the first census (1838) attempts were made to obtain data on the occupational structure of the population for – as we read in the instructions sent to the provincial committees – "it is known that the condition of a country differs according to the proportion of the people who live off the land, or capital, or industry."[71] However, in the censuses of both 1838 and 1848 the table that was supposed to report the figures on occupations remained empty: census enumerators had not applied uniformly the occupational categories suggested by the Commissione and the data collected were consequently unreliable. There were also attempts to draw up a statistics of agriculture, while Cavour was Minister of Agriculture and Commerce in 1850–1852; but in this case too the data collected were never published because of their recognized lack of accuracy.[72]

After 1848 the Commissione practically ceased to function. As the head of the office for the census, Filippo Cordova, explained in the proposal he sent to Cavour for the organization of the statistical services of the new nation (May 1860), the role of the Commissione Superiore had become somewhat obsolete after the granting of the constitution, the *Statuto*.[73] Earlier its existence had been justified by the absence of representative institutions: the Commissione was to provide a "kind of guarantee of truthfulness and check on the administration." Now that principles of liberty and constitutional guarantees informed the government, there was no longer a need for an institution that was not organically part of the administration. The ministries

70 According to Filippo Cordova ("Rapporto al Conte di Cavour ultimo ministro di Agricoltura, Industria e Commercio del Regno Sardo", in *I discorsi parlamentari e gli scritti editi ed inediti preceduti dai ricordi della sua vita* [Rome, 1893], vol. IV, p. 365). Seven volumes of statistics on the *Movimento commerciale* were published by the Ministry of Finances for the years 1851–1857. R. Luraghi, *Agricoltura, industria e commercio in Piemonte dal 1848 al 1861* (Turin, 1967), p. 103, mentions an industrial census ordered by Cavour in 1852, for which, however, no archival evidence remains.

71 See *Informazioni statistiche . . . Censimento della popolazione*, p. lxiv.

72 Romeo, *Cavour e il suo tempo*, pp. 6–7, footnote 11. The Ministry of Agriculture and Commerce was actually abolished at the beginning of 1852 by Cavour himself, then prime minister, and its duties distributed to other ministries. Statistics (which had been assigned to the Ministry of Agriculture in 1850) became then a jurisdiction of the Ministry of the Interior (see ISTAT, *Dal censimento dell'Unità*, p. 14).

73 "Relazione al Ministro di agricoltura e commercio," in *I discorsi parlamentari*, pp. 349–373. Cordova (1811–1868), an expert lawyer who had defended many municipalities against the encroachment of the feudal aristocracy in pre-1848 Sicily, was a Sicilian émigré who entered the "party" of Cavour. He held various ministerial appointments in the Italian government.

had already directly assumed the task of gathering data on their activities and fields of concern. Cordova, however, argued that there continued to be a need for an institution to coordinate all the work necessary to assemble information of the state of the country. To this end he proposed the creation of a central directorate of statistics within the Ministry of Agriculture, Industry and Commerce. The eventual creation of this institution in October 1861, and the simultaneous closing of the offices of statistics in Palermo, Florence, Naples, Parma, and Modena, sanctioned in yet another domain the continuity between Piedmont and its successor, the Kingdom of Italy.

THE STRUGGLE FOR PUBLIC NUMBERS IN THE GRAND DUCHY OF TUSCANY

In the Restoration years the administration of the Grand Duchy of Tuscany routinely collected a great deal of information on the object that, since the eighteenth century, had become an important state concern, namely population. Indeed, in the field of demographic statistics Tuscany stood out among the other Italian states.[74] Beginning in 1817 each community of the Grand Duchy had to send monthly returns of births, marriages, and deaths to a central office (Ufficio dello Stato Civile) modeled after the institution established by the French. At the local level, however, parish registers were the basis for counting, and parish priests were therefore the functionaries of the Stato Civile. With this material, the functionaries of the central administration, under the direction of the Ministro dello Stato Civile (Minister of Civil Registration), had to draw up quarterly reports describing population trends in each community, in each provincial district, and in the Grand Duchy as a whole.[75] They did so by diligently filling in hundreds of neat tables which, accompanied by scholarly commentaries and references to the vast European literature on political arithmetic and political economy, were presented to the Grand Duke in a handsome leather binding.[76] In these works, besides presenting the actual figures and making comparisons over time as required by the 1817 in-

[74] See P. Bandettini, ed., *La popolazione della Toscana alla metà dell'Ottocento* (Turin, 1956), p. 1.

[75] See *Motuproprio*, June 18, 1817, and the attached instructions in *Raccolta delle leggi del Granducato della Toscana*, vol. IV (Florence, 1817), pp. 135–148. A brief account of the organization of the Tuscan Stato Civile is in P. Bandettini, *L'evoluzione demografica della Toscana dal 1810 al 1889* (Turin, 1960), pp. 59–67.

[76] These works are preserved in the Archivio di Stato, Florence (henceforth ASF), Segreteria di Gabinetto Appendice, especially nos. 103, 104, 106 and 107.

structions, they also proceeded, in some cases, to work out demographic projections.[77]

This remarkable monitoring of the population and its movements had an eighteenth-century precedent in the research of the clergyman Marco Lastri (1731–1811), agronomist and member of the prestigious agronomic society, the Accademia economico-agraria dei Georgofili. His interest in population figures stemmed from his participation in the Enlightenment culture of the time and from his support of the reforming policies of Grand Duke Peter Leopold.[78] Author of a historical study on the population of the Tuscan capital, he had created a coefficient to estimate the population of the Grand Duchy based on his record-keeping of demographic events and following contemporary models of political arithmetic.

Parish enumerations and the calculations of political arithmetic continued to be a fundamental point of reference for Tuscan administrators of the first half of the nineteenth century in their attempts to identify population trends, and mortality in particular.[79] This did not prevent them from trying to take also a complete, nominative census of the population in order to correct the "mistakes to which [were] inevitably subjected annual enumerations."[80] This attempt, however, was not a success: the results of the 1841 census were not considered reliable and were not published, except for some general figures.[81]

77 See for example ibid., no. 106, manuscript entitled "Tavole progressive del rapporto esistente fra la popolazione e la mortalità in ciascuna comunità della Toscana calcolate dietro l'esperienza del lasso di 17 anni dal 1808 al 1824." Mortality figures were calculated for each community and in relation to forty-one different hypotheses of total population and for each quarter of the year. The author was probably the Ministro dello Stato Civile himself, Giuseppe Gasbarri.

78 See M. P. Paoli and R. Graglia, "Marco Lastri: aritmetica politica e statistica demografica nella Toscana del '700," *Annali della Fondazione Luigi Einaudi* 12 (1978), pp. 117–215.

79 Generally in the nineteenth century, political arithmetic was seen as a corpus of calculations and ratios developed in order to evaluate the resources of a state in the absence of a census, and therefore as an alternative to the method of complete enumerations: see for example A. Quadri, *Storia della statistica dalle sue origini sino alla fine del secolo XVIII per servire d'introduzione ad un prospetto statistico delle provincie venete* (Venice, 1824), p. 236, and G. Boccardo, *Dizionario dell'economia politica e del commercio così teorico come pratico utile non solo allo scienziato ed al pubblico amministratore, ma eziandio al commerciante, al banchiere, all'agricoltore ed al capitalista*, vol. 1 (Turin, 1857), pp. 190–191.

80 G. Piccinetti, "Censimento nominativo della popolazione del Granducato di Toscana eseguito nell'anno 1841. Memoria letta dal Socio ordinario Giovanni Piccinetti nell'Adunanza del dì 5 Maggio 1844," in *Atti dell'I. e R. Accademia economico-agraria dei Georgofili* 22 (1844), pp. 105ff.

81 On this census see also Bandettini, ed., *La popolazione della Toscana*, who tends to give a more positive evaluation of its results.

Along with demographic statistics, public order was the other major object of close monitoring in the Grand Duchy. Starting in the 1820s frequent statistical reports classifying crimes and criminals as well as the "productivity" of the various courts and the punishments meted out were included in great synoptic tables sent to the Duke. Comparisons between different years and different districts were made, and comments on the meanings of the figures offered for his attention always accompanied the tables. These data were mostly for internal use and not for publication.[82]

Given the strong reformist tradition that developed in this state in the second half of the eighteenth century, and Tuscany's persistent links with the Habsburg Empire (where statistics enjoyed an important position among administrative practices, as we are going to see), we could expect a greater investigative involvement of the Tuscan administration in areas other than population and public order. Indeed, the Restoration grand dukes Ferdinand III (1814–1824) and Leopold II (1824–1859) maintained the principal tenets of eighteenth-century reforming policies – a *laissez-faire* attitude in the grain trade, for example – and chose their administrators from among tolerant and politically moderate individuals, some of whom had collaborated with the French administration.[83] Between 1819 and the early 1830s, they continued and completed the cadastral survey begun under that administration,[84] and, as we have seen, continued to devote much attention to demographic statistics. As a whole, however, the Restoration rulers had no grand reforming ambitions, the only exception being the works of land reclamation pursued in several areas of the state and in particular in the Maremma Grossetana. There the physical conditions of the population were closely watched to find out whether the environmental changes improved living conditions in the area: in the 1840s one of the members of the committee on public health for the province of Grosseto, the physician Antonio Salvagnoli Marchetti, compiled medical statistics for the area, as well as a few studies of its economy.[85]

[82] Several of these reports are preserved in ASF, Segreteria di Gabinetto, especially nos. 190, 241–263.

[83] See C. Rotondi, ed., *I Lorena in Toscana* (Florence, 1989), esp. essays by R. P. Coppino ("Restaurazione e ceti dirigenti in Toscana," pp. 81–105) and Z. Ciuffoletti ("I moderati toscani e la tradizione leopoldina," pp. 121–138).

[84] On the circumstances of the cadastral survey in Tuscany see G. Biagioli, *L'agricoltura e la popolazione in Toscana all'inizio dell'Ottocento. Un'indagine sul catasto particellare* (Pisa, 1975), esp. pp. 14ff.

[85] See A. Salvagnoli Marchetti, *Saggio illustrativo le tavole della statistica medica delle Maremme Toscane compilata per ordine del Granduca di Toscana*, (Florence, 1844), and *Memorie economico-statistiche sulle Maremme Toscane*, (Florence, 1846). On the policies

But all these were projects of a sectoral kind. The drawing up of a general statistics for the state was not part of the vision of the grand duke and his administration. In fact, the very idea of making a full statistical description of Tuscany appeared as a challenge to the existing authority when it was promoted in the mid-1820s by some well-established intellectuals and enlightened aristocrats. The leading lights of the Tuscan intelligentsia and landowning aristocracy (all members of the Accademia dei Georgofili), did not (or did not want to) know that their project would meet such institutional opposition when they gathered to form the "Società toscana di geografia, statistica e storia naturale" in November 1826. Among them were the indefatigable cultural organizer Giovan Pietro Vieusseux, the astronomer V. Antinori, the chemist Antonio Targioni-Tozzetti (nephew of the more famous eighteenth-century naturalist Giovanni Targioni Tozzetti), the physician Pietro Betti, the educator and amateur geographer Attilio Zuccagni Orlandini, the agronomist Marquis Cosimo Ridolfi, the mathematician Guglielmo Libri, and the economist Marquis Gino Capponi.[86]

In the pages of the *Antologia* Vieusseux explained that they aimed at collecting "all the elements of a good geographical, statistical and physical description of Tuscany," and linked their initiative to the scholarly and reforming tradition of the eighteenth century, which, in the person of Giovanni Targioni Tozzetti, had begun to widen the boundaries of the naturalist enterprise by including a description of human society in its economic and demographic aspects.[87] At the same time, Vieusseux emphasized the novelty of their project, insisting that the science of statistics occupied a higher position than geography and natural history, because it provided a ground for the study of political economy, "the only secure basis for good legisla-

of the grand dukes in this period see R. P. Coppino, *Il Granducato di Toscana dagli anni "francesi" all'Unità* (Turin, 1993).

86 See G. P. Vieusseux's account of the inaugural meeting of the association – along with a list of the founders – in "Società toscana di geografia, statistica e storia naturale patria," *Antologia* 24 (1826), pp. 184–193.

87 Vieusseux opened the previously mentioned article with an epigraph taken from Tozzetti's *Relazioni d'alcuni viaggi fatti in diverse parti della Toscana per osservare le produzioni naturali e gli antichi monumenti di essa* (Florence, 1751). On the eighteenth-century Tuscan naturalists see F. Rodolico, *La Toscana descritta dai naturalisti del Settecento. Pagine di storia del pensiero scientifico* (Florence, 1945), and F. Venturi, *Settecento riformatore. Da Muratori a Beccaria* (Turin, 1969), pp. 337ff. Another Tuscan economist, Aldobrando Paolini, a few years later extolled the Tuscan tradition of statistical investigation and especially the researches developed under Peter Leopold: see his "Studio e progressi della statistica in Toscana," *AUS* 36 (1833), pp. 209–230.

tive decisions." He also lamented its present neglect "in spite of the fact that the Italians have two masters in this science (Gioia and Balbi) and that for some years the administrative needs of a conquering army have made us collect the elements of those prospectuses and tables that were sent to the ministers of the great empire."[88] Implicit in this reference to Napoleonic statistics was a critique of the lack of initiative in this field on the part of the Restoration rulers.

The founders of the Società intended to mobilize the whole of Tuscan civil society in the compilation of this information. To this end they envisaged meetings in the provincial towns where

> the professor, the student, and the landowner from Florence will sit alongside the worthy magistrate, the honorable landowner, the respectable priest, and the simple but estimable farmers from the provinces, who have contributed with zeal and care to our archives and our museum by writing essays and sending samples of natural products.[89]

This project never went beyond a preliminary stage, since the Società fell victim to the repressive course taken by Leopold II's policies in the early 1830s (when publication of the *Antologia* was also prohibited).[90] It stimulated, however, the activity of two learned members of the Società, Emanuele Repetti and Attilio Zuccagni Orlandini, who privately, and separately, undertook the task of accurately describing Tuscany, the former with the *Dizionario geografico fisico storico della Toscana*, the latter with the *Atlante geografico, fisico e storico del Granducato di Toscana*, a title from which the word "statistical" had been excluded – as the author later complained – by the intervention of grand ducal censorship.[91] The statistical merits of Zuccagni Orlandini's *Atlante* received recognition through the prestigious pen of Gian Domenico Romagnosi, who remarked on the difficult circumstances in which the

88 Vieusseux, "Società toscana," p. 185.

89 Ibid., p. 190. This project of involving provincial notables in the overall description of the state bore many analogies with the provincial committees on statistics in the Kingdom of Sardinia.

90 The dissolution of the association is mentioned by Castiglioni, who however does not refer to a specific date ("Introduzione storica," p. 48). It probably took place around 1830, as appears from a letter of Vieusseux to A. Balbi dated March 6, 1830 (preserved at the Biblioteca Nazionale in Florence).

91 Zuccagni Orlandini's *Atlante* was published in Florence in 1832, Repetti's *Dizionario* appeared in installments between 1833 and 1846. The story of grand ducal censorship prohibiting the term "statistical" in the title of Zuccagni Orlandini's work is reported by the same author in a memoir attached to the *Compte-rendu des travaux du congrès général de la statistique réuni à Bruxelles les 19, 20, 21 et 22 septembre 1853* (Brussels, 1853), pp. 219–220.

author had to work, and expressed the hope that in the future those limitations would be overcome.[92]

The creation of statistical institutions in the Kingdom of the Two Sicilies and in the Kingdom of Sardinia found a warm supporter in Tuscany in the governor of Siena, Count Luigi Serristori (author of the first statistical essay on the whole of Italy), who proposed in the *Annali universali* the creation of a similar body in the Grand Duchy.[93] Throughout the 1840s members of the Accademia dei Georgofili never ceased to praise the value of comprehensive and "absolutely public" statistics for a good administration of men and things, as well as for economic improvement.[94] However, it was only after the moderate liberals entered the government at the end of 1847 that a Tuscan statistical office was instituted with the purpose of implementing a full-fledged program of official statistical research. Count Cosimo Ridolfi, a moderate liberal who became Minister of the Interior in the new government, appointed the experienced Zuccagni Orlandini as director of this office (which was a department of the Ministry of Finance, Commerce, and Public Works) in May 1848.[95] The office was divided into four sections which were supposed to direct investigations on physical topography, population, industry, and economic–political administration. Among the initiatives undertaken by Zuccagni Orlandini in his new position was the immediate publication of the data that were being collected: in the first volume of his *Ricerche statistiche sul Granducato di Toscana*, which appeared that same year, he linked the institution of official statistics in Tuscany to the teachings of Gian Domenico Romagnosi who

> was able to discover a practical function in the social sciences, drawing them away from the abstract meditations of the philosopher and proclaiming them to be skills which [can] teach [us] to act ... From the need to study them, that famous economist derived the immediate task of statistics, a light that could help us to understand fully and to act confidently in every branch of public administration: this is why this science became the logical agent of every well-regulated government: the search for eco-

92 See G. D. Romagnosi, review of *Atlante geografico, fisico e storico del Granducato di Toscana*, *AUS* 37 (1833), pp. 238–242.

93 See his "Progetto di una commissione di statistica in Toscana," *AUS* 64 (1840), p. 315.

94 The quote comes from Piccinetti, "Censimento nominativo della popolazione." See also "Della utilità di una statistica che pubblicasse i prezzi degl'Immobili, congiuntamente alle loro principali qualità. Memoria del socio ordinario Felice Francolini, letta nell'Adunanza ordinaria del dì 7 Gennaio 1844," *Atti dell'I. e R. Accademia dei Georgofili* 22 (1844), pp. 20–34.

95 See Castiglioni, "Introduzione storica," pp. 49ff. The official establishment of the office took place in January 1849.

> nomic, moral, and political data was recognized as an administrative task; therefore the results of the coordinated researches will have to be presented with full publicity.[96]

Aside from being the "agent" of a "well-regulated government," in the novel political situation statistics had also wider politico-pedagogical implications: in the "Constitutional Regime," the Tuscan statistician went on, the publication of figures had to be accompanied by the willingness on the part of the citizens to collaborate with the authorities by providing "adequate answers to the questions that [were] already circulating" and overcoming their diffidence towards the diffusion of information on their economic activities.[97] The organization of official statistics in Tuscany did not involve the notables: the central office directed by Zuccagni Orlandini communicated directly with the *cancellieri comunitativi*, the administrative employees of the 246 communities of the Grand Duchy, or with their mayors.

The liberal climate and the events that led in October 1848 to the formation of a government under the chairmanship of a democratic politician, Giuseppe Montanelli, saw an increased mobilization of "statistical activists" in Tuscany, confirming that statistics in all political quarters had become a paramount concern as well as a symbol of where one stood. An example is a certain Gaetano Landi, the author of a "Statistica civile della Toscana." Landi addressed his manuscript to the Minister of the Interior, Giovanni Guerrazzi, thanking him for the confidence he had shown in giving him the commission of this work, a source of relief from his state of "humiliating unemployment." This work was supposed to present the "moral, political and economic conditions of Tuscany"; exhibiting concerns analogous to those expressed in the Napoleonic departmental statistics,[98] it was to be "the Inventory of all Tuscany such as it was in the new Era of the Democratic Ministry. It reveals the evils of the past and the needs of the present and will be a point of departure for future progress."[99] In the same month of December 1848, the local government of Livorno – at the time the most important democratic stronghold in the Grand Duchy – affirmed its intention to give all possible help for the development of statistics in Tuscany in order to favor the "well-being of the People." To this end

[96] The quote comes from the preface, p. vii. The volume was published in Florence.

[97] Ibid., p. viii.

[98] The prefects' statistics were supposed to measure the progress made since 1789: see M.-N. Bourguet, *Déchiffrer la France. La statistique départementale à l'époque napoléonienne* (Paris, 1988), pp. 73–75.

[99] The manuscript volume (dated December 1848) is in ASF, Ministero dell'Interno, no. 3143; the quotes are from the dedication to the Minister (which has no page numbers).

it instituted a "Comitato di statistica patria" which was to collaborate with the office of statistics in Florence and encourage anyone who intended to operate along the same lines.[100]

The constitutional regime, however, was short-lived. The Tuscan statistical office survived the defeat of the liberal government in April 1849, only to be incorporated within the registry office, a shift which reduced its importance quite dramatically.[101] Its director, Zuccagni Orlandini, could no longer count on the help of employees exclusively devoted to statistical researches, since the business of registration was now their main task; he often complained about the uncooperative attitude shown by the local administrators to whom he addressed his requests for information.[102] However, as his political outlook was moderate, he kept his position, and pursued several investigations, the results of which he made public in the five volumes of his *Ricerche statistiche sul Granducato di Toscana*, a title, as has been observed, reminiscent of M. Lastri's *Ricerche sull'antica e moderna popolazione della città di Firenze.*[103] In this case as well, population statistics represented the largest part of the enterprise, which also included substantial data on the state of primary education, convicts and their crimes, the prices of foodstuffs, the merchant navy, the army, mineral waters, foundlings, annual fairs and weekly markets, and lunatic asylums, along with a model for a "community statistics," that is a full description of a town or village.[104] These were only a few of the topics that Zuccagni Orlandini had wished to explore in his researches: attempts to study other topics such as grain production in the Grand Duchy, or industry, had been frustrated, he said, by the lack of cooperation among peripheral administrators and authorities.[105]

The diverse range of topics on which Zuccagni Orlandini collected data reflected the encyclopedic ambition of a statistical enterprise rooted in the tradition of eighteenth-century scholarship and natural

[100] See the announcement in *Monitore Toscano*, no. 36, 14 (December 1848), p. 1.

[101] See Castiglioni, "Introduzione storica," p. 49. The decree was issued on December 6, 1849.

[102] See especially the introductions to vols. III and IV of his *Ricerche statistiche sul Granducato di Toscana* published in Florence in 1852.

[103] Paoli and Graglia, "Marco Lastri," p. 212.

[104] Vol. I appeared in 1848, vol. II in 1850, vols. III and IV in 1852, vol. ff in 1854. In 1856 Zuccagni Orlandini started a new series of publications, which he called *Annali statistici*; however he published only a short volume of demographic data for 1854.

[105] See his complaints in the introductions to vols. III and IV. In his *Elementi di statistica* (Florence, 1869), pp. 122–133, he mentions a failed attempt to produce an industrial census of the province of Florence.

history. But the objects of the *Ricerche* were the expression of the exposure of the author to the new trends in social–statistical research which were developing elsewhere both in Italy and in Europe, as appears from his sensitivity to so-called "moral statistics." This consisted of the figures for crime, suicides, foundlings, and illegitimacy, subjects which could not fail to attract Zuccagni Orlandini's catholic and paternalistic outlook.[106] The models of the Tuscan statistician were mostly Italian and French, ranging from eighteenth-century Italian writers on political economy such as Genovesi and Verri to contemporary statistical authorities such as Romagnosi and Gioia, to French social economists such as De Gerando, and even to Quételet.

In a pattern that we will find recurrent throughout this story, official statistics was able to enter fully only certain areas of investigation. Demographic and moral statistics dominated the picture (although no new census was taken after that of 1841, and data came from parish registers and *stati animarum*), while economic statistics, actively sought by Zuccagni Orlandini, remained largely elusive. The former were easier to obtain because they depended on the existence of institutions (the parish, the hospital, the prison) which customarily kept records of their charges and had developed various techniques of monitoring as well as systems of classification. Even though Zuccagni Orlandini himself expressed doubts on the meaning of some of the figures which referred to the populations inhabiting these institutions, he nevertheless always published them, firmly believing that a crucial aspect of the statistical enterprise was to make the numbers public.

In the administrative reorganization which occurred under the new Tuscan government of 1859, there was, however, no place for the old statistician. He was appointed to a newly established chair of statistics in an institution of higher education in Florence, a change about which he complained bitterly because of the loss of his functions and the lower salary that came with it.[107] Others, probably more "deserving" from a political standpoint, took up the responsibility of directing the statistical office of Tuscany, until this too was eliminated by the decree of October 1861 which created a central statistical office for the whole of Italy in Turin.

106 On moral statistics in France see M. Perrot, "Premières mesures des faits sociaux: les débuts de la statistique criminelle en France (1780–1830)," in *Pour une histoire de la statistique* (Paris, 1977), pp. 125–136.

107 See his letters preserved at the Biblioteca Nazionale in Florence, especially the one dated May 26, 1860 and sent to his political patron T. Corsi. See also A. Maccioni Anguillesi, "Un geografo dell'Ottocento: Attilio Zuccagni Orlandini," *Rivista geografica italiana* 55 (1948), pp. 114–115.

FROM SECRECY TO OPENNESS: IMPERIAL NUMBERS IN LOMBARDY–VENETIA

In his well-informed historical introduction to the censuses of the ancient Italian states published in four sumptuous volumes by the new Kingdom of Italy in 1862, the Milanese doctor and statistician Pietro Castiglioni reserved much praise for the population statistics of the Habsburg Monarchy. In spite of the "painful memories" that the foreign domination had left in that part of Italy, observed Castiglioni,

> it would be unjust to deny whatever little good we gained. And I do not hesitate to say that the statistical reforms inaugurated by Maria Theresa and by Joseph II and their development under Austria, although not imbued with the spirit of good government and openness, are a matter of praise that the scholar cannot ignore. Perhaps in no other European country will we be able to find over a period of a century a series of documents that offer more evident characters of certainty, and more regularity.[108]

Echoing an admiration for the enlightened government of eighteenth-century monarchs shared by other Lombard writers (all at the same time harsh critics of the contemporary Austrian Monarchy), Castiglioni's praise for the uniform methods of investigation and for the accuracy of the Habsburg bureaucracy was accompanied by a complaint about the secrecy which surrounded the results. This secrecy made the information useless for the Lombard people who so intelligently contributed to the enterprise by providing the original data.[109] Once in Vienna, Castiglioni contended, the immense amount of data collected was buried among all others of the same kind sent from the other provinces of the Empire.

In many respects Castiglioni's views were accurate and balanced. His praise of the continuity and quality of the demographic data collected in the Austrian Empire was certainly justified, although, being based essentially on the work of parish priests, the Empire's method of data gathering was not without problems. Nevertheless, the systematic collection of annual data on the state of and changes in the population

[108] Castiglioni, "Introduzione storica," p. 193. Castiglioni was born in Cremona in 1824. After obtaining a degree in medicine from the University of Pavia in 1847, he practiced the profession for a short time in Milan before migrating to Piedmont following the defeat of the 1848 revolution. He then became a member of the Piedmontese parliament and worked as a *medico condotto* and political journalist. Politically he was always very close to Cavour. After unification he debated and wrote on the issues of sanitary reform and the control of prostitution. See A. Calani, *Il Parlamento del Regno d'Italia* (Milan, n. d.), vol. 1, pp. 114–120.

[109] Ibid., p. 196.

which began in 1769 under the government of Maria Theresa represented an enterprise with few analogs in the rest of Europe.[110] In the mid-eighteenth century the Austrian bureaucracy had also been involved in another important investigative experience, although only indirectly of a statistical kind, namely the survey undertaken for the drawing up the land register in Lombardy, a survey highly praised by enlightened intellectuals such as Pietro Verri.[111] In 1750 the Giunta del Censimento, the committee appointed to supervise and direct the works for the survey, requested from every Lombard community accurate answers to a series of forty-five questions on the juridical status of all the lands included in their territory, on all their charges and debts, their sources of revenues, and so on.[112]

In his polemic Castiglioni exaggerated the degree of secrecy surrounding the production of Austrian statistics, whose results, although not published until 1846 (and even then, only partially), were distributed in small quantities to the press. But he also expressed a feeling quite diffused among the educated elites of Lombardy–Venetia under the last decades of Austrian domination, the feeling of being excluded from the process of production of very important information that served to represent the condition of their country, and thus ultimately of themselves. But this is not the place to tell in any detailed fashion a story which is of great interest in itself (and to my knowledge has not been fully narrated by anybody),[113] namely the story of the compiling of the statistics of that huge multi-ethnic state, the Habsburg Empire. I will limit my account to those aspects of the development of official statistics in the Empire that concerned Lombardy–Venetia more directly, and that can allow for a comparison with the cases previously described.

As the Austrians lost their possessions in Italy to the French, Lombardy too was invested by the remarkable activism in the field of administrative statistics that characterized all lands occupied by the

[110] See S. Zaninelli, *Un secolo di vita economica lombarda 1748–1848* (Milan, 1950).

[111] On Verri's evaluation see C. Capra, "Il Settecento," in D. Sella and C. Capra, *Il Ducato di Milano dal 1535 al 1796* (Turin, 1984), p. 350.

[112] Ibid., pp. 310–323.

[113] Brief overviews in R. Meyer, "The History and Development of Government Statistics in Austria," in J. Koren, ed., *The History of Statistics. Their Development and Progress in Many Countries* (New York, 1908), pp. 85–122; S. Zaninelli, "Una fonte per la storia dell'economia del Lombardo-Veneto nella prima metà del secolo XIX: le 'Tafeln zur Statistik der Oesterreichischen Monarchie,'" *Archivio economico dell'unificazione italiana* 12 (1963); Oesterreichischen Statistischen Zentralamt, *Von der Direktion der Administrativen Statistik zum Oesterreichischen Statistichen Zentralamt 1840–1990* (Vienna, 1990).

French, resulting in new systems of data collection, and in the first, fairly extensive publication of the material collected by the prefects and their assistants. We have already mentioned how intellectuals who sided with the revolution and approved of its eventual moderate outcome, such as Melchiorre Gioia, gave their energetic collaboration to the enterprise. Besides producing remarkable descriptive works, the practice of statistics in the public apparatuses of the Kingdom of Italy generated much debate and controversy, which expressed conflicting views about the scope of state intervention in society and the balance between different sectors of the apparatuses of the state. As in Napoleonic France, the proliferation of statistics was part of the process of state building, and as such it involved the creation of new economic and political opportunities for the members of the educated classes who had embraced the values of Enlightenment culture.[114]

After the fall of the Napoleonic Kingdom of Italy in the spring of 1814, the restored Austrian administration immediately proceeded to the political and administrative reorganization of the Italian provinces. These – Lombardy and the territories formerly belonging to the Republic of Venice – were made into the Kingdom of Lombardy–Venetia, proclaimed in 1815 to meet the autonomistic aspirations of the Italians, but in fact deprived of any autonomy and fully subordinated to Vienna. The new rulers purged the state apparatus and many members of the educated classes lost the positions they had gained. To be sure, the Austrians did not discard all the administrative innovations introduced by the previous rulers.[115] In particular, the restored Austrian rulers did not interrupt the collection of statistical information, although the French institutions were dismantled and secrecy became the rule again.

In 1815–1816 the new provincial *delegati*, who replaced – although not endowed with the same kind of power – the prefects, compiled statistical descriptions of their provinces following virtually the same schemes as those used formerly by the prefects in the compilation of their departmental statistics.[116] These works were compiled for internal

[114] See A. Cova, "Osservazioni sulle origini delle statistiche del Regno Italico," *Annali dell'Istituto Storico Italo-Germanico in Trento* 5 (1979), pp. 117–140; F. Sofia, *Una scienza per l'amministrazione. Statistica e pubblici apparati tra età rivoluzionaria e Restaurazione* (Rome, 1988), chs. 3–4. A review of this recent literature in R. Romani, "Quale sapere sociale per il Risorgimento? Alcuni libri recenti sulla statistica napoleonica," *Società e storia* 54 (1991), pp. 893–911.

[115] For a recent account see M. Meriggi, *Il Regno Lombardo-Veneto* (Turin, 1987).

[116] See Archivio di Stato, Milan (henceforth ASM), Studi. Parte moderna, no. 1137. The information had been required by the Central Aulic Commission on Organization in Vienna.

use, and were not supposed to be published. There were however exceptions. In the case of the Venetian provinces, the central government in Vienna, which intended to have very detailed information about this new area annexed to the Monarchy, assigned the task to Antonio Quadri, a zealous functionary who had started his career in the previous regime. Quadri managed to produce a semi-official statistical publication that described the Venetian provinces in eighty-six tables and a volume of expository prose.[117] In a work which showed a remarkable mastery of eighteenth-century political arithmetic, the author claimed the primacy of the Venetians in the invention of statistics.

But beyond collecting information on the individual provinces and the newly acquired territories, the ambition of the Monarchy was that of setting up an apparatus which would assure the regular transmission of statistical information from the periphery to the center of the Empire, and that would produce a picture of the whole structure. Already in 1819 the Aulic Commission for Commerce in Vienna drew up a scheme of a general statistics of the Empire whose elements were to be presented in five tables devoted essentially to human and animal populations and agricultural and industrial production. In the same year two *giunte centrali statistiche* (central statistical committees), one in Milan and one in Venice, were established: they were supposed to provide the Aulic Commission in Vienna with advice on the changes to which the survey plans devised at the center ought to be subjected to best fit local realities. The members of these committees were appointed by the governors of Milan and Venice and chosen from very loyal members of the local aristocracy and the bureaucracy. Presiding over the works of the *giunta* of Milan for example was Alfonso Castiglioni, a prominent patrician who had strongly opposed the Napoleonic regime.[118] The advisory work of these committees lasted for a few years, generating more than anything else much correspondence between the various layers of the Austrian administration in Lombardy–Venetia and Vienna.[119]

In 1828 Vienna created a statistical office within the general office of state accounting. In Lombardy–Venetia the actual work of gathering together the data collected by communal and provincial

[117] See A. Quadri, *Prospetto statistico delle provincie venete* (Venice, 1826) and *Atlante statistico delle provincie venete* (Venice, 1827). On the author and the writing of these works see M. Berengo, "Antonio Quadri e le statistiche venete della Restaurazione," in *Studi veneti. Offerti a Gaetano Cozzi* (Vicenza, 1992), pp. 391–407.

[118] On A. Castiglioni see *Dizionario biografico degli italiani*, s. v. by C. Capra.

[119] This correspondence is today at the ASM, Studi. Parte Moderna, nos. 1140, 1142–1147.

authorities and sending them to Vienna was assigned to the functionaries of the accounting offices of Milan and Venice. It was only in 1840 that the statistical office was transformed into a full-fledged statistical service. As head of this office was appointed in 1841 Baron Carl von Czoernig (1804–1889), an aristocrat from Bohemia with a passion for social analysis. In previous years, while secretary to the governor of Lombardy, he had undertaken on his own a statistical investigation of the state of Lombard agriculture and the condition of the peasantry. Czoernig was an intelligent observer of the society of Lombardy–Venetia although he filtered everything through the lenses of an organicist ideology. He extolled the social role of the nobility and lamented the atomization in which social relations had fallen in the Italian *Land*. Yet he was also at the same time a supporter of a moderate socio-economic modernization, promoting especially the creation of railroads and the development of trade and international exchange. All these matters he studied relentlessly, collecting and publishing statistical data.[120]

Although participating in the European elite culture of the time, the work of Austrian administrators such as Czoernig was also informed by "indigenous" bodies of knowledge: by the formidable traditions of cameralism and *Polizeiwissenschaft*, which became a matter for study at university by prospective civil servants of the Habsburg Empire in the mid-eighteenth century with Johann H. G. von Justi and then the extremely influential Joseph von Sonnenfels;[121] and by the discipline of *Statistik* systematized by G. Achenwall at the University of Göttingen in the mid-eighteenth century, taught in the universities of Vienna and Prague since the late eighteenth century, and, as we have seen, introduced in the universities of Lombardy–Venetia in 1817.

In spite of the fact that figures were not all-dominant in the texts used in the statistical courses taught in the universities of the Empire, teaching the statistics of the Austrian Empire did require some access to them, and this could be singularly difficult given the secrecy by which they were surrounded. In fact some professors of statistics, such as Johann Springer, who taught in Vienna in the 1830s and 1840s, were allowed to look at the tables compiled by the administration for in-

[120] For a brief biographical profile of Czoernig see the obituary published in *Bulletin de l'Institut International de Statistique* 4 (1889 [but 1890]), pp. 281–286. See also L. Faccini, "Carl Czoernig e la statistica agraria in Lombardia," *Società e storia* 21 (1980), pp. 931–950, and Meriggi, *Il Regno Lombardo-Veneto*, pp. 107ff.

[121] On German cameralism see P. Schiera, *Dall'arte di governo alle scienze dello stato. Il cameralismo e l'assolutismo tedesco* (Milan, 1968); on cameralism and *Polizeiwissenschaft* in the German states and Austria see K. Tribe, *Governing Economy. The Reformation of German Economic Discourse 1750–1840* (Cambridge, 1988), ch. 4.

ternal use. Eventually in 1846, a few years after the appointment of Czoernig to head the statistical services in Vienna, the Austrian authorities changed their policy and began to publish these tables, under the title of *Tafeln zur Statistik der österreichischen Monarchie.*[122] At the meeting of the international statistical congress which took place in Vienna in 1857, Czoernig extolled the virtues and effects of the publication of statistics:

> The *indisputable* power of numbers acted forcefully. One saw at a glance [*d'un coup d'oeil*] the immense wealth of the not yet fully exploited resources of the country; one perceived the satisfactory regularity of a financial administration which had been the object of unfavorable *opinions*, fed or strengthened by the secrecy which surrounded it; finally all patriotic hearts felt uplifted by the awareness of belonging to a state which could put in the balance of European interests such a considerable material and moral power [original emphases].[123]

Besides uplifting the hearts of the Austrian patriots, a knowledge of official statistics, Czoernig observed, was required by an "enlightened public," not to mention that "scientific works that are not destined to be brought out into the full glare of public inspection are exposed to the danger of stagnating or of becoming more and more superficial."[124] The creation of special chairs of statistics in the universities and of a special office of statistics in the administration was one of the signs, in Czoernig's view, of the Habsburg Monarchy's serious commitment to progress and the improvement of the administration.

Already before the decision to make public some of the numbers collected by the bureaucracy, the Habsburg Monarchy was interested in promoting the image of the country's rapid development. The *Gazzetta privilegiata*, the official daily of Lombardy–Venetia, emphasized the progress of civilization in the Kingdom and provided figures about this improvement even before the establishment of official statistics.

The volumes of Austrian official statistics which began to appear in 1846 were monumental compilations in strictly tabular form, packed with numbers in a characteristically dense way. Data – for the Monarchy as a whole and for its provinces – were generally classified into three main categories: land and inhabitants, state administration, and "civilization." Civilization included diverse kinds of headings from agricultural production, industry, and commerce, to roads and rail-

122 The series of the *Tafeln* was published, although not very regularly, from 1846 to 1868; afterwards it was replaced by statistical yearbooks.

123 *Compte-rendu de la troisième session du Congrès international de statistique réuni à Vienne les 31 août, 1, 2, 3, 4, 5 septembre 1857* (Vienna, 1858), p. 12.

124 Ibid.

roads, to hospitals, to schools, in other words all those headings that signaled the progress of the country, indicators of development, we could call them, signs of its modernization. The category "state administration" was divided into subsections illustrating the number of state personnel and the activity of the administration with regard to justice, police, revenues and expenditure, and so on. There were a few variations over the years with reductions or increases in the number of tables presented and some shifting emphasis on topics, but as a whole the structure of the publication remained fairly stable. Other, more convenient, summary publications complemented this voluminous series.

One should not exaggerate, however, the degree to which the figures of the Monarchy became really open to scrutiny, even leaving out the issue of the size and composition of the reading public which could have access to them. Not all the figures collected and processed by the bureaucracy were published: those that would allow the "public" to evaluate the way the Austrian rulers used the revenue were not published; for example the internal budgets of the *Länder* continued to be kept secret, thus making it easy for the anti-Austrian polemicists, as Meriggi has noted, to accuse the Habsburg Monarchy of plundering the Italian provinces through rapacious taxation.[125] This selective attitude was certainly not limited to Austria: other states which at the time had begun to publish collections of official statistics did not necessarily publish everything they knew; yet in the Habsburg Monarchy this attitude was bound to be exploited politically by increasingly dissatisfied subjects who had espoused the language of nationhood.

The compilers of the *Annali universali di statistica* kept their readers informed about the numerical data released by the Austrian Monarchy. For them it was often a matter of defending the image of the Italian minority in the Monarchy: as in the case of the publication of crime statistics which generated in Viennese journals remarks about the ability of the Italians to escape justice. The publication of these figures prompted the compilers of the *Annali* to publish also a note observing that the statistical returns offered themselves to different interpretations.[126] Sometimes figures were published that showed that Lombardy–Venetia was the most industrialized province of the Monarchy as measured by numbers of people occupied in manufactures, trades, and professions, and in terms of value produced.[127] As the events of

[125] *Il Regno Lombardo-Veneto*, p. 275.

[126] "Statistica criminale dell'Impero austriaco," *AUS*, 2nd s., 26 (1850), pp. 279–285.

[127] See especially "Nuova statistica dell'Impero austriaco," *AUS*, 2nd s., 25 (1850), pp. 273–278; 30 (1852), pp. 297–315; 31 (1852), pp. 65–86 and 295–321.

the 1848 revolution and the political repression that ensued widened the gap between the Monarchy and many sectors of the Lombard intelligentsia, these data on the economy – unaccompanied by any comment – appeared to provide evidence for the distinctiveness of Lombardy and its claims to greater autonomy and even independence.

Thus after having committed itself to the publication of official statistics, the Habsburg Monarchy found the same statistics being used as weapons against itself or being rejected by emerging nationalist critics, who were sensitive to what the numbers said about them, and to what the rulers made of the numbers. We will have occasion to return to this later. Here it is enough to recall Cesare Correnti's most famous pamphlet *L'Austria e la Lombardia*, a work first published in 1847 in a fictitious cover which bore the title *Cenni statistici di Melchiorre Gioia*. In it Correnti indicted the imperial administration for allegedly looking at things Italian with contempt, and for making its employees into machines who spent their time filling "statistical tables as if this were the supreme and most tangible result of public institutions."[128] Here the target became the very statistical zeal of which the Monarchy was so proud. While criticizing most aspects of Austrian rule in the Italian provinces, Correnti denounced particularly the excessive fiscal drain imposed by the Austrian rulers in Lombardy and solemnly declared that "if we will show by means of the numbers that from their aggregation to Austria the Italian provinces do not get a minimal advantage, we will have also shown that their situation is illegitimate even from a material standpoint."[129] After the defeat of the 1848 revolution, Correnti migrated to Piedmont and there he became one of the authors who engaged in a kind of statistics whose purpose was no longer the representation of existing Italian states, but of Italy as a whole, as a viable new national entity. We will see in the next chapter the characters and outcomes of this particular use of the language of numbers.

[128] Page 34. Correnti (Milan 1815–1889) studied law at the University of Pavia and subsequently worked in the Austrian administration. There he acquired a knowledge of the socio-economic conditions of Lombardy that he used in his collaboration with various periodicals of the time (among these the *AUS*). After 1846 he became one of the most vocal opponents of Austrian rule in Lombardy and was among the leaders of the 1848 insurrection. For a biographical profile see *Dizionario biografico degli Italiani*, s. v. by L. Ambrosoli. For further details on his life and work see following chapters.

[129] Correnti, *L'Austria e la Lombardia*.

CHAPTER 5

"PATRIOTIC" STATISTICS

> Not long ago in the preface to our *Abrégé de Géographie* we did not hesitate to indicate the wide and numerous deficiencies that we still see in the very geography and in the statistics of Italy considered as a whole. We do not yet know any work both satisfying the needs of the enlightened time in which we live and the progress made by these two sciences ... we see the best geographies and the most renowned itineraries describing minutely or outlining in detail a statue, a painting, a medal, or the remains of an ancient building, while not mentioning an immense tunnel bored in the bowels of a mountain to open a new road in order to promote industry and commerce ... or [the gigantic structures built] to connect two manufacturing and commercial places ... or to save large tracts of countryside from flooding.[1]

Thus in 1834 the Venetian geographer and statistician Adriano Balbi expressed the irritation felt by many Italian intellectuals towards the conventional descriptions of Italy, products of an erudite gaze which looked at the country as a repository of antiquities, and of endless travelers' accounts portraying the Mediterranean peninsula as a picturesque realm to be exploited for tourists' delight.[2] The authors of those

1 A. Balbi, "Dell'Italia e del saggio statistico del conte Luigi Serristori," in *Scritti geografici, statistici e vari pubblicati in diversi giornali d'Italia, di Francia e di Germania da Adriano Balbi raccolti ed ordinati per la prima volta da Eugenio Balbi*, vol. I (Turin, 1841), p. 12 (orig. publ. in *Gazzetta privilegiata* [Milan, 1834]).

2 On the images of Italy abroad in this period and travelers' literature see F. Venturi, "L'Italia fuori d'Italia," in *Storia d'Italia*, vol. III, *Dal primo Settecento all'Unità* (Turin, 1973), pp. 1187ff. Another harsh critic of the antiquarian approach was Carlo Cattaneo: see especially his "Osservazioni sulle guide di Pisa, Torino, Firenze e Padova pubblicate in onore del congresso degli scienziati," in *Opere edite ed inedite*, ed. A. Bertani, 2nd edn. (Florence, 1948), pp. 106–121 (orig. publ. in *Il politecnico* 6 [1843], pp. 471–483). A. Balbi was born into a noble Venetian family in 1782. After a brief experience as a Napoleonic official, he went to Paris and became a prolific and well known author of geographical textbooks (translated into several languages), and geo-statistical compilations. In 1833 he moved to

traditional depictions were always happy to remind the reader about the birthplace of an antiquarian or of a mediocre poet, but would only vaguely mention institutions of public education or libraries, and would be totally silent about "important facts" such as, wrote Balbi, "the state of population, industry, commerce, navigation and the important topic of moral statistics based on the number of paupers, foundlings, students, suicides, and crimes."[3]

Echoes of the polemic waged by both enlightened and romantic intellectuals against a bookish and classical culture which impeded the advancement of Italy resounded in these words uttered by a scholar who after several years spent in France had just become the official adviser on matters of geography and statistics to the Emperor Francis II.[4] Balbi's complaint was also a rhetorical device to introduce a review of one of the new works he called for, Luigi Serristori's *Saggio statistico dell'Italia.*[5] Full of tables with numbers of people, schools, ships, exported goods, periodicals, suicides, and so on, the work of this Florentine economist responded fully to the new urge to collect signs of civilization, and to make that scientific inventory of the institutions and resources of society which Balbi contrasted to the old antiquarianism. But Serristori's *Saggio* was also the first specimen of a new genre, which became popular in the 1830s–1850s, namely the genre of the statistical description of Italy in its entirety. Repeatedly in that same period collaborators to statistical journals wrote expressing their wishes that from the descriptions of many places a picture of the whole of Italy would eventually materialize; and participants at the meetings known as Congresses of Italian Scientists tried to assemble a statistical knowledge of the whole peninsula.

To construct a quantitative portrayal of the country was the task of what came to be known, after the unification of the country, as "patriotic statistics." As a collaborator on the *Annali universali di statistica,*

Vienna where he was appointed imperial adviser on geography and statistics at the court. He left Vienna in 1839 and returned to Venice where he died in 1848 (see *Dizionario biografico degli italiani*, s. v.).

3 Balbi, "Dell'Italia," p. 13.

4 The "classicists" themselves, however, espoused some of the values of their adversaries: see M. Guglieminetti, "'Decadenza' e 'progresso' dell'Italia nel dibattito fra classicisti e romantici," in *La Restaurazione in Italia. Strutture e ideologie. Atti del* XLVII Congresso di storia del Risorgimento italiano (Cosenza, 15–19 settembre 1974) (Rome, 1976), pp. 251–307.

5 Serristori's *Saggio statistico* appeared in Vienna in 1833. It had been preceded – the same year – by a shorter version entitled *Saggio di un Atlante Statistico dell'Italia.* In *Annali di statistica. Saggio di bibliografia statistica italiana*, ed. L. Bodio, 2nd edn. (Rome, 1885) there is also a reference to a *Statistica d'Italia* (1825) by the Venetian Gian Aostino Carli, a work which it has been impossible to locate.

Enrico Monnier, observed in 1869, while it was unpatriotic in the years before unification to believe in the figures released by the governments, "private" statistics became a weapon of the nationalist opposition:

> there was a *patriotic statistics* which, by studying the whole peninsula, *associated the northern provinces to those of the South through numbers*, attacked the artificial and unnatural policies of the past, and thus by overturning the barriers which formerly existed, on paper reconstituted the national unity. Italy was made *scientifically* [emphases added].[6]

Monnier was referring here in particular to the works of the 1850s in which the term "patriotic" really meant "nationalist". To the people and groups engaged in the struggle for national independence, naming, counting, and portraying the nation-to-be were vital tasks, indissolubly part of their struggle. For these people patriotic statistics was no surreptitious discursive operation, nor an unwitting complicity of knowledge and politics. Earlier works were patriotic in another sense, in the sense that the term had acquired since the eighteenth century when it began to be employed to indicate a kind of civic patriotism, a concern for the condition of one's *patria* (fatherland) and a willingness to work for its improvement.[7] In this sense one's *patria* could be simultaneously one's town of origin, the state in which this was located, or Italy as a whole, the fatherland that increasingly commanded the allegiance of men of letters and science.

This urge for description recalls another historical conjuncture marked by a voracious descriptive appetite: that of the early 1800s in Napoleonic France when Minister of the Interior Chaptal designed a grand program of departmental statistics in order to make a portrait of the French nation at the start of a new era.[8] As every revolution, being

6 E. Monnier, "Le popolazioni italiane: nuovi studi statistici," *Annali universali di statistica* (henceforth *AUS*), 4th s., 37 (1869), p. 243. To find such a clear awareness among the statisticians themselves of the *constitutive* role of representation provides strong support for the perspective informing my analysis, a perspective whose main kernel can be found in Roger Chartier's words: "The representations of the social world themselves are the constituents of social reality. Instruments of power, the stakes of struggle as fundamental as economic struggles, the systems of classification or images of social order are all transformers of this very order": from "Intellectual History or Sociocultural History? The French Trajectories," in D. LaCapra and S. L. Kaplan, eds., *Modern European Intellectual History: Reappraisals and New Perspectives* [Ithaca, 1982], p. 41).

7 On this kind of patriotism in Lombardy–Venetia see G. Rutto, "Riforme e patriottismi nell'Austria di Maria Teresa," in A. De Maddalena, E. Rotelli, and G. Barbarisi, eds., *Economia, istituzioni e cultura in Lombardia nell'età di Maria Teresa*, vol. II, *Cultura e società* (Bologna, 1982), pp. 903–923.

8 D. Nordman and J. Revel, "La Formation de l'espace français," in A. Bourguière

a new appropriation of a territory, requires a new mapping of it,[9] this enterprise expressed the need on the part of the new French elites to bring the French territory under their gaze. During the Italian Risorgimento something similar to this descriptive appropriation of the territory took place although, necessarily, on a different scale and with a lack of unified direction. The statistical descriptions of Italy produced before unification were an overt rhetorical weapon in the hands of the Italian reformers and patriots: by responding to the "false" images produced by foreign observers and to the "useless" representations of the antiquarians, they aimed at establishing Italy's "true" picture, and in some cases also constituted a precise political statement about the legitimacy and viability of the future nation.

If patriotic statistics – that "aggressive arithmetic," to use another of Monnier's expressions – truly constituted a way of *making* a new nation, what objects made up a statistics of Italy? What was considered appropriate and included in it? What was excluded? What kind of space did statistics create? What effects and implications did they have from a cognitive standpoint? These are the questions that we will try to answer in the following pages by looking at some versions of this "scientific" building of the nation. That the Italian national space was produced statistically is an important matter. While in the case of France or Britain the representation, and identity, of the nation was predicated on a centuries-long national history, no political unity was to be found in the past of the peoples inhabiting the Italian peninsula, unless one wanted to go back to Roman times. In this context, therefore, where unity had to be created, the work of representation was bound to affect, perhaps more than elsewhere, the very identity of the new political organism.

BUILDING THE PORTRAIT OF THE ITALIAN "NATION": A CALL FROM THE CONGRESSES OF ITALIAN SCIENTISTS

In 1835, in a review of a work of local statistics, Defendente Sacchi, one of the collaborators on the *Annali universali di statistica*, urged that similar works should be written in all the localities of Italy, so that "in a few years there would be precise and reliable material for composing the *Corografia* [Chorography], the Statistics of Italy, for composing a

and J. Revel, eds., *Histoire de la France*, vol. 1, *L'Espace français* (Paris, 1989), pp. 88–89. See also D. Nordman, "L'Espace objet: le département," *Annales: économies, sociétés, civilisations* 45 (1990), pp. 445–462.

9 G. Dematteis, *Le metafore della Terra. La geografia umana tra mito e scienza* (Milan, 1985), p. 19.

work in which, by putting together all the partial elements of the nation, a portrait of the entire nation may be prepared."[10]

In this view a uniform descriptive scheme, applied to the patchwork of states, regions, and local communities that severally formed the territory of the peninsula, could eventually provide the overall description of Italy that its political division made impossible. This call for a portrait of the Italian "nation" came from an author who was not directly involved in nationalist politics, but exercised his patriotic militancy in the sphere of culture and education.[11] In 1828 the *Annali* had begun to publish a feature called "Bollettino statistico italiano" which brought together specifically numerical information about the Italian states. In 1832 G. P. Vieusseux expressed the intention to publish in the *Antologia* a monthly column presenting the "intellectual, moral, and economic progress of our Italy" and requested the collaboration of the *Annali*. The latter responded enthusiastically:

> since our experience has proven that statistical information about our peninsula can be found easily only in nearby countries where they publish a Journal of Statistics, thus we, like the compilers of the *Antologia*, should collect and publish news concerning the modes of being and doing of the populations of central and southern Italy, of which we are usually deprived, and we could for our part spread the statistical information that concerns northern Italy.[12]

In the various calls for a multiplication of local statistics in order to build a general picture of Italian society, the assumption was that the gathering of similar data for all the individual parts would eventually produce a harmonious whole. This assumption had a parallel in the imagining of the future of Italy which was prevalent in the 1830s and 1840s, namely the idea of a confederation of Italian states. A confederation would draw together political entities that shared common cultural traits and economic interests while at the same time leaving

10 Review of *Calendario lunese per l'anno 1835* and *Lo spigolatore novarese. Almanacco per l'anno 1835*, *AUS* 43 (1835), p. 7. The same ideas were expressed in a review (by an unidentified "L. Z.") of the Tuscan *Almanacco aretino per l'anno bisestile 1836*, which appeared in the *Nuovo giornale de' letterati* 31 (1835), pp. 96–101, as well as in other places.

11 On this author see E. Gabba and D. Zanetti, eds., *Defendente Sacchi: filosofo, critico, narratore* (Milan, 1992). Born in the vicinity of Pavia in 1796, he held a law degree from the university of Pavia, and, after attempting unsuccessfully an academic career in that university, spent the rest of his life collaborating with various journals and publishing houses in Milan. He died there in 1840.

12 The passage comes from *AUS* 31 (1832), p. 89 and is quoted in R. Ariotti, "Primi apporti di Carlo Cattaneo all'introduzione delle tecniche statistiche nella ricerca economica e sociale. La collaborazione agli Annali di Statistica," *Statistica* 29 (1969), p. 740.

unharmed the specific individuality of each member, which geographical and historical factors had forged throughout the centuries.[13]

The idea of collecting the same data for all the Italian states in order to make a picture of the whole of Italian society inspired the scholars, investigators, and public officials who met at the scientific conferences known as Congresses of Italian Scientists. These meetings were held in eight different Italian cities between 1839 and 1847, following a tradition first established in 1816 in Switzerland, and since then rapidly spreading elsewhere in Europe. The congresses made possible intellectual exchanges and collaborations not only among savants involved in the field of the hard sciences, but also among social researchers scattered throughout the whole peninsula.

Since, for political reasons, there was no section devoted exclusively to statistics or social science, participants with an interest in the "positive" study of society met in the section on agronomy and technology. There they presented the results of their studies and set the agenda of inquiries on various topics (from agriculture to foundlings, from savings banks to prisons, from literacy to child labor and industry) to be made in all the Italian states.[14] These investigations had several purposes. On the one hand, they were to "map the land," in both a literal and metaphorical sense. At the first congress, for example, a committee was created to coordinate the work of several local researchers for an investigation into the state of land ownership and agrarian practices.[15] Given the widespread preference of the Italian elites for a model of development primarily based on agriculture, this interest should not be surprising. Congress participants were also interested in studying the conditions of the silk-threading sector, which the supporters of the "agricultural vocation" of Italy saw as a "natural" complement of agriculture, and was at the time the engine of industrial development in the northern regions.[16]

13 For an account of major politico-ideological developments in these decades see S. J. Woolf, *Il Risorgimento italiano*, vol. II, *Dalla Restaurazione all'Unità* (Turin, 1981), pp. 415–527.

14 On the history of these meetings see F. Bartoccini and S. Verdini, *Sui congressi degli scienziati* (Rome, 1952). For an analysis of their ideological components see G. C. Marino, *La formazione dello spirito borghese in Italia* (Florence, 1974). For an examination of their more properly scientific contribution see G. Pancaldi, ed., *I congressi degli scienziati italiani nell'età del positivismo* (Bologna, 1983).

15 *Atti della Prima Riunione degli scienziati italiani tenuta in Pisa nell'ottobre del 1839* (Pisa, 1840), pp. 35–39.

16 A permanent committee on the silk industry was formed to coordinate investigations by the individual researchers: see *Atti della Settima Adunanza degli scienziati italiani tenuta in Napoli dal 20 di settembre al 5 di ottobre del 1845* (Naples, 1846), pp. 544–545. On the role of the silk sector in the context of nineteenth-century

Besides mapping the land, the proposed investigations aimed at collecting a "positive" knowledge of the "facts" in order to propose appropriate intervention. For example, one of the participants at the congress held in Turin in 1840, the Piedmontese reformer Lorenzo Valerio, called attention to the "moral" conditions of the workers in the silk industry.[17] At a subsequent meeting, he proposed the compilation of a "general statistics, if not of all Italian industry, at least of all the most important Italian manufacturing plants, aiming less at its value than at the economic and moral state of the *industrianti* [industry workers]."[18] The state councilor of the Kingdom of Sardinia, Ilarione Petitti di Roreto, was also an active proponent of an inquiry into the conditions of children working in factories. As was underlined in a session of the Fourth Congress in 1842 (in which the proposal was first discussed), the inquiry was to establish whether this phenomenon represented a real problem in Italy, as it did elsewhere in Europe.[19]

In these projects it is evident how the reference model was the new social statistical investigation into the dimensions of industrial and urban growth which had developed in France and Britain in the 1820s and 1830s. For example, both Valerio and Petitti knew well the studies that Louis-René Villermé was at that time pursuing in France into the question of children's work in the new factories,[20] and those on the

Italian industrialization see L. Cafagna, *Dualismo e sviluppo nella storia d'Italia* (Venice, 1989), especially the essay on "Il modello di sviluppo dell'ultimo Cavour e l'opposizione dei primi interessi industriali," pp. 223–255.

17 *Atti della Seconda Riunione degli scienziati italiani tenuta in Torino nel settembre del 1840* (Turin, 1841), p. 295.

18 See G. Sacchi, "Progetto di una statistica generale dell'industria italiana in relazione al miglior essere della popolazione operaia," *AUS* 71 (1842), p. 187. The expression "industry workers" does not render the sense of the awkward and archaic sounding *industrianti*, a word (destined to disappear quite soon) coined to refer to a reality which in Italy was barely incipient.

19 *Atti della Quarta Riunione degli scienziati italiani tenuta in Padova nel settembre del 1842* (Padua, 1843), p. 502. Two years earlier Petitti had published "Del lavoro dei fanciulli nelle manifatture. Dissertazione," *Memorie della reale Accademia delle scienze di Torino*, 2nd s., 3 (1841), pp. 209–306.

20 Valerio explicitly recalled the example of the French hygienist Villermé when he proposed a general statistics of industry: "let me express a wish, namely that soon a scientific association will rise in Italy, an association of friends of the country, and that it will do what the French Institute did recently with the mission conferred on Doctor Villermé. The question of industry becomes increasingly important every day and acquires the size of a giant in modern society, therefore there is an urgent need to let in the light of investigation": G. Sacchi, "Progetto di una statistica generale dell'industria italiana," p. 188. On Valerio see G. Bravo, *Torino operaia. Mondo del lavoro e idee sociali nell'età di Carlo Alberto* (Turin, 1969), pp. 268–289. On Villermé see W. Coleman, *Death is a Social Disease. Public Health and Political Economy in Early Industrial France* (Madison, 1982).

question of public health that were to lead to Chadwick's report on the sanitary condition of Britain's laboring classes in 1842.[21] Some medical doctors too tried to persuade their colleagues of the validity of the statistical method in their practice. A physician from Milan, Giuseppe Ferrario, engaged in a battle to build a "national clinical statistics" based on the adoption of a uniform system of data collection.[22] Italian researchers intended to bring Italy not only economically, but also scientifically closer to the centers of economic and cultural development in Europe. This is evident also in the invitation they sent in 1839 to Adolphe Quételet, then one of the most famed European savants, to participate in their first congress.[23]

A sentiment of national pride and a desire to bring about economic and social change so that the Italian states would not miss the "train of civilization"[24] informed many of the plans of the congress participants, as is apparent in the words of the secretary of the agriculture and technology section at the Fourth Congress, Bonaiuto Paris Sanguinetti, who thus commented on the projected statistics of Italian agriculture:

> A statistics aiming at measuring the forces and the resources of our agriculture will greatly benefit the whole community. It will become an element for the compilation of an agronomic map of Italy, which unfortunately we still lack. Therefore let us all make all possible efforts to propagate the kinds of knowledge which honor the Peninsula; the best way to make oneself respected is to make oneself known![25]

Although not possessing that unity of patriotic sentiments which an old hagiographic tradition of Risorgimento historiography attributed to them, nor having that instrumental role in the making of an organic Italian intelligentsia on which G. C. Marino has insisted at length,[26] the congresses had a remarkable symbolic value as reified instances of a possible confederation, of a combination of parts into a harmonious whole, the "unity of Italian science" to which many of their participants ardently aspired.[27]

21 E. Chadwick, *Report on the Sanitary Condition of the Labouring Population of Great Britain 1842*, ed. with an introduction by M. W. Flinn (Edinburgh, 1965).

22 *Atti della Prima Riunione*, pp. 227ff.

23 On this episode see M. Battistini, "Le relazioni di Adolfo Quetelet con i dotti italiani," *Rivista delle scienze mediche e naturali di Firenze* 20 (1929), pp. 115–127, and "Documenti italiani nel Belgio. La corrispondenza del conte Ilarione Petitti di Roreto con Adolfo Quetelet," *Rassegna storica del Risorgimento* 23 (1936), pp. 1024–1080.

24 *Atti della Quarta Riunione*, p. 560.

25 Ibid., p. 552.

26 *La formazione dello spirito borghese.*

27 The expression is in *Atti della Quinta Unione degli scienziati italiani tenuta in Lucca nel*

Virtually all the investigative projects concocted and proposed at the congresses remained, however, at the initial stage and did not achieve the desired outcome. They stimulated the activity of individual researchers, but did not produce the complete and uniform statistics of Italian society they called for. Still more decisive in those years in drawing a picture of the nation – of the national society, that is, since there was not yet a national body politic – was the individual initiative of scholars and researchers.

STATISTICAL CHOROGRAPHY: A STATIC REPRESENTATION OF ITALY

Since the early 1830s, responding to the demand – coming from the ranks of a patriotic readership – for a comprehensive description of Italy, several works appeared which offered an overall picture of the peninsula. In 1832, a Milanese scholar, G. B. Rampoldi, published a *Corografia dell'Italia*, with places described in alphabetical order.[28] Between 1835 and 1845 the erudite Tuscan doctor and member of the prestigious Accademia dei Georgofili, Attilio Zuccagni Orlandini, published in Florence the *Corografia fisica, storica e statistica dell'Italia e delle sue isole*. In 1845 the newly founded Ufficio di corrispondenza geografica (Office of Geographic Correspondence) of Bologna released an *Annuario geografico italiano* which, according to the editor, was intended to stimulate fruitful exchange and collaboration among Italian scholars for the development of geographical studies and the making of the "much desired general description of Italy."[29] In the same year, Adriano Balbi collected and published his several geographical and statistical essays on Italy;[30] a few years later appeared a description of Italy by the geographer F. C. Marmocchi, which was to begin a larger series of volumes comprising a "popular Italian chorography;"[31] another

settembre del 1843 (Lucca, 1844), p. 496, and was uttered by a certain Maiocchi, professor of physics from Milan and compiler of the *Annali di fisica chimica e matematica.*

28 3 vols. (Milan, 1832).

29 *Annuario geografico italiano* (Bologna, 1845), p. x.

30 *Miscellanea italiana. Ragionamenti di geografia e statistica patria. Raccolti e ordinati da Eugenio Balbi* (Milan, 1845). In the introduction to this work (which is dedicated to the Seventh Congress of Scientists) "the author and the compiler" claim that for a long time they had been "thinking about a *Description of Italy in Its Natural Boundaries*, which for several reasons [they] were unable so far to complete. Therefore in these reasonings on the geography and statistics of the fatherland they intend to offer almost a harbinger of that work" (p. v).

31 *Geografia d'Italia. Libri due. Nel primo de' quali si discute della geografia fisica e nel secondo della geografia politica e storica d'Italia da' più remoti tempi a' dì nostri* (Italia [but Bastia], 1850), and *Geografia politica, storica, etnografica e statistica d'Italia da' più remoti*

Corografia d'Italia by the Milanese publicist Massimo Fabi was published in 1854.[32]

Among these works, Zuccagni Orlandini's *Corografia fisica, storica e statistica dell'Italia e delle sue isole* occupied a respected position, and not only because of its imposing size.[33] In it the author – who, as we have seen, would become in 1847 the director of the statistical office of the Grand Duchy – extended the chorographical and statistical investigation he had previously carried out on Tuscany to the whole Italian peninsula.[34] The work on Tuscany, he claimed, grew out of his love for the Tuscan "fatherland," while his wide travels throughout Italy and the lack of any comprehensive description of the whole peninsula made him decide to go on to the larger enterprise despite the many difficulties that such a task entailed.[35] Perhaps less high-minded motivations also played their role: namely the rivalry between Zuccagni Orlandini and a fellow Tuscan scholar, Emanuele Repetti, who, like Zuccagni Orlandini, was a member of the Accademia dei Georgofili. Repetti's recently begun geographical and statistical dictionary of Tuscany was rapidly gaining a reputation as a thorough and accurate work, thus threatening to overtake the reputation of Zuccagni Orlandini's own work on Tuscany.[36]

Zuccagni Orlandini's *Corografia* constitutes a paradigmatic example of an approach to the description of the country which took the form of an assemblage of geo-historical–statistical descriptions of all the states (one by one) existing within the "natural boundaries" of Italy,[37] including all those areas under "foreign domination," such as Corsica, Malta, Monaco, besides of course the Kingdom of Lombardy–Venetia and other Italian-speaking areas which were part of the Austrian Empire. At the same time, this work embodies a mode of representa-

tempi a' dì nostri (Italia [but Bastia], 1851); the quotation is from the latter, p. iv. The same author had already published the *Prodromo della storia naturale generale e comparata d'Italia* (Florence, 1844), which dealt exclusively with physical geography.

32 *Corografia d'Italia. Gran dizionario storico-geografico-statistico delle città, borghi, villaggi, castelli, fiumi, monti ecc. della Penisola* (Mantua, 1854).

33 Twelve volumes published in Florence between 1835 and 1845.

34 See his *Atlante geografico, fisico e storico del Granducato di Toscana* (Florence, 1832). On this author see ch. 4.

35 *Corografia*, vol. 1, p. 174.

36 On this rivalry see A. Maccioni Anguillesi, "Un geografo dell'Ottocento: Attilio Zuccagni Orlandini," *Rivista geografica italiana* 55 (1948), pp. 104–105. This essay provides biographical information.

37 Zuccagni Orlandini shared with A. Balbi a generous, so to speak, interpretation of the extent of these natural boundaries; see Balbi's "Sunto topografico dell'Italia," in *Miscellanea italiana*, pp. 9–34.

tion which may be called transitional between an older, antiquarian/erudite tradition of description and the more matter-of-fact statistical descriptions that became popular in the first half of the nineteenth century.[38] To be sure, descriptive changes had already begun to emerge in the late eighteenth century within the erudite tradition, in particular a widening of the subject matter with an increasing inclusion of information on economic and scientific topics, forming a specifically "chorographic" tradition of description.[39] These transformations converged with the changes taking place at the same time in the practice of geographers. The latter, too, as we mentioned earlier, began at this time to include numerical information (the Venetian Adriano Balbi, whose words are cited as the epigraph to this chapter, was one of the main representatives of this type of statistical geography).

In the absence of clear boundaries between statistics and geography, reviewers and commentators classified these hybrid works according to their taste and ideological convictions: usually in the category of statistics, but sometimes in that of geography. It is not accidental that in Zuccagni Orlandini's work the adjective "statistical" followed "physical" and "historical." As the author explains in the volume that summarizes information on Italy as a whole (the last to appear, but, significantly, numbered the first in the whole collection), the primacy of the physical and historical treatment derived from the need to establish the intrinsic unity of the object described: "Physics [i.e. the description of the territory] and history, very sublime sciences and invulnerable to the violence of the strongest, will give us indisputable rights to deal with Italy as an undivided state."[40] In spite of these intentions, Zuccagni Orlandini's work followed conventional practice subordinating the organization of the description to existing state boundaries.

Relying mainly on published materials (in spite of the author's frequent reference to his direct search for new information which he tried to obtain by sending questionnaires to the governments of the Italian states),[41] Zuccagni Orlandini's work appears at first scarcely in-

38 Examples of this older erudite tradition of description are F. Ughelli, *Italia sacra* (Venice, 1722); G. A. Zannetti, *Nuova raccolta delle zecche e delle monete d'Italia* (Bologna, 1745); G. R. Carli, *Antichità italiche* (Milan, 1795).

39 See R. Bordone, "Spunti archeologici nelle descrizioni erudite fra Sette e Ottocento," in V. Fumagalli and G. Rossetti, eds., *Medioevo rurale. Sulle tracce della civiltà contadina* (Bologna, 1981), pp. 139–154.

40 *Corografia*, vol. 1, p. vi.

41 A few times he had some success, significantly in the case of Piedmont, where the Minister of the Interior told the intendants to fill in Zuccagni Orlandini's very detailed questionnaire. Reference to this is in *Corografia*, vol. 1, pp. 177–179; copies of the returned questionnaire can be seen at the Archivio di Stato, Turin, Intendenza

novative *vis-à-vis* the large numbers of erudite descriptions of Italy or parts of it – namely those dealing with its history, fauna and flora, literature, antiquities, illustrious men, and so on – on which it relies. The long sections describing the physical configuration and the historical vicissitudes of each state necessarily summarized existing knowledge. It was not always easy to distinguish the *Corografia* from those works that Balbi accused of being entirely turned towards the Italian past, unable to reveal any positive sign of progress.

Yet, Zuccagni Orlandini's *Corografia* was not merely an assemblage of old and rehashed erudition. In fact it contained elements that did not appear in his erudite sources, and that his contemporaries would classify as "useful knowledge," namely detailed information on government institutions and the administrative subdivisions of each state, as well as their population and economic activities, all components of what Zuccagni Orlandini called "statistical chorography." As in most works of this kind, a qualitative mode of description predominated. This meant again listing things in a mainly expository prose form. Numbers were used for illustrative purposes and, as Zuccagni Orlandini maintained, only when they could be relied upon, in order to avoid the illusory sense of accuracy communicated by those "strikingly handsome tables"[42] which filled out many contemporary works. Since Zuccagni Orlandini mistrusted in particular the figures then available on "industry" (a term used in its traditional meaning to indicate any kind of economic activity), and was not able to obtain many new ones from his informants, he preferred to use adverbial expressions or quantitative adjectives – such as "much," "little," "few," "many" – in his effort to give an accurate depiction of the state of things.

Needless to say, Zuccagni Orlandini's *Corografia* paid due homage to the teaching of the two "masters" of statistics in Italy, Gioia and Romagnosi: the primacy of topography satisfied Gioia's prescriptions, while Romagnosi's emphasis on history and the institutions of government was reflected in the long sections on these subjects which followed the description of each state's physical configuration.[43] Yet if

di Susa, no. 174. More often Zuccagni Orlandini had to rely on the existing literature from which he claims he excluded all "doubtful" figures (*Corografia*, vol. I, p. 157).

42 *Corografia*, vol. I, p. 157. The principal objects of his polemic were Serristori's *Statistica dell'Italia* (on which see the next section of this chapter), and a work by the British J. Bowring, *Statistica della Toscana, di Lucca, degli Stati Pontifici e Lombardo-Veneti e specialmente delle loro relazioni commerciali*, n. p., n. d. [London, 1838]: see *Corografia*, vol. I, pp. 155–156.

43 Zuccagni Orlandini refers to Gioia in *Corografia*, vol. IV, p. 6, and to the "immortal Romagnosi" in the introduction to vol. I, p. 155.

Gioia had seen Zuccagni Orlandini's work, he would probably have complained about it – and not only because of his notoriously bad temper. While Gioia had insisted on the investigation of the *relations* among the phenomena described, on the determination of "influences" and "causes" which linked different phenomena, Zuccagni Orlandini operated more within the logic of the collector. As in this logic the territory was conceived as a repository of things, Zuccagni Orlandini's description took the form of a linear and well-ordered inventory: topography amounted to a list of the mountains, plains, rivers and lakes, coasts, minerals, plants and animals, and a description of the climate, as well as of the characteristics of the inhabitants (considered naturalistically as part of the landscape) enclosed in a given administrative unit.[44] The relation between topographic inventory and the rest was lost, relations of influence and causality were difficult to see, and thus what was considered a major merit of statistics, that of making a *colpo d'occhio* possible, was missing. Moreover, not content with giving general depictions of whole states, Zuccagni Orlandini entered their minute subdivisions for thousands of pages.[45]

Statistical information was thus weighed down by erudite description, and smothered – to use the words of a reviewer – by an "unappeasable curiosity."[46] The result was a ponderous text, which did not always favorably impress the public and the critics. As the geographer and economist Gerolamo Boccardo observed some years after the publication of the *Corografia*, Zuccagni Orlandini's work had not received the attention it deserved because of its huge size and its high cost, which ran against the new demand of the market, increasingly oriented as it was towards more convenient and less expensive texts.[47] Significantly, six years after the appearance of the last volume of the *Corografia*, a work summarizing Zuccagni Orlandini's several thousand pages in a mere thousand appeared in the Florentine collection *Biblioteca dell'italiano*. Under the title of *Geografia politica dell'Italia*, it gave proportionally more space to numerical information, the best indicator, according to the author, of the "prosperity or lack of strength of peoples."[48]

[44] D. Nordman ("L'Espace objet," p. 449) speaks of a topographical representation "by places, lines, figures" with reference to French departmental statistics.

[45] The case in point is represented by the Kingdom of Sardinia, whose statistical chorography amounted to 2,700 pages, followed by the Kingdom of the Two Sicilies with "only" a thousand pages of statistical information.

[46] Review (signed C. M.) in *Biblioteca italiana* (February 1837), pp. 153–172.

[47] "Degli studi geografici e del loro stato presente in Italia," *Archivio storico italiano*, n. s., 5 (1857), pp. 72–73.

[48] M. Bianchi, *Geografia politica dell'Italia* (Florence, 1851), p. 8.

In spite of Zuccagni Orlandini's patriotic intentions, the *Corografia* appeared as a static picture of the existing order. The heritage of the past lay heavily on the long sections minutely describing the administrative partitions of the existing states, which certainly could not but appear to some readers as information soon to be made obsolete by an impending new order. Then there were the hundreds of pages of historical narratives for each state which seemed to defeat the purpose of the work as a whole – "to deal with Italy as an undivided state" – by reifying instead historic divisions and consequently localism. Zuccagni Orlandini's predominantly expository mode of description and the endless repetitions certainly did not favor a perception of the country as a whole, nor allow for any easy comparison among the described states. After thousands of pages of local descriptions, the volume on Italy as a single unit – containing a summary of geographical data, a brief history of the peninsula, and statistical information on industry – appeared as a rather stiff and perfunctory supplement, not up to the symbolic function it was supposed to perform.

The static quality of Zuccagni Orlandini's description resulted not only from his subjection to existing political and administrative boundaries, but also from the actual descriptive conventions within which he operated, namely the subordination of the material to the primacy of topography, the inclusion of historical narratives and the mainly verbal style of presentation. A more dynamic quality in description could be acquired by abandoning these conventions: this is what appears in a work published around the same time, the *Statistica dell'Italia*, by another Tuscan scholar, Luigi Serristori.[49] In this work the language of numbers acquired full dominance and accompanied the disappearance of topography as a controlling form.

THE DYNAMICS OF NUMERICAL COMPARISON

Luigi Serristori (1793–1857), the son of a prominent aristocratic family from Florence, graduated in mathematics at the University of Pisa with a thesis on steam engines. A firm believer in the great promise of technology, he contributed to the short-lived liberal journal *Conciliatore* in the late 1810s, and then after a period spent abroad first in Russia as

49 The full title is *Statistica dell'Italia contenente le statistiche del Regno di Sardegna, Principato di Monaco, Isola di Corsica, Ducato di Parma, Modena e Lucca, Repubblica di S. Marino, Granducato di Toscana, Stati Pontifici, Regno delle Due Sicilie, commercio dell'Italia con alcune delle principali nazioni* (Florence, 1835–1839, in installments). This was an extended version of his brief *Saggio statistico dell'Italia* (Vienna, 1833). A new and revised edition of the *Statistica* was published in Florence in 1842.

officer in the army of the Emperor Alexander I, and then in Austria, he returned to Florence in 1833. There he joined the Accademia dei Georgofili and began a long-lasting collaboration with the *Annali universali di statistica* and the *Giornale agrario toscano*.[50]

Like Zuccagni Orlandini's *Corografia*, Serristori's *Statistica* distributed its information separately for each one of the Italian states as well as for areas belonging to other countries and yet included within what he and the geographers of his time generally understood as the geographical boundaries of the peninsula (e.g. the island of Corsica). While in an earlier and shorter version of this work, the *Saggio statistico dell'Italia*,[51] he had included Lombardy–Venetia, the *Statistica dell'Italia* did not include a distinct section of material on these provinces of the Habsburg Empire although data referring to them were often used for comparative purposes. This exclusion was clearly motivated by political discretion since the work was published by the official publisher of the Grand Duke of Tuscany (the Stamperia Granducale), which certainly did not intend to displease the Habsburg dynasty. Needless to say, this glaring omission attracted consistent criticism among readers and reviewers.[52]

In contrast to Zuccagni Orlandini's verbose style, Serristori allowed himself only very limited room for comments concerning the reported facts and numbers, which he exhibited in an abundance of tables. And even more in contrast to the *Corografia*, the *Statistica* avoided almost any reference to topography, reducing it to measurements of area. In this it resembled the work of the Belgian J.-H. Hassel, the *Statistique de l'Europe*,[53] which summarized in long, narrow columns the numerical information considered essential for knowing the state of a country. There the numbers for the Italian states (starting with the

50 See a brief biography by A. Sapori, *Luigi Serristori* (Florence, 1925). His activities are also mentioned by U. Carpi, *Letteratura e società nella Toscana del Risorgimento. Gli intellettuali dell'"Antologia"* (Bari, 1974), and by R. P. Coppino, *Il Granducato di Toscana dagli anni "francesi" all'Unità* (Turin, 1993). Serristori participated also in the process of moderate liberal reform and then revolution which took place in Tuscany in 1847–1848, but left politics in 1849. Some accused him of having supported the restoration of the Grand Duke.

51 (Vienna, 1833). The following year he updated his work by publishing a *Primo supplemento al Saggio statistico dell'Italia* (Vienna, 1834).

52 See for example M. Montezemolo's review in *Giornale agrario toscano* 14 (1840), p. 60, which also lamented the "silence observed on the island of Malta." See also the review in *AUS* 78 (1843), pp. 3–5: the reviewer (signed F. L., namely Francesco Lampato) invites Serristori to add Lombardy–Venetia in a new edition of his work since Balbi described it as within the Austrian Empire while in fact it was "entirely located ... in the Peninsula."

53 Brussels, 1827. Serristori praised this author in his *Saggio statistico*, p. 6.

Kingdom of the Two Sicilies) all fitted in three pages: areas, populations, number of houses and cities, armed forces, finances. Economy of space was necessary for this kind of reference work.

In Serristori's *Statistica* the description of each state was almost entirely numerical. It opened with figures on the so-called *ramo ecclesiastico* (ecclesiastical branch), followed by the administrative branch (which included population), then the military, the judicial, the educational, the industrial (which included agriculture), and finally the commercial "branches." There was, however, little consistency from one state to the other, a trait picked up by some reviewers who also criticized Serristori for the lack of order in the sequence of states: they were not presented in a geographical order, nor "by analogy of political, industrial, or commercial condition."[54] Sensitive to these criticisms, in the second edition of the work he tried to meet them by presenting the states in a roughly north-to-south sequence and opening all the sections with data on territorial size and population.

The kind and quantity of numbers displayed for each state varied remarkably. Clearly the availability of data determined the space devoted to each state; for example in the case of the Grand Duchy of Tuscany the large amount of population statistics reflected the efficient system of demographic monitoring developed in that state which had resulted in a great availability of this kind of data.[55] The sheer display of data and figures was made sometimes at the expense of their direct comparability. However, behind the apparent subjection to what was available, Serristori pursued a clear argument and deployed the figures to support it.

As he explained in the introduction to his work, the Florentine aristocrat intended to establish the "relative degree of present civilization" in the several states of Italy.[56] To this purpose he selected a few quantitative indicators to represent each state.[57] To rank the Italian states in their order of material prosperity (a fundamental component of "civilization"), the most telling "symptom" was to be found in population density, an outcome of such factors as property relations in the coun-

[54] Montezemolo, p. 61.

[55] Population statistics on Tuscany occupied eighteen pages, as against, for example, the two pages devoted to demographic figures on Piedmont.

[56] *Statistica* (1842 edn.), pp. 3–5.

[57] He never mentions the sources of his figures, claiming this would weigh his work down (and this was another element about which his critics complained). The data seem to come mostly from published sources, although we cannot exclude the possibility that he tried to get information directly from state authorities (he complains about his unanswered requests in the case of the Kingdom of the Two Sicilies: see *Statistica* [1842 edn.], p. 253).

tryside, the circulation and availability of land, the conditions of communications, and the ease of commerce. The ranking of the Italian states by this measure showed that the Duchy of Lucca occupied the top position, and the island of Sardinia the bottom one. Without excluding the role of "physical" causes, Serristori made a point of stressing that they were not pertinent to his concerns.[58] The prosperity of Lucca was primarily related to the diffusion of small landownership. Similarly, the poverty of certain areas, although not independent of their topographical configuration, depended really on variables of a social and political nature. These were the factors – factors which could be modified by human intervention – upon which Serristori focused in order to plead the causes that he cared about, such as the elimination of customs barriers among the Italian states and the improvement of elementary education.[59]

As for the degree of culture, Serristori provided – following a common practice – the numbers of schools and students per inhabitants. He regretted not being able to compare levels of "morality," since its indicators (the ratio of crimes and criminals to the population) were rarely available. Instead he put a certain emphasis on the ratio between dioceses and population so as to show the "density" and income of the clergy, an indicator whose meaning was not explained, but left to the sagacity of the reader.[60] The indicators chosen by Serristori allowed him in particular to plead the cause of the extension of public education, which he considered an essential requisite for the economic progress of any country. The model situation in this respect was represented by the Kingdom of Lombardy–Venetia, which although not the object of a specific treatment was brought in for comparative purposes; in comparison to Lombardy–Venetia almost no state for which data were available could exhibit a good record; the worst educational condition was exhibited by the Kingdom of the Two Sicilies, whose populations were "*among the less advanced of the peninsula* [original emphasis]."[61] It is worth reflecting for a moment on this last statement by Serristori. While perceptions about the condition of southern Italy had certainly been negative in previous times (they abounded in the travelers' literature), the first half of the nineteenth

58 Ibid., p. 3.

59 Ibid., pp. 6–7.

60 Serristori considered the proportion of clergy to general population to be an indicator of *incivilimento* (the higher the proportion, the lower the degree of civilization) and observed how Italy was surpassed only by Spain in this respect: see *Primo supplemento*, p. 24. The polemic against the clergy is of course a recurrent theme in Italian patriotic discourse.

61 *Statistica*, p. 252.

century witnessed an important change. The new indicators devised by statisticians such as Serristori to compare and rank states translated these perceptions into hard, "positive" evidence, endowed with an authority superior to that of travelers' descriptions. They made differences immediately defined, visible at a glance.

The widespread use of percentages, ratios, and rank ordering gave a distinct character to Serristori's work. His comparative purpose and the systematic search for the "causes" capable of "influencing" the state of a country distinguished it not only from descriptions such as Zuccagni Orlandini's *Corografia*, but also from later works which mainly aimed at counting and displaying the resources of the future nation. But the search for causes and the comparative aim were linked also to a specific feature of the *Statistica* that made it stand out in the midst of chorographical works, namely the absence of topographical information. It would be wrong to think this feature of Serristori's work was the necessary outcome of the larger scope of his statistics. It is true that, in contrast to a local statistics that was often the result of a *direct*, visual knowledge of places and things (a knowledge obtained for example by notables, doctors, naturalists through actual travel within the territory of a province), the statistics of larger units required a more abstract procedure. But the work of Zuccagni Orlandini is there to show that the larger scope of an investigation did not necessarily imply that topography had to be eliminated from a description. Yet while Zuccagni Orlandini's work was a genre destined to become obsolete, Serristori's was the way forward.

We might recall that around the time of the publication of Serristori's *Statistica* a French author we have already met, P.-A. Dufau, who thought that statistics should deal exclusively with "numerical facts," gave a theoretical justification to the "deterritorialization" of statistics.[62] He claimed that it was not the territory which produced the population, as topographic statistics implied, but it was the population which by appropriating and transforming the territory through human labor actively produced it itself.[63] What we witness here is actually part of a more general trend occurring in the nineteenth century: the subordination, if not abandonment, of physical space in empirical social science and social theory. Beginning in the early nineteenth century liberal political economists denied space any conditioning role in the

[62] I am borrowing a term used by A. Desrosières and quoted by Nordman and Revel, "La Formation de l'espace," p. 95.

[63] *Traité de statistique ou théorie de l'étude des lois d'après lesquelles se développent les faits sociaux; suivi d'un Essai de statistique physique et morale de la population française* (Paris, 1840), p. 93.

process of economic development (and this in spite of the emphasis that many put on a development that ought to follow the "natural vocation" of a country). Then space disappeared from social theory altogether – although at different times and to a different extent in different countries – for reasons that are very complex and that still await full investigation. The rise of historicism in the early part of the century has been considered a major intellectual component of this trend.[64] In light of the history of statistics, it seems very likely that a process of reciprocal reinforcement took place. On the one hand, the assumption of a uniform space[65] – namely a space deprived of any physical-historical characterization – began to inform discourses on society and the practice of representation. On the other hand, the very increase in the employment of numbers in descriptive practices contributed to bring about a "dot-like world," to use Pierre Dockès's expression.[66]

Although addressing the question of the difference among states, and therefore of a difference taking place *in space*, numerical statistics reduced the territory to a simple numerical expression (its area) which served essentially to produce other numerical indicators, such as the much valued ratio of population to area, namely "population density." Serristori's deterritorialized statistics accompanied rather well the socio-political agenda of the author, who at the time was directly engaged in a plan to promote a customs union among the Italian states following the model of the German *Zollverein*.[67] As Serristori's comparison of the degree of civilization among the Italian states clearly aimed at an indictment of those practices which hindered the improvement of their economies, the "immutable" elements of the picture – such as the physical landscape – lost their

64 E. W. Soja, *Postmodern Geographies. The Reassertion of Space in Critical Social Theory* (London and New York, 1984), pp. 31–35.

65 According to Nordman and Revel, "La Formation de l'espace," p. 95, this was a necessary assumption, or, better, a condition of possibility, of entirely numerical statistics.

66 The expression is from *L'Espace dans la pensée économique du XVIe au XVIIIe siècle* (Paris, 1969), p. 9.

67 He published two articles on this subject in the *AUS*: "Sulla possibilità di un'unione doganale tra gli stati italiani," *AUS* 75 (1843) pp. 292–300; and "Di una unione doganale tra gli stati italiani," *AUS* 78 (1843), pp. 197–202. On these plans see C. Rainone, "Liberisti e Liberali: pensiero economico e pensiero politico in Italia avanti il 1861," in *Nuove questioni di storia del Risorgimento e dell'unità d'Italia*, vol. 1, (Milan, 1976), pp. 513–564. It is interesting to note that from the time of its constitution in 1833 the German *Zollverein* began immediately to collect statistics on all its member states: see J. and M. Dupâquier, *Histoire de la démographie. La statistique de la population des origines à 1914* (Paris, 1985), pp. 262–263.

relevance. Not by accident the demise of the topographical tradition and the emergence of an abstract approach to space in the statistical descriptions of Italy were related to a reform project, namely the overcoming of custom barriers and the "equalization" of all Italian states from the standpoint of the laws of the market.[68] The more numericized statistics added to the credibility of this project. An abstract idea of space sustained the utopia of the unhindered market; only thus could the same recipe (free trade) be suggested for universal improvement.

That Serristori's *Statistica dell'Italia* challenged strongly held convictions in the practice of Italian statisticians is evident from the reactions it excited. The reviewers were critical of the order of exposition he adopted. Giuseppe Anastasio, one of the collaborators on the *Giornale di statistica* published in Palermo, charged Serristori with not knowing the proper order of presentation of the statistical objects, precisely because he had ignored, among other things, the primacy of topography.[69] Others lamented the choice of a "state by state" descriptive scheme (instead of a thematic scheme with the states intermingled),[70] a feature of Serristori's and Zuccagni Orlandini's works that was to become a major target for criticism in the 1850s: then Cesare Correnti defined the work of the Tuscan statistician as "a sequence, or rather a superimposition, of booklets bound together for the sake of the title page."[71] By that time Serristori's representation of the country was no longer satisfying, and in effect no longer patriotic enough when, after the defeat of 1848, many supporters of the national cause had moved from a federalist to a unitary stance. However, the main innovations of his work, namely the dynamics of comparison and the creation of an "abstract space" through numbers, became part of the subsequent and more unitary statistical descriptions of the country. But before moving to this new phase of the description of Italy, let us explore further the issue of the relations between conceptualizations of space and descriptive conventions.

68 There are analogies here with the work of the French statisticians of the 1820s and 1830s in which the division – drawsn for analytical purposes – of the French territory bore no relation to any geographical characteristics, and served and an ideological battleground: see R. Chartier, "Science sociale et découpage régional. Notes sur deux débats 1820–1920," *Actes de la recherche en sciences sociales* 8 (1982), pp. 27–36.

69 *Giornale di statistica* 1 (1836), pp. 60–70.

70 Montezemolo, p. 70. In fact the *Saggio* contained precisely this different approach.

71 [C. Correnti, ed.], *Annuario statistico italiano. Anno I 1857–58* (Turin and Milan, n. d. [1858]), p. 364.

THE HOLISTIC ALTERNATIVE OF REGIONAL DESCRIPTION

My previous observations on Serristori were not meant to suggest the existence of a simple relation of causality between economico-political agendas, paradigms of causation, and descriptive practices. Nor did I mean to imply that the dismissal of the geographical causality implicit in topographical statistics was derived from the embracing a free-trade agenda. In fact the relations between politics and epistemologies are always extremely complex. The Milanese Carlo Cattaneo, another convinced supporter of policies inspired by free-trade convictions and, like Serristori, a believer in technology and economic development, maintained a clearly "territorialized" approach. In his case, however, space was not the simple list of mountains, rivers, and plains to which the territory was reduced in many topographical statistics (and which it was easy to eliminate from the new statistics). The territory was conceptualized as a domain of relations able to stamp its imprint on society and history, and in its turn always shaped by their influence.

Carlo Cattaneo (Milan 1801–1869), the most illustrious student of Romagnosi, was the best representative of the Lombard intelligentsia engaged in the work of cultural modernization. Extremely versatile in his interests, he wrote extensively on diverse subjects – philosophy, history, economics, anthropology, geography, psychology, technology – always applying his vast knowledge to issues of present concern in the social and political debate. From 1828 to 1838 he contributed to the *Annali universali di statistica*, which he left in 1839 in order to found a more specialized journal (*Il politecnico*) devoted to the development of a modern scientific culture and the integration of science and technology in society. In contrast with Gioia and Romagnosi, Cattaneo did not write on the theory of statistics. Yet he was unrivaled in developing his arguments in an inductive manner with the help of statistical data.[72]

From a political standpoint, Cattaneo was the most important representative of a liberal and democratic federalism, opposed to the unitarian perspective and to the moderates' ideas of a confederation of Italian states. Cattaneo never specified exactly what he thought the basic components of a future Italian federal state should be; in fact until 1848 he was primarily concerned about the acquisition of greater

72 See Ariotti, "Primi apporti di Carlo Cattaneo all'introduzione delle tecniche statistiche," p. 751. After the revolution of 1848, Cattaneo went into exile to Switzerland, and, in contrast to other democrats, never accepted Piedmontese hegemony in the process of national unification. For a biographical profile see *Dizionario biografico degli italiani*, s. v. by E. Sestan.

autonomy for Lombardy–Venetia within the Habsburg Monarchy, and about the progress of federalism on a European scale. His perspective shared some of the geo-historical arguments for a confederation characterizing many moderate federalists, namely, that given the long history of political division on the one hand, and the geographical diversity of the peninsula on the other, political unification was not desirable.[73]

But besides this emphasis on diversity, what made his federalist perspective distinct was its methodological and ideological underpinnings. The fundamental ideological motive consisted in the idea that liberty cannot be safe within a centralized political structure. Indeed, Cattaneo's federalism fed on the tradition of municipal autonomy in Lombardy and especially on an idealized representation of the degree of municipal autonomy attained by this area under the government of Maria Teresa. This had to be maintained in order to avoid the inevitably despotic consequences of state centralization.[74]

By the "methodological" component I mean Cattaneo's strong anthropo-geographical perspective in the analysis of human societies. Specific human societies are the product of the interaction of multiple factors – among which population and natural environment play a dominant role – in specific places. These places do not coincide necessarily with the boundaries of states although states of course influence their making. Cattaneo's native Lombardy was an example of such a place, the best he could think of. To the description of Lombardy Cattaneo devoted his most famous work, the *Notizie naturali e civili sulla Lombardia*, an ambitiously conceived, but never completed, geo-statistical monograph.[75] Lombardy was an area distinct "from a natural and civil standpoint" from the rest of Italy, bearing the mark of a historical development which had transformed its southern part into the most advanced area of the plain of the Po, dominated by an agriculture which paralleled that of capitalist England.[76] The rich Lombard agri-

73 On the various components of Italian federalistic thought see an old work by A. Monti, *L'idea federalistica nel Risorgimento italiano. Saggio storico* (Bari, 1922); for a more recent contribution which covers only authors from the northern regions of the peninsula see F. Della Peruta, "La federazione nel dibattito politico risorgimentale 1814–1847," in Della Peruta, *Conservatori, liberali e democratici nel Risorgimento* (Milan, 1989), pp. 309–339.

74 This is a view stressed particularly by N. Bobbio in his essay "Stati uniti d'Italia," in Bobbio, *Una filosofia militante. Studi su Carlo Cattaneo* (Turin, 1971), pp. 3–55.

75 Milan, 1844.

76 These arguments are developed by Cattaneo in his introductory essay to the *Notizie naturali e civili sulla Lombardia*, pp. xi–cxii, now in C. Cattaneo, *Scritti storici e geografici*, ed. G. Salvemini and E. Sestan, vol. 1, (Florence, 1957), pp. 325–433.

culture of the plain was the result of the commercial wealth accumulated in the cities, and especially in the capital city, Milan;[77] although this rich capitalistic agriculture contrasted with the poorer northern areas of the hills and the mountains, these had also enjoyed the same enlightened civil legislation.[78]

Understood in these terms, as the result of a complex historical process, the region constituted an organic whole; as such, and only as such, it made sense to describe it; indeed it was the only unit that justified description. The regions of the Italian peninsula which thus deserved to be illustrated should not – or not always – overlap with the existing political units. Such was the case for example of the island of Sardinia (to which Cattaneo devoted a study that essentially followed the lines of the *Notizie naturali* on Lombardy), then a part of the Kingdom of Piedmont.[79]

If, for Cattaneo, only the regional space constituted the proper scale for the illumination of economic and social problems and the search for solutions, it is evident that description of the whole country apart from its regional partition was not even considered by him; nor was it worth the effort simply to accumulate local descriptions of towns or districts, for the sum of all those pictures would not provide a satisfactory depiction of the whole:

> Moreover such narrow and minute partitions of the country involve too many resemblances and endless repetitions. And *few are the districts that within their limited area include the principal sources of their natural and civil state* [emphasis added], making it possible to account for them in a reasoned manner without drawing information from the surrounding lands ... Instead of writing every year here and there in Italy a volume on the hundredth or three-hundredth particle of the *bel paese*, it seemed suitable resolutely to adopt [as unit of analysis] a whole region, so that it could be considered under a certain conceptual unity, such as Venetia, or Tuscany.[80]

77 This nexus has been the object of many analyses, among which perhaps the most enlightening is L. Gambi, "I valori storici dei quadri ambientali," in *Storia d'Italia*, vol. 1 (Turin, 1972), p. 34.

78 This account of the making of a region bears some analogies with the model elaborated in the 1820s by the precursor of modern "location theory," the German economist Johann H. von Thünen; however, it never gets as specific as the latter in showing the effect of a city on the usage of rural land. On von Thünen see the introduction by P. Hull to *Von Thünen's Isolated State*, (Oxford and New York, 1966), pp. xi–xliv.

79 "Della Sardegna antica e moderna," in C. Cattaneo, *Scritti storici*, pp. 189–254 (orig. publ. as "Di varie opere sulla Sardegna," *Il politecnico* 4 [1841], pp. 219–273).

80 C. Cattaneo, "Avviso al lettore," in *Notizie naturali e civili*, now in C. Cattaneo, *Scritti storici*, p. 326.

The quote comes from the preface to the *Notizie naturali e civili* and shows clearly how the link between city and countryside worked both ways: as it did not make sense to describe a province aside from the sources of its economic and civil life (usually an urban reality), it did not make sense to describe a city without the territory from which it derived its nourishment. To adopt the city as the whole unit of description resulted, in Cattaneo's view, from a narrowly antiquarian and municipalist attitude that he strongly opposed. The *Notizie naturali e civili* itself was written following a conflict which significantly pitted Cattaneo against the proponents of a quite different vision. It is worth recalling the episode.

The municipal Congregazione (the organ, partly elected by the city council, which ran the city) decided in 1843 to sponsor a publication on the city for the Sixth Congress of the Italian Scientists to be held in Milan the following year, and Cattaneo was given the task of coordinating the enterprise, assembling the contributions from various authors, and writing the final text. But after a few months he resigned citing a conflict between his own conceptions of the scientific nature of the work and those of the Congregazione and of some of the collaborators. He in fact was adamantly opposed to the recommendations of the Congregazione, which wanted the publication to be essentially a guide to the city.[81] As a result, two publications were presented to the Congress of Scientists of 1844: the official one, sponsored by the Congregazione, collected in a rather disorderly way diverse information on Milan and its immediate surroundings;[82] the one compiled under the direction of Cattaneo, the *Notizie naturali e civili sulla Lombardia*, consisted of the first two sections (defined as "inorganic and organic nature") of a proposed systematic depiction of the whole region around Milan which ought to include additional sections on population, public health and assistance, agriculture, industry, commerce, public order, language, history, legislation and culture, and

[81] The letters he addressed to the Congregazione Municipale are in Cattaneo's *Epistolario*, ed. R. Caddeo, (Florence, 1949), vol. 1, pp. 128–148, 162–172, 427–428; the letters addressed by the Congregazione to him are conserved in the Museo del Risorgimento in Milan, Carte Cattaneo, no. 30.

[82] *Milano e il suo territorio*, 2 vols. (Milan, 1844). The first volume, entitled "The People," covered the following: history, church and rituals, nomography and statistics, education, hygiene, charity, festivities, theaters and pastimes; the second volume, entitled "Nature and Art," consisted of the following sections: physical geography and geology, waterways, roads, commerce and industry, agriculture, archives, libraries, museums, collections, painting and galleries, buildings, surrounding area.

finally the city and its remarkable places.[83] Clearly no compromise had been possible.

Being founded on a notion of region defined on an essentially geographical–historical rather than political or legal basis,[84] what we could call Cattaneo's project of "regional statistics" took the territory and its description seriously. And in contrast to the static dimension that characterized topography in works such as Zuccagni Orlandini's *Corografia*, the territory in Cattaneo's holistic perspective was an inherently dynamic element: it interacted with population and history in accounting for the present conditions of a given society.[85] Space here was truly *integrated* in social analysis.

Cattaneo's project, however, was never carried out. Not even the carefully planned description of Lombardy was completed, and the Milanese scholar lamented the indifference met by his proposal for the "establishment in every region of Italy of a perpetual collection of local studies, similar to the one that we began with the title of *Notizie naturali e civili sulla Lombardia*."[86] Both political and epistemological reasons accounted for Cattaneo's lack of success. Aside from the virtual political defeat of democratic federalism in the national revolution, the notion of region remained difficult to define satisfactorily by the very people who intended to make use of it. The concept of region is notoriously vague, lacking a widely accepted definition even among geographers who range from attempting to find "natural regions" to constructing regions on the basis of purely external characteristics.[87] It is significant that Cattaneo never set out to try to chart a geography of Italian regions.

Those who did found the enterprise tricky. In an attempt to devise a regional partition of Italy that would not reflect existing state divisions,

[83] For a detailed listing of all the objects see "Prospetto d'una raccolta di notizie naturali e civili sulla Lombardia, proposta da alcuni studiosi per l'occasione del Congresso scientifico di Milano," in Cattaneo, *Scritti storici*, vol. I, pp. 309–324 (orig. publ. in *Il politecnico* 7 [1844], pp. 212–222). Several unpublished manuscripts which should have been included in vol. II of *Notizie naturali e civili* are preserved among the Carte Cattaneo, nos. 31–32 in the Museo del Risorgimento in Milan. Lack of financial support was probably the cause of the failure to publish vol. II.

[84] On the territorial dimension in Cattaneo's works see L. Ambrosoli, "Cattaneo e i problemi del territorio," *Nord e Sud* 21 (1974), pp. 80–94.

[85] Cattaneo defined the territory an "immense deposit of toils" (quoted by C. De Seta, "Città e territorio in Carlo Cattaneo," *Studi storici* 16 [1975], p. 454, from another famous essay by Cattaneo, *La città considerata come principio ideale delle istorie italiane* [Milano, 1858]).

[86] "Annuario Geografico Italiano," in Cattaneo, *Scritti storici e geografici*, p. 80 (orig. publ. in *Rivista Europea. Giornale di scienze morali, letteratura ed arti* [1845], p. 452).

[87] See T. W. Freeman, *A Hundred Years of Geography* (London, 1961), ch. 6.

another Milanese student of statistics, Cesare Correnti, concocted in the 1850s a scheme of sixteen regions, in which, however, he essentially gave new names (based on their most salient geographical traits) to historical–political units: thus Lombardy became "Transpadana," Tuscany "Val d'Arno," the Marches "Costiera Adriatica," and so on.[88] He later partially revised the list observing that:

> This partition question is ticklish and several people rebuked me on this. As far as I am concerned I think that when the geographical distinctions, articulations, and names prevail over the usual divisions of subnationalities and provincialities, it will be a great advantage.[89]

While stating that it would be necessary "in these partitions to follow the indivisible rights of the land [*inviolabile diritto della terra*] and the impartial witness of rivers and mountains," Correnti affirmed that "in order to avoid excessive complications" he had to carve out of the Italian peninsula nineteen regions which were "nearer to the present political and administrative divisions."[90] Indeed the political and administrative partitions of the existing states limited the attempts to rethink the territory of the peninsula.

As a unit of description, the "region" had to face competition both in the political arena and in collective imagery. It had to face the reality of existing states, the persistence of strong municipal identities, the ideal of national unity, and later, after the constitution of the Kingdom of Italy, the administrative existence of provinces. Yet, as we will see in Chapter 7, the statisticians of the new Italian state (Correnti among them) were eventually to be successful in making "regions" a unit of observation and analysis even more important than the other entities. But in the new national statistics regions had no longer the same rich texture that they had in Cattaneo's geo-anthropological approach.

THE PRIMACY OF POPULATION

In the 1850s the movement to make Italy *known to herself* and show that Italians as well "keep track of the progress of the sciences ... nor are [they] behind any other nation with regard to useful institutions

88 C. Correnti, "Fisionomia delle regioni italiane," in *Scritti scelti di Cesare Correnti in parte inediti e rari*, ed. T. Massarani, vol. II (Rome, 1890), pp. 372–399 (orig. publ. in *Il Nipote del Vesta Verde* 5 [1852], pp. 42–61).

89 "Ancora delle nostre regioni," in C. Correnti, *Scritti scelti*, vol. II, pp. 446–447 (orig. in *Il Nipote del Vesta Verde*, 8 [1855], pp. 144–163).

90 Ibid., p. 449. On the difficulties faced by the supporters of local autonomy (but after 1861) in defining the notion of "region" see R. Ruffilli, *La questione regionale dall'unificazione alla dittatura* (Milan, 1971), pp. 22–23.

and good intentions"[91] became more incisive. The *Annali universali di statistica* signaled this new phase by starting to publish in 1852 a column called "Cronaca statistica italiana." This was included in the section of the journal called "Bulletin of Italian and foreign statistics and of the most important inventions and discoveries, or progress of industries and useful knowledge" which first appeared in 1833. Mindful of censorship problems, the introduction to the first article published under the new heading specified that it would "contain only statistical facts and documents. Political facts are not part of our studies."[92] In the same years appeared several works called statistical and/or political-economic *annuari* (yearbooks). The choice of the word "yearbook" well expresses the purposes of this kind of texts. A yearbook purported to be, by definition, a reflection of objects in constant movement, a mirror of a rapidly changing situation, while statistical monographs of the topographical kind were based on a relatively static perception of society. In the 1850s for the intellectuals engaged in the struggle for national unification the only accurate representation of Italy was one that could be offered by a yearbook, namely a temporary one, reflecting the precariousness of the post-1848 order and the expectations that it would soon be overcome by armed action and diplomacy. Several yearbooks were explicit tools of patriotic propaganda, whose aim was "to give a body to the limbs torn apart and dispersed," to gather "an inventory of our forces, and ... to present, if not the muscles, at least the skeleton of the nation's body."[93]

For the Milanese Pietro Maestri, author of the *Annuario economico e statistico dell'Italia per l'anno 1853* – from which these words are taken – and future head of the Statistical Office of the Kingdom of Italy,[94] the

91 *AUS* 79 (1844), p. 7.

92 *AUS*, 3rd s., 30 (1852), p. 73.

93 [P. Maestri], *Annuario economico e statistico dell'Italia per l'anno 1853, contenente studi e statistiche riguardanti la popolazione d'ogni stato e provincia, le istituzioni comunali, l'agraria, l'industria, le amministrazioni politiche, la marina militare d'Italia*, (Turin, n. d. [1853]), pp. vi–vii. For obvious reasons, the name of the author does not appear on the cover, but Maestri's authorship was acknowledged by contemporary reviewers: see for example G. B. Michelini, "Degli studi economici e statistici in Italia," *Rivista contemporanea* 16 (1859), pp. 152–156. Maestri was also the author of the *Annuario economico-politico* (Turin, 1852).

94 On P. Maestri (1816–1871) there are no biographical studies. For a brief outline of his life see ISTAT, *Decennale 1926 IV – 1936 XIV* (Rome, 1936), pp. 71–74; some information also in F. Della Peruta, *I democratici e la rivoluzione italiana. Dibattiti ideali e contrasti politici all'indomani del 1848* (Milan, 1974), pp. 166–170. Born in Milan, he was the son of a civil servant employed as deputy director of a bookkeeping office in the Austrian administration; he received a degree in medicine from the university of Pavia. Initially active in the democratic milieu of the Lombard capital, he had to leave Milan after the revolution of 1848. After a brief Proudhonian inter-

skeleton of the nation's body was constituted by its population, municipal institutions, agriculture, industry, political administration, and armed forces. It was indeed a basic framework when compared to the richer proliferation of objects and categories that characterized local statistics. In the genre of "national" statistics, those objects that were immediately relevant to military confrontation and to the construction of the new state acquired a dominant place in the description. It is thus not surprising to find an increased interest in the armed forces and in state structures at a time when the supporters of the national cause were getting closer to reaching the goal of independence, and the same applies to population which now acquired a primacy among the objects of description, while the topographical introduction virtually disappeared, or was transferred elsewhere. Population above all was the foundation of power, and people were the power of a nation.[95]

Consider another significant example of this propaganda literature, the *Annuario statistico italiano* (1858) compiled by Cesare Correnti as a continuation of the series of yearbooks initiated by Maestri.[96] This former employee of the Austrian administration of Lombardy–Venetia and collaborator on the *Annali universali di statistica* had – like Maestri and many others – migrated to Piedmont after the failure of the 1848 revolution in Milan and had embraced a unitarian perspective on the national question.[97] The title of the work clearly stated its agenda: as if Italy were already a political entity, this was its first statistical yearbook (in fact the new Italian state would not begin the regular publication of its official statistical yearbook until 1878). After a long and revealing introduction with data concerning the other European countries, population occupied the opening section on Italy with a rich exhibit of figures and categories: total number of people; distribution by provinces, districts, and municipalities; distribution by sex (proudly the compiler announced that the male population exceeded the female

lude, he embraced more moderate views in the mid-1850s, when, after a period of exile in Piedmont, he moved to France.

95 Meanwhile the description of the territory became increasingly the object of a distinct, more properly geographical genre of writing.

96 The name of the author, however, does not appear on the title page.

97 In Piedmont he became a member of the so-called Commissione per i lavori statistici (Committee on statistical works) which had the purpose of coordinating the political propaganda in Lombard cities in preparation for a new insurrection in 1849 (see *Dizionario biografico degli italiani*, s. v.). Perennially oscillating between moderate and democratic political positions, Correnti was also to have, like Maestri, a primary role in the compilation of statistics in the Kingdom of Italy. See also chs. 7 and 8.

STATI		POPOLAZIONE			SUPERFICIE del territorio in chilom. quadr.	POPOLAZ. specifica o abitanti per ogni chil. quad.
		Data dell'ultimo censimento	Abitanti verificati	Presunti al 1 genn. 1857		
Due Sicilie	Continente	1 genn. 1854	6,843,355	6,986,906	79,233 00	88 18
	Sicilia	1854	2,231,020	2,294,373	25,393 50	90 35
Regno L. V.	Lombardia	1855	3,009,505	3,057,765	21,585 45	141 66
	Venezia	1855	2,493,968	2,526,606	23,881 59	105 80
Stati Sardi	Continente	1848	3,785,160	3,997,607	40,161 09	99 54
	Sardegna	1848	547,112	568,098	24,096 06	23 58
Stati Romani		1849	3,019,359	3,127,027	41,434 63	75 17
Toscana		1856	1,779,338	1,794,658	22,082 76	81 27
Modena		1855	609,139	616,883	6,019 66	102 47
Tirolo Italiano		1851	538,524	551,882	15,741 65	35 06
Trieste, Istria, Gorizia . .		1851	527,539	549,311	8,524 46	64 44
Parma		1854	508,784	514,083	6,201 13	82 90
Corsica		1852	236,251	243,982	8,746 91	27 89
Malta		1851	123,496	129,207	374 67	433 80
Ticino		1850	117,759	119,955	2,675 05	44 84
Grigioni Italiani		1850	14,506	15,037	853 91	17 61
Monaco		1848	7,627	7,915	23 15	341 90
S. Marino		1852	5,700	5,844	57 15	102 26
Totale per l'intiera regione italica			26,398,142	27,107,139	327,085 82	82 87

Figure 2 Italy unified by numbers: from [C. Correnti, ed.], *Annuario statistico italiano. Anno I 1857–58* (Turin and Milan, n.d. [1858])

which meant that "there [was] a prevalence of the most active forces"),[98] by domestic condition (married or not), by profession (again with a particular attention to the number of priests, the message being the fewer the better), by religion; number of births, deaths, marriages; average and probable duration of life; average annual growth and estimated time for the population to double; population density; number of foreigners; migration movements.[99] The section on topography did not disappear in this text, but it followed the one on population, and consisted mainly of a bibliographical essay on cartography and a proposal for a regional partition of the Italian peninsula aiming, as we read, at "studying the differences in order to find the similarities [*concordanze*]." It was in other words an attempt to posit the basis for the administrative geography of a new political entity.

If we look at nineteenth-century works which ranked states in order of importance – for example Xavier Heuschling's *Manuel de statistique*

[98] *Annuario statistico italiano. Anno I 1857–58*, p. 371. The same observation is in C. Cattaneo, "Sulla densità della popolazione in Lombardia e sulla sua relazione alle opere pubbliche," in *Scritti economici*, ed. A. Bertolino, vol. II (Florence, 1956), pp. 131–165 (orig. publ. in *Il politecnico* I [1839], pp. 29–52).

[99] *Annuario statistico italiano. Anno I 1857–58*, pp. 366–384, and 427–457. The first half (pp. 5–365) of this small but thick volume was taken up by considerations on the history and theory of statistics (with a quite Romagnosian tone), and by statistics on the states of Europe and America, included, as the author says, for comparative purposes. The data on Italy follow on pp. 366–589.

ethnographique universelle – the criterion of population appears to be dominant in spite of the acknowledgment that both territory and population are the "primary elements of every nation."[100] In claiming the necessity of a state for their nation, Italian patriots resorted to an argument that could be heard and fully understood by European liberals of their time, what I would call the argument of "large numbers." As Eric Hobsbawm has observed, in the first half of the century size and population were the most important criteria for establishing the legitimacy of a claim to nationality: according to the *New English Dictionary* a nation was an "*extended* aggregate of people,"[101] a definition we find in almost the same terms in Italian publications of the time such as the *Nuova Enciclopedia Popolare*:[102] "a nationality is constituted when in a *large aggregate* of people the ideas, the material and moral interests, and principally the purpose of their activity are almost the same [emphasis added]." This latter quote echoes the voluntarist conceptions of that Italian "apostle" of democratic nationalism, Giuseppe Mazzini, who did not think that in Europe there was room for more than twelve nations.[103] Let us recall that once unified in 1870, the Kingdom of Italy acquired a very respectable position: it was the fifth among European states in terms of inhabitants (almost 27 million).

Moreover, in the absence of other signs of progress, another demographic indicator, that of population growth, could offer some consolation. Serristori, certain that the "natural increase of the absolute population of a state is always a sign of progressive civilization," claimed that his figures disproved the "generally established opinion that population growth in Italy is inferior to most other European states."[104] Pietro Maestri, in his *Annuario economico e statistico dell'Italia per l'anno 1853*, took pains to measure the average demographic increase for each of the Italian states and for Italy as a whole, and found out that the latter was at least higher than that of France, but lower than those of Belgium, England, and the United States.[105]

The advantage of relative numbers such as the rate of population growth is that they offered an assessment of a country's potential

[100] Brussels, 1847. The quote is from p. 9. The classification of states by importance is on pp. 37–40. Heuschling was the secretary of the Commission central de statistique of Belgium.

[101] See E. J. Hobsbawm, *Nations and Nationalism since 1780. Programme, Myth, Reality* (Cambridge, 1990), p. 30. The emphasis is added by the Hobsbawm.

[102] Turin, 1847.

[103] Hobsbawm, *Nations and Nationalism*, pp. 31–32.

[104] L. Serristori, *Saggio statistico dell'Italia*, p. 9.

[105] Page 99.

independently of its size. Cesare Correnti, for example, in his polemic against Austrian domination in Lombardy–Venetia warned against those "crude and materialist statistics which make a confusion between strength and corpulence" and went on arguing about the viability of a country formed by the union of Piedmont-Sardinia and Lombardy–Venetia. He then offered data on the Kingdom of Sardinia and Austria showing that in spite of the smaller population of the former, it was sounder from a financial standpoint and better equipped militarily.[106]

Thus the great stress on "population density" that we find in all these authors becomes fully understandable. Serristori already used this relative number as a primary indicator of prosperity to rank the Italian states on the scale of "civilization." And in his very well-known essay "Sulla densità della popolazione in Lombardia" published in *Il politecnico* in 1839, Carlo Cattaneo stressed that what really counted as an indicator of prosperity was not the absolute number of people so much as the ratio of people to the area they occupied. Population *density* was the real "representation of *civilization* [original emphasis]," which went beyond the mere "material strength" of a nation.[107] "Making possible in a given area the prosperity and civilized existence of the maximum number of living beings" was, in Cattaneo's view, the "ultimate problem of every nation."[108] Pietro Maestri did not hesitate to call this ratio "the most evident and certain measure of the good and evil of universal conditions."[109] Demographic density, along with population growth, not only made possible the evaluation of the performance of a government in an "objective, scientific" way, it also indicated whether the productive potential of a nation had been fully put to work; it expressed the ideal of a rationalization and optimization of the relation between population and territorial resources that economists define as "optimum population."[110]

We have underlined so far how population figures (absolute and relative) were deployed in pro-development and pro-national arguments in the genre of patriotic statistics. But the display of demographic data in the opening section of the *Annuari* had also a precise rhetorical function: it was an effective way of evoking the unity of the country, of

106 *L'Austria e la Lombardia* (Italia [but Milan], 1847), p. 124.
107 "Sulla densità della popolazione," *Scritti economici*, vol. 1, p. 134.
108 Ibid.
109 *Annuario economico e statistico dell'Italia per l'anno 1853*, p. 45.
110 On this notion see E. P. Hutchinson, *The Population Debate. The Development of Conflicting Theories up to 1900* (Boston, 1967). In Cattaneo's view Lombardy had already attained a "balanced concentration [*addensamento*] of its population" ("Sulla densità della popolazione," *Scritti economici*, vol. 1, p. 134).

representing a unified nation. As Cesare Correnti explained in 1858, the exposition of Italian statistics was to follow an order of presentation going from the most to the least uniform object beginning "with the objects in which there most clearly prevails a similarity [*conformità*] among the various Italian areas;" population was precisely one of these most uniform objects.[111] As a matter of fact population could be categorized everywhere in the same way. Its density exhibited analogous variations and trends over the territory of the whole peninsula: if it varied from state to state, it also exhibited variations *within* each state which depended on similar causes.[112] As Pietro Maestri observed, high rates of demographic growth occurred in areas as different as Sicily and Tuscany indicating that the natural wealth of the peninsula prevailed over government misdeeds.[113]

Many other numbers filled the yearbooks of the 1850s, numbers painting a portrait of an "intellectual," an "economic," a "political," a "military" Italy. Several of them were more "prophetic,"[114] namely conjectural, than anything else. Yet Correnti felt confident enough to declare, in an outburst of patriotic optimism, that the figures so added together were the expression only of a present condition, not of the country's "latent capacities."[115] By adding numbers together, the yearbooks of the 1850s inevitably played down the diversity of situations to be found in the various Italian states. It was a rather mechanic or, better, arithmetical unity that they constructed and imposed on a heterogeneous reality. In this respect they bore resemblance to the sudden unification of the country which would occur only a few year later. While stressing the virtue of the whole, they paid less attention to the deficiencies of its parts, and produced an overly optimistic picture of the future ahead.

But in spite of the different descriptive conventions and the diversity of political projects behind the varieties of patriotic statistics that we have examined in this chapter, all these works had something in common: they aimed at establishing a positive identity for Italy and its inhabitants, an identity such as that possessed by the nations of north-western Europe, the leaders of political and economic progress. By stressing the quantity of "things" present within the territorial boundaries, they also gave a peculiar content to the idea of nation: that of being a repository of resources – of people, land, institutions –

111 *Annuario statistico italiano. Anno I 1857–58*, pp. 588–589.
112 [P. Maestri], *Annuario economico e statistico dell'Italia per l'anno 1853*, pp. 45ff.
113 Ibid., p. 99.
114 As even Correnti recognized: see *Annuario statistico italiano Anno I 1857–58*, p. 386.
115 Ibid., p. 365.

to be managed and governed. A far cry, one can observe, from that conception of nationhood – stressing the will of a people, who share the same history, culture, and language, in forming an independent political community – that Mazzini had been promoting since the 1830s.

CHAPTER 6

THE IDENTITY OF THE ITALIANS

We have seen how Italian patriotic statisticians, in order to escape the stereotyped images of the peninsula which stressed the torpor and misery of its inhabitants, highlighted by means of numbers and "facts" the signs of progress, the dynamic elements that made the Italian people worthy of the place they were claiming among the industrious and civilized countries of Europe. Yet there were statistics that could be used to support exactly the kind of stereotyping that Italian patriots were trying to avoid; this was particularly the case of "moral statistics," namely the counting of crimes, suicides, foundlings, and all sorts of signs of disorderly, "immoral" lives, in order to evaluate the morality of a society. From the time of their emergence in the late 1820s moral statistics promised to be a powerful new instrument for ranking peoples and states, or for drawing boundaries within established ones. Lower or higher percentages of violent crimes or of suicides seemed to represent the moral fabric of a society in a way which was both evocative and authoritative, and spoke to the concern about the nature of modern society that gripped the nineteenth-century elites.

Although crime was monitored and counted both in the pre-unification states and then in the Kingdom of Italy, the first extended essay on moral statistics bearing on the Italians was published as late as 1865–1866, in Venice. This city was at that time still part of the Habsburg Empire, although by the end of 1866 it would be annexed to the Italian state. The long title of the essay ("Esposizione critica delle statistiche criminali dell'Impero austriaco, con particolare riguardo al Lombardo-Veneto, secondo i resoconti uffiziali del quadriennio 1856–1859, e col confronto dei dati posteriori")[1] was indicative of its

[1] It appeared in *Atti dell'I. R. Istituto Veneto di scienze, lettere ed arti*, 3rd s., 11 (1865–1866), pp. 151–211, 331–409, 483–510, 599–652, 993–1051, 1237–1258; 12 (1866–1867), pp. 227–268. Subsequently the essay was also published as a separate volume entitled *Le statistiche criminali dell'Impero austriaco nel quadriennio 1856–59 con particolare*

content: a critical examination of the official crime statistics of the Habsburg Monarchy which paid particular attention to the figures concerning the Italian provinces, namely Lombardy–Venetia. A remarkable exercise in statistical exegesis, this essay was also the first rigorous application of the language and methodological procedures of Quételetian social statistics by an Italian scholar. The author of this analysis, Angelo Messedaglia, is already a familiar figure: professor of political economy and statistics in Padua and at the same time a savant who pursued a variety of different scientific and cultural interests from mathematics and astronomy, to classics, to contemporary American poetry (he was a translator of Longfellow and himself a composer of romantic verses). We should bear in mind his 1851 critique of the way statistics was taught in the universities of Lombardy–Venetia and his call for a radical reorganization of this teaching. The change sought by Messedaglia did not come, but the young scholar was assigned a full teaching position in Padua in 1858. By the mid-1860s he had published several works, among which a critical analysis of the work of Malthus.[2] It was thus with the authority of an already well-known scholar that Messedaglia presented his critical report on Austrian statistics to the academicians gathered in the Imperial Regio Istituto Veneto di scienze, lettere e arti, the most prestigious scientific institute of Venetia which had just elected him member.[3]

In this essay Messedaglia set out to refute the official Austrian interpretations of crime statistics and especially the Austrian views of the character of the inhabitants of the Italian provinces, views that portrayed them as people driven by southern passions, *passioni meridio-*

riguardo al Lombardo-Veneto e col confronto dei dati posteriori fino al 1864 inclusivamente. Esposizione critica (Venice, 1866–1867).

2 *Della teoria della popolazione principalmente sotto l'aspetto del metodo. 1. Malthus e dell'equilibrio della popolazione colle sussistenze* (Verona, 1858).

3 After having studied law at the University of Pavia, Messedaglia taught statistics, political economy, and finance at the Universities of Padua and Rome. In 1866 he was elected representative to the Lower House, where he sat in the center–right, and was appointed senator in 1884. From 1869 until his death he also sat on the central advisory committee on statistics of the Kingdom of Italy. He played a fundamental role in the design of a new land tax system in the mid-1880s. There are no full-fledged biographies of this important representative of positivist culture in liberal Italy; recent works which provide some information are A. Pellanda, "Angelo Messedaglia parlamentare e accademico nel contesto socio-politico italiano del secondo Ottocento," in A. Tagliaferri, ed., *I ceti dirigenti in Italia in età moderna e contemporanea* (Udine, 1984), pp. 457–467; R. Romani, "Il Messedaglia dimezzato," *Venetica. Rivista di storia delle Venezie* 2 (1984), pp. 109–130; Romani, "Romagnosi, Messedaglia e la 'scuola lombardo-veneta': la costruzione di un sapere sociale," in R. Camurri, ed., *La scienza moderata. Fedele Lampertico e l'Italia liberale* (Milan, 1992), pp. 177–210.

nali. The target of Messedaglia's critical observations was not new: I mentioned earlier how in 1850 the compilers of the *Annali universali di statistica* had rejected the conclusions drawn by Viennese commentators about the character of the Italian subjects of the Empire as shown by criminal statistics. The interpretation offered by Austrian authorities was questioned, but no attempt was then made to deconstruct the numbers and the procedures that had generated them. In the absence of a systematic critique or alternative data, the issue remained inconclusive, and thus it was left to the preference of the reader how to interpret the data. Messedaglia's innovation was that a new statistical *methodology* was mobilized for the purpose of refuting a perceived misrepresentation of the inhabitants of Lombardy–Venetia. The alternative interpretation proposed was to be based on technical arguments, although, as we will see, these did not involve much more than a systematic comparison – together with some reworking – of means and proportions summarizing various aspects of the judicial process and of criminality across the various provinces of the Austrian Empire.

There are thus several reasons why the issue of moral statistics and, more particularly, Messedaglia's work deserves attention. First of all, moral statistics were an important vehicle for the debate on the nature of *incivilimento* and the meaning of *italianità* in which the educated and liberal elites of the Italian states were engaged in the first half of the century. Since the formulation of a "climatic" theory of people's character in the eighteenth century, the inhabitants of southern Europe had been endowed with a set of characteristics which as a whole amounted to a specific stereotype of the *meridionale*: passionate, hot-tempered, drawn to music and poetry, but impatient of long and disciplined meditations fundamentally lazy, spoiled by too gentle a climate, and finally, closer to savagery than the more sober and resolute dwellers of the North.[4] This image recurred very frequently in the travelers' literature and made its way also into "scientific" publications. The rejection, or the qualification, of this image was therefore part of the effort to define a new identity for a people who were claiming a place in "civilized" Europe.

Another reason for considering here moral statistics in general and

[4] The obligatory reference for the eighteenth century is, of course, Montesquieu who, however, was not at all as rigid as some of his interpreters and followers. For an example of a text packed with stereotypes written in the early nineteenth century see Ch.-V. de Bonstetten, *L'Homme du Midi et l'homme du Nord ou l'influence du climat* (Geneva, 1824), a work which was criticized by M. Gioia in *Annali universali di statistica* (henceforth *AUS*) 5 (1825), pp. 245–301.

Messedaglia's work in particular is that this will allow us to confront the issue of methodological innovation and one of the contexts of its emergence and consolidation. Thanks to his knowledge of mathematics, Messedaglia was probably the first in Italy who not only in theory, but in actual practice, went beyond the boundaries of a descriptive conception of statistics. He was not alone; as we will see more extensively in the next chapter, in the 1860s the creation of the new state coincided with a shift in the conceptualization and practice of statistics. The methodological questions associated with descriptive statistics ceased to be central to the debate among statisticians – although that genre of writing continued to exist. Statistics became increasingly concerned not only with counting resources but also with finding the "laws" of society. Significantly, Messedaglia applied the new analytical tools – and even experimented with new ones such as the technique of adjustment – on a distinctly political issue.

Finally this is a story that we must not omit because of the importance of the issue of law and order in nineteenth-century Italy. Recent scholarship has emphasized that questions of crime and public order occupied a central position in the preoccupations of the Italian elites both before and after unification. As John Davis has argued, before unification sectors of the bourgeois classes were driven to support political change because of, among other reasons, their conviction that the existing rulers were not able to assure their security.[5] Then in the difficult socio-political context of post-unification Italy, the ruling elites tended to look at social and political tensions, especially in parts of the South, through a "vocabulary of crime and public order."[6] Not surprisingly Italy was the country that contributed most to the creation of the "science" of criminal anthropology. The tendency to translate social questions into a vocabulary of crime noted by Davis was not, however, an Italian peculiarity, as witnessed by the category of "dangerous classes" elaborated in the Paris of the first half of the nineteenth century.[7] It is thus necessary to look more closely at what the quantification of crime itself came to mean in a society that increasingly interrogated statistics to find its own identity.

5 This is John Davis's argument in his *Conflict and Control: Law and Order in Nineteenth-Century Italy* (London and Atlantic Highlands, N. J., 1988).

6 Ibid., p. 290.

7 On the Parisian "dangerous classes" see the classic study by L. Chevalier, *Laboring Classes and Dangerous Classes in Paris during the First Half of the Nineteenth Century* (Princeton, 1973).

MORAL NUMBERS IN RISORGIMENTO ITALY

The expression "moral statistics" as used in the nineteenth century refers to a particular object of investigation, the statistics of "deviancy." For historians of statistics it has also become the epitome of the new statistics of the first half of the nineteenth century, a fully numerical social science which abandoned the conventions of the older description of states. This numerical social science covered a great variety of objects and fields of investigations, but was particularly interested in quantifying people's social behavior. Impressed by the regularities he noticed in the statistics of population and of crime from year to year, in the 1830s Quételet came to endow average numbers with a capacity to truly represent a whole society. He made of the mean a regulative notion, the ideal and "center of gravity" of the body social. From the observation of moral numbers he also derived the fateful idea of the "penchant for crime" of a society (equal to the number of criminal acts committed divided by the population) precipitating a long quarrel about the role of free will in human action.[8]

In the first half of the nineteenth century, however, many practitioners of statistics did not associate moral statistics with the new statistics of Quételet. Indeed Quételet's methodological lessons were lost on most practitioners, and they limited themselves to collecting and analyzing moral statistics as yet another way of measuring civilization, of judging the health or sickness of the social body, of debating policy issues. This was true also in the case of the country that invented moral statistics, namely France. The material prerequisites, so to speak, of moral statistics were the numerical data on crime and the administration of justice that the French Ministry of Justice began to publish in the mid-1820s. These data were reported and discussed in various journals and publications. The director of the department of statistics at the Ministry of Justice, the jurist André-Michel Guerry (1802–1866), processed them in sophisticated ways in a study that became the classic work of reference in the debate, the *Essai de statistique morale de la France* (1833). Although Guerry was undoubtedly methodologically innovative (he was among the first to use colored maps to represent the geography of crime), the impact of his work must be seen in the context of an ongoing discussion on the consequences of economic development in France that had been opened a few years earlier by a staunch advocate of industrialization, Baron Charles Dupin.

8 Quételet's early contributions to social statistics date from the late 1820s. On statistics as a numerical social science and on Quételet see T. M. Porter, *The Rise of Statistical Thinking 1820–1900* (Princeton, 1986), chs. 1–2.

In his analysis of the industrial and commercial wealth of France, Dupin "dualized" France, so to speak, arguing that the more economically developed North (defined roughly as the part of France north of an imaginary line going from Saint-Malo or Cherbourg to Geneva) fared better than the South in many respects, such as school attendance: a France *éclairée*, he argued, stood against a France *obscure*.[9] Publication of the crime statistics, however, complicated the matter for the supporters of industrial "civilization," who had the problem of explaining to their critics why, as these figures showed, rates of crimes were not lower in the regions with higher rates of school attendance. They found a solution by disaggregating the data by type of crime; to them we owe the idea that modern societies are less *violent* than peasant ones: that crimes "against property" characterize more developed societies, while crimes "against the person" are a mark of the less developed ones. In what appears to be a modified version of the "doux commerce" argument, for example, the French lawyer and subsequent inspector of prisons, Charles Lucas, argued that the immense growth of wealth that accompanies a civilized society naturally produces also an increase in the "opportunities for being harmful" (he was referring to the lure of riches that can induce people to behave in an unlawful fashion), and that one should compute these in order to assess fairly the increase of crime. Should one, he asked rhetorically, give up all the increased "opportunities for rejoicing" brought about by civilization in order to avoid its negative consequences? Clearly the answer was an emphatic "no": lack of civilization did not guarantee a lower *degree* of criminality, but was in fact characterized by a more violent *kind* of criminality, as appeared from the examination of the criminal record of Spain, "one of the countries less advanced in modern civilization," where the number of crimes against the person was far greater than that of crimes against property.[10] In the pages of the *Bulletin des sciences géographiques*, the organ of the French Society for the Propagation of Scientific and Industrial Knowledge, Adriano Balbi too warned against those interpretations which saw a simple inverse correlation between crime and education, and proposed a multicausal explanation of the

9 See R. Chartier, "Les Deux Frances. Histoire d'une géographie," *Cahiers d'histoire* (1978), pp. 393–415 (there is an English version in *Social Science Information*, 17 [1978], pp. 527–54). The work in which Dupin speaks of a "France éclairée" is *Forces productives et commerciales de la France* (Paris, 1827).

10 Review, signed "D.", of "Influence de l'instruction et de la civilisation en général sur la diminution des délits et des crimes (extrait de l'ouvrage ci-après cité; par C. Lucas)," *Bulletin des sciences géographiques, économie publique, voyages. 6e section du Bulletin universel publié . . . par la Société pour la propagation des connaissances scientifiques et industrielles* 14 (1828), pp. 106–116.

patterns of crime.[11] Balbi did not doubt that education, although not enough to defeat crime, was certainly needed to overcome "savagery and superstition" and the worst types of crime that came with them. More ambiguous in contrast was Guerry in his *Essai* of 1833. By carving out an alternative territorial partition of France and using different indicators of literacy, he showed that the least literate regions were not necessarily the most criminal ones, while crimes against property were consistently more numerous in the more industrial and literate areas. He thus provided evidence for those inclined to question the positive picture of industrialism, and cast a shadow on the nature of the development experienced by France *éclairée*.[12]

Most participants in these discussions used numbers indiscriminately. They often made comparisons between countries and regions resorting to absolute numbers instead of using, as we would expect, proportions and percentages. But this was common practice and does not prove bad faith at a time when statistical manuals did not specify what kind of numbers were acceptable. Nevertheless, it is interesting to notice, as Porter aptly observes, that "the bewildering complexity of statistics was invariably rediscovered whenever a result arose that contradicted expectations or prejudices."[13] Most of the time statisticians completely subordinated figures to the arguments they desired to make.

The Italians could not be insensitive to this European debate concerning, as it did, the nature of the modern civilization they aspired to join, although this aspiration was not unqualified. In fact they entered the discussion almost at its very beginning. In 1829 Romagnosi published some "statistical observations" on the *Compte général de l'administration de la Justice criminelle*, the official annual publication of the French Ministry of Justice.[14] In this work he questioned the very assumptions of the participants in the debate, taking issue with those who considered it inevitable that since civilization "creates more things to be stolen"[15] it will bring about more crimes motivated by cupidity. In Romagnosi's view those who equated civilization with the mere increase in material wealth had a very narrow idea of it; on the contrary, inherent in its definition was not only economic, but also moral and political progress. Similarly, the positive effects of education could be

[11] "Rapport du nombre des crimes à l'état de l'instruction publique en France," Ibid., 20 (1829), pp. 252–264.

[12] Chartier, "Les Deux Frances," p. 400–402.

[13] Porter, *Rise of Statistical Thinking*, p. 174.

[14] "Osservazioni statistiche al proposito del libro intitolato: *Conto generale dell'Amministrazione della giustizia criminale in Francia durante l'anno 1827 presentato al Re dal Guarda-Sigilli. Parigi 1828. Stamperia Reale*," *AUS* 19 (1829), pp. 1–25.

[15] Ibid., p. 11.

expected only in a society where means of subsistence were equitably distributed. Without a substantial discussion about the very meaning of the terms used in the debate, Romagnosi insisted, people tended to draw foolish conclusions from criminal statistics.

Gioia's participation in the discussion in the late 1820s was limited to a brief note on the suicides that had occurred in Lombardy in the period 1817–1827.[16] For suicides too were an integral part of moral statistics and generated vast interest at the beginning of the century. Early statistics on suicide were inscribed in the debate on the nature of civilization, but also served to feed or reject national stereotypes, as in the case of the squabble between French and English commentators on whether suicide was a *mal anglais* connected to a specifically English propensity to melancholy.[17] In Gioia's *Filosofia della statistica* (1826) suicides belonged in a panoply of "symptoms" of the conditions of a society and were listed with several other "physical quantities" which served to represent the "moral qualities" of a society.[18] But already in his earlier work, Gioia had paid close attention to crimes, constructing elaborate models of tables in which the activities of judicial magistrates and the characteristics of criminals would be faithfully represented.[19] He observed that the proportion of crimes to a population was not a sufficient indicator of morality; a more accurate way of proceeding would be that of calculating the ratio between criminal impulses (whose levels depended on a host of objective and subjective circumstances) and actual crimes committed.[20] In full accordance with his productivist and utilitarian ideology, morals were a "branch of economics," and statistics were "to determine the influence [of morals] on production and consumption."[21]

By the 1830s and 1840s reports on French moral statistics were a regular feature of the *Annali*: Defendente Sacchi, for example, reported extensively on the work of Guerry, expressing however some doubts about the negative conclusions reached by the author on the relation between education and crime.[22] In introducing a summary of the work of the French statistician P. Fayet, an anonymous compiler took care to observe that similar researches should be done also "da noi" (here), in order to know "the effect of literacy on people's morality,"

16 "Nota sui suicidi avvenuti nelle Provincie Lombarde," *AUS* 17 (1828), pp. 67–68.

17 I. Hacking, *The Taming of Chance* (Cambridge, 1990), ch. 8.

18 *Filosofia della statistica* (Milan 1826), p. 377.

19 *Tavole statistiche ossia norme per descrivere, calcolare, classificare tutti gli oggetti d'amministrazione privata e pubblica* (Milan, 1808). The following quotes come from an edition published in Chieri in 1854.

20 Ibid., pp. 218–219.

21 Ibid., p. 220.

22 *AUS* 36 (1833), pp. 193–200.

but without "making too much of mathematical operations" that might produce false conclusions. The same interest – and warnings – were expressed in *Il progresso* of Naples and the *Giornale di statistica* of Palermo, where several authors contributed well informed articles that showed a full knowledge of Guerry's and Quételet's works. Like the authors writing in the *Annali*, these writers too recommended interpretative caution, arguing that the positive effects of education were wrongly questioned on the basis of often unreliable numbers. A follower of Romagnosi and collaborator on the *Giornale di statistica*, the Sicilian jurist and economist Emerico Amari, expressed the frustrations felt by many who saw the same statistics used to support opposite standpoints, and asked for a "norma direttrice" (guiding principle) in the making of judicial statistics so that one would be more confident in drawing conclusions from them,[23] and would avoid the kind of "sophisms" uttered by Guerry.[24] Amari's denunciation of the defects of crime statistics was part of a running debate, more theoretical than methodological in nature, but it was also a covert polemic about the crime statistics which the Ministry of Justice of the Two Sicilies had begun to publish in 1833 and upon which writers close to the government, such as the Neapolitan prosecutor Pietro Calà Ulloa, had commented extensively in very favorable terms.[25]

It is clear from the foregoing discussion that the debate on moral statistics in the Italian states had strong ideological overtones. It was yet another means of defining an "Italian way to modernity"[26] in which the advocacy of modernization went together with a great deal of moralizing and the proclamation of interpretive and practical precautions: to avoid the excesses both of a "civilization" understood in a purely materialistic fashion, and of a statistical approach transformed into *cifro-*

23 See "Difetti e riforme delle statistiche de' delitti e delle pene," *Giornale di statistica* 3 (1838), pp. 270–334.

24 "Società statistica di Londra, e i suoi lavori," Ibid., 3 (1838), p. 52.

25 The first official publication of the Kingdom of the Two Sicilies in this field was the *Quadro statistico generale sull'amministrazione della giustizia penale ne' reali domini di qua del Faro per l'anno 1833 rassegnato a Sua Maestà il Re dal Ministro Segretario di Stato di Grazia e Giustizia* (Naples, 1835). Another volume appeared in 1851. Data on Sicily were published separately. P. C. Ulloa's articles on the topic of crime statistics are in *Il progresso delle scienze, delle lettere e delle arti* 10 (1835), pp. 161–173 ("De' reati e della civiltà"), and pp. 240–259 ("Dell'esposizione dei reati in Inghilterra e in Francia, e del quadro statistico dell'amministrazione della giustizia penale nel Regno di Napoli"); 13 (1836), pp. 211–235 (review of the official crime statistics [civil justice] for 1833); 14 (1837), pp. 39–54 (review of the official crime statistics [penal justice] for 1833).

26 I am borrowing this expression from G. Bollati, "L'Italiano," in *Storia d'Italia*, vol. 1, *I caratteri originali* (Turin, 1972), pp. 949–1022.

mania (obsession with figures). Italian publicists generally posed the need for a more philosophically based investigation against the making of quick generalizations on important issues merely on the basis of ambiguous numerical indicators. In this manner one could avoid drawing injudicious conclusions, and statistics could continue to provide the guiding light of true knowledge for statesmen and legislators.

While moral statistics from early on attracted a great deal of attention among Italians as yet another arena for the ideological debate on the nature of civilization, and the desirability of economic and political development, there were no significant works in this genre before the original contribution of Messedaglia,[27] who was also the first to fully apply the lessons of Quételetian social physics. Although earlier Italian statisticians had known the work of Quételet,[28] they lacked some of the prerequisites which were necessary for the development of the same kind of studies; as we know, they were not mathematicians but humanists who never paid much attention to questions of probability theory. Moreover, a work of moral statistics *à la* Quételet required the availability of data, large quantities of data of a serial kind on crimes and suicides spanning several years. Italian statisticians were aware of this need. As the rapporteur of the Piedmontese committee on judiciary statistics, Pasquale Stanislao Mancini, observed in 1852, according to the "law of large numbers" one could not "attribute any value to statistical facts of various kinds unless in a series of long and uninterrupted observations."[29] Although some states like the Kingdom of the Two Sicilies (in the 1830s) and Piedmont (in the early 1850s) had

27 An exception is C. I. Giulio, "Osservazioni sui fatti principali e sulle leggi del movimento della popolazione negli Stati di S. M. il Re di Sardegna," in *Informazioni statistiche raccolte dalla Regia Commissione Superiore per gli Stati di S. M. in Terraferma. Movimento della popolazione* (Turin, 1843), pp. 635–736.

28 A summary translation, by the physician Andrea Bianchi, of the *Physique sociale* was published in installments in the *AUS* between 1837 and 1838. See *AUS* 53 (1837) pp. 9–33, 129–154, 241–258; 54 (1837) pp. 9–28, 157–169, 301–307; 55 (1838) pp. 191–203; 56 (1838), pp. 37–47; 57 (1838), pp. 156–166; 58 (1838), pp. 39–52. Some Italian scholars also enjoyed a personal relationship with the Belgian savant. They had occasion to meet him at the first Congress of Italian Scientists held in Pisa in 1839, and even before then during his first trip to Italy in 1830, as well as in Brussels where he befriended several Italian patriots who went into exile after the failed revolutionary attempts of the early 1820s and 1830s. Their correspondence is in the archives of the Académie Royale de Belgique. On his trips to Italy and his relations with some Italian scholars there is an old article by M. Battistini, "Le relazioni di Adolfo Quételet con i dotti italiani," *Rivista di storia delle scienze mediche e naturali di Firenze* 20 (1929), pp. 115–127.

29 [Commissione di statistica giudiziaria], *Statistica giudiziaria degli Stati Sardi*, vol. 1, *Statistica giudiziaria civile, commerciale, e del contenzioso amministrativo degli Stati Sardi per gli anni 1849–50* (Turin, 1852), p. xv.

started to assemble and publish their own judiciary statistics,[30] official numbers, as we have seen, had a fairly unstable existence in the pre-unification monarchies, and only rarely did they form the long and homogeneous series needed for the exegetical exercises of moral statistics. The Habsburg Empire in contrast had been publishing data on crime since 1846. Besides providing a reliable series of data on crime, imperial official statistics offered Italian scholars a political motivation to work on moral statistics.

THE IDENTITY OF THE ITALIANS: MESSEDAGLIA AND THE CRIME STATISTICS OF THE AUSTRIAN EMPIRE

Since definitions of morality and moral standards differ not only among different societies but also within the same society, moral statistics as an attempt to measure the morality of a society is bound to be an arena of contention. Besides, by using counts of unlawful and/or immoral actions as means to assess the moral fabric of a society, moral statistics is almost by definition a realm of paradoxes. Consider for example the paradigmatic case of the relation between gender and crime. Statisticians constantly confronted the issue of whether women's participation in crime was a sign of a more or less civilized social order. The answer of course depended on one's view of what the relationships between the sexes should be. If commentators agreed that a more civilized social order was one in which women behaved more like men, and applied this reasoning to crime, then the paradox of the moral nature of society being measured by a behavior that otherwise society condemned was well apparent.

An almost intractable problem arising from crime statistics is to be found also with respect to cross-national comparisons: the comparison of crime statistics across states is an extremely difficult enterprise, if not altogether impossible, given the differences in the systems of penal legislation, in the definitions of what constitutes a crime, in the systems and extent of application of policing, and so on, all of which make crime a construct which varies greatly from society to society, and also within the same society through time. However, avoiding cross-national or temporal comparisons when data are available is almost impossible, since statistics acquire meaning only through comparison.

[30] On the Kingdom of the Two Sicilies see note 25. On Piedmont see the work cited in note 29 and vol. II, *Statistica giudiziaria penale degli Stati Sardi per l'anno 1853 e ragguagli comparativi con gli anni 1854 e 1855 e con altri anni anteriori* (Turin, 1857). The committee on judiciary statistics was created in 1850.

To be sure, the most alert nineteenth-century statisticians were aware that comparisons are appropriate only when made between homogeneous realities. Guerry warned against cross-national comparisons in a large work on the moral statistics of France and England which he published in 1864, thirty years after the first *Essai*, and which received the prestigious "Prix Montyon de Statistique" from the Academy of Sciences of Paris. In spite of its title (*Statistique morale de l'Angleterre comparée avec la statistique morale de la France*), the author claimed that he did not compare the two countries, but only constructed two distinct statistics.

It is to Guerry's work of 1864 that Messedaglia dedicated a review essay a year before publishing his own work on Austrian crime statistics.[31] While praising Guerry's imaginative use of maps and in general his skillful graphic representation of statistical differentials within the boundaries of a state, as well as his invention of new kinds of averages, Messedaglia took issue with many of Guerry's standpoints on theoretical and methodological grounds. He first of all criticized the lack of a statistics of the penal process in all its aspects without which no full sense of the meaning of the figures on crime could be made. This was an important point to which was linked Messedaglia's own understanding of the meaning of the measure of crime:

> crime itself cannot be assumed to be fully an expression of morality in the sense of moral habit [*costume morale*]. It powerfully reflects the ensemble of moral and civil, economic and political conditions; it is the outcome not only of depravity but also of need, of cunning as well as of the system and the variable circumstance of repression; in a word of the political state and the state of social relations as a whole. To be sure, as an *effect*, crime appears to be a disorder of moral reason in the proper meaning of the word; but its primary *causes* can be many and diverse, and we need a very subtle study difficult [sic] to extract the real and specific element of morality."[32]

Although not completely original, this statement contained a much needed and clear definition of the limits of moral statistics which those who dealt with moral numbers too often forgot. The more conservative Guerry had been eager to distance his work from the theory of probability which was accused of being a denial of free will when applied to moral or social matters – a question which had been an object of heated debate since the 1830s and continued to be so in the 1860s and 1870s.

31 "Relazione critica sull'opera di M. A. [sic] Guerry: Statistica morale dell'Inghilterra comparata alla statistica morale della Francia," *Atti dell'I. R. Istituto Veneto di scienze, lettere e arti*, 3rd s., 10 (1864–1865), pp. 1068–1085, 1135–1168.

32 Ibid., pp. 1141–1142.

Messedaglia in contrast sided strongly with Quételet on all points, stating that statistics was "nothing more than the calculus of probability" and that the mean was "the most probable value of the true nature of a thing."[33] This statement is important since it gives us a full sense of the importance of the mean: being supposed to tell the truth of a thing, the calculation of means ought to be handled with rigor and expertise. For Messedaglia, Guerry had actually made a mistake by calculating means on the basis of exceedingly long series of figures. He had certainly obtained "large numbers" that could eliminate the effects of what were called in Quételetian jargon "accidental causes," but had thus hidden the working of "regular" causes. For Quételet, accidental or "perturbing" causes amounted to the free will of individuals. Regular causes were those found operating once the effects of the former had been eliminated; they included geographical factors, economic and religious institutions, sex, age, profession, and so on.[34]

As Messedaglia himself recognized,[35] the 1864 review of Guerry's work served him as a kind of theoretical and methodological introduction to the essay he presented to the Istituto the following academic year. The statistics that were examined in this work had been published by the central statistical agency of the Austrian Empire in 1861 and 1862 for the years 1856–1857 and 1858–1859 respectively; data for 1861–1864 had appeared only in summary form.[36] In the volumes containing the data for 1856–1859 the numerical tables were introduced by a fairly long interpretive essay in which the data were summarized and commented upon. The tables presented in these introductions were organized to include summary figures on the results of judicial procedures and the collective profile of criminals (through means and percentages) for the Empire as a whole and for all its provinces. Comments highlighted the diverse geographical patterns shown by the figures and tried to account for them. The anonymous commentators pointed out the provinces whose statistical means represented an excess or "transgression" (*Überschreitung* is the word used) of the general mean, erected into the model or normative value.

[33] Ibid., p. 1150.

[34] For Quételet's systematic treatment of causes see his *Lettres à S. A. R. le Duc Régnant de Saxe-Coburg et Gotha sur la Théorie des probabilités appliquée aux sciences morales et politiques* (Brussels, 1846), part 3.

[35] "Relazione critica sull'opera di M. A. Guerry", p. 1073.

[36] See *Tafeln zur Statistik der Oesterreichischen Monarchie. Neue Folge. III. Band. Die Jahre 1855, 1856 und 1857 im fassend* (Vienna, 1861), and *Tafeln zur Statistik der Oesterreichischen Monarchie für die Jahre 1858 und 1859* (Vienna, 1863); *Übersichtstafeln zur Statistik der Oesterreichischen Monarchie für das Jahre 1861 und 1862* (Vienna, 1863); *Statistisches Jahrbuch der Oesterreichischen Monarchie für das Jahr 1863* (Vienna, 1864).

The main task Messedaglia set for himself in the essay of 1865–1866 was to challenge the indictment pronounced by the Austrian authorities on their Italian subjects as a people ruled by *passioni meridionali*. The Austrians read the high number of cases dismissed during the phase of preliminary investigation in the Italian provinces as the result of a predisposition to *vendetta*, a most typical "southern passion."[37] Unwillingness on the part of the population to collaborate with official justice and the cunning of the delinquents were in addition adduced as reasons for the frequent failure to convict.[38] The proportions of violent crimes among the accused in the Italian provinces also fitted the stereotype of a people less able to keep passions in check.

Through a closer examination of the types of such crimes in which the preliminary investigations were not able to find sufficient evidence (*indizi*) to bring a culprit to the next stage of prosecution, Messedaglia argued that these consisted mainly of those crimes of which there were frequent complaints but for which the identification of the offender was objectively more difficult to prove, such as arson, abandonment of infants, abortion, and those crimes which depended on the level of public surveillance such as larcenies and burglaries.[39] Before accusing the moral predisposition of a whole people, one should – this was the kernel of Messedaglia's initial argument – first carefully disaggregate the total figures into their various components and consider a host of possible factors accounting for them.

The next step in the analysis of the Veronese statistician focused on what happened in "special" investigations, namely investigations against specific suspects, as distinct from general preliminary investigations: here the figures showed that the subjects of Lombardy–Venetia thus targeted were more likely to be brought to trial than those of almost all the other provinces of the Empire.[40] Lombardy–Venetia stood out also with regard to the proportion of suspects kept under arrest during the investigative process.[41] Messedaglia argued even further that the relatively lower proportions of people accused of crimes in Lombardy–Venetia were a statistical artifact that hid more than they revealed: in fact in order to be meaningful these numbers had to be adjusted to take into account the variations from province to province in the proportion of the actual indictments to the preliminary investigations. The adjusted figures showed remarkably different results: the proportion of accused people in Lombardy–Venetia in 1858 rose from 1 per 3,984 inhabitants to 1 per 978 by this index,

37 *Tafeln . . . 1855, 1856 und 1857*, p. 84, and *Tafeln . . . 1858 und 1859*, p. 78.

38 *Tafeln . . . 1858 und 1859*, p. 84.

39 "Esposizione critica," p. 201.

40 Ibid., pp. 332–336.

41 Ibid., p. 353.

much closer to the proportion of 1 accused to 655 inhabitants that was the general average for the Empire;[42] thus in this respect too Lombardy–Venetia did not exhibit very abnormal averages. In order to make meaningful comparisons – this was the clear lesson, as Messedaglia insisted – one had to produce first of all *homogeneous* numbers. Ironically the Italians, in order to be more "normal," namely closer to the overall mean, needed to be more criminal.

Through his own calculations, Messedaglia was also able to show that the inhabitants of the Italian provinces were more likely to be condemned, once brought in front of a court, than other subjects.[43] Thus what the figures really showed, in this analysis, was the inefficiency of the system of public surveillance as shown in the high number of thefts which went unpunished, and the discriminatory treatment to which Italians were subjected once brought to trial. As for the types of crimes, Messedaglia observed that it was true that Lombardy–Venetia had a higher proportion of violent crimes, namely crimes against the person, but (and this was an important "but") this figure referred only to the people who had been judged and condemned. Now, from the preceding analysis, it was clear that these were only a small part of the total since a large volume of lesser crimes escaped the reach of the law. Therefore it was more accurate to evaluate the frequency of certain types of crimes on the basis of total complaints: once this was done, the whole claim about the relative predominance of violent crimes in Lombardy–Venetia was shown to have no basis whatsoever.[44] The relatively low proportion of property crimes was thus not the sign of a lower level of "civilized" criminality, but the result of ineffective policing, especially of the *furti campestri* (thefts in the countryside) which, Messedaglia observed, "had become a real *flagello* [scourge]."[45] Thus, once corrected, the averages of the "southern" Lombardy–Venetia were quite similar to those of the northern provinces of the Monarchy.

Wide variations in the pattern of crime from year to year also enabled Messedaglia to refute firmly explanations of statistical differentials reflecting ethnic prejudices: as a matter of fact this variability pointed to contingent, variable causes which were probably at work as opposed to – in Quételet's typology – the "constant" ones, namely those that exercised regular influence, such as climate, time of the day,

42 Ibid., p. 363. 43 Ibid., pp. 376–377. 44 Ibid., p. 614.

45 Ibid., p. 621. The high number of *furti campestri* is to be linked to the aggressive process of privatization of common lands which took place in the 1840s and 1850s in the Italian provinces: on this process see for example P. Brunello, *Banditi, questuanti, ribelli: proteste contadine in Veneto e in Friuli 1814–1866* (Venice, 1981).

season, and so on.[46] Moreover, as the law of large numbers clearly taught, small numbers did not warrant firm conclusions: thus for example not much could be made of the sometimes relatively higher percentages of rapes to be found in the Italian provinces, which could be taken to "legitimate the accusation or the suspicion that we (we *southerners*!) are very inclined to the sins and crimes of the flesh [original emphasis]."[47] The numbers here were too small to allow for meaningful conclusions to be drawn especially if they were compared with the very low percentages of another indicator of morality in this field, namely the proportions of illegitimate births. The very low level of illegitimacy in the Italian provinces in contrast to the very high ones found in the German provinces struck a positive note about the morality of the former, a difference of which, Messedaglia insisted, Italians should make more.[48]

Messedaglia made other significant remarks when examining the sex variable: the low proportion of women criminals in Lombardy–Venetia (and in other southern areas) as a mark of primitive cultural condition could not go unquestioned. Considering that the kind of crime in which women were more involved in "civilized" countries were those against property, and that these precisely went particularly unpunished in the Italian provinces, one would have obtained more accurate results about the criminal behavior of women in the latter if thefts were excluded from the totals for all the provinces.[49] In this case too Lombardy–Venetia was probably closer to the imperial norm than the figures seemed to show. Here, paradoxically, Messedaglia was obliged to argue for the existence of a higher percentage of crimes than that shown by official figures, this being the symptom of a more "moral" condition because it approached the average, which in its turn approached the figure for the regions that were by definition more "civilized". The "primitive" factor therefore was imperial justice, not Italian criminality.

Faithful to his appeal for caution in cross-national comparison, Messedaglia made a very moderate use of comparison with other countries, or, to be more precise, with one other country, France. He relied, however, on the French data once more to indict the Austrian Monarchy because of its more ineffective, and yet at the same time, exceedingly severe administration of justice. Thus the high percentages of cases dismissed during preliminary investigations[50] – a sign of a lack of sufficient surveillance – were combined with the absence of the jury

[46] Ibid., p. 637. [47] Ibid., p. 643. [48] Ibid., p. 645.
[49] Ibid., p. 1007. [50] Ibid., pp. 342–343.

system, adopted by most civilized nations,[51] and with the persistent use of corporal punishment, abandoned by civilized nations,[52] to paint a rather grim picture of the administration of justice in the Habsburg Empire.

Considering the negative attributes attached to the representation of the *meridionale* in general, it is not surprising that Messedaglia resented the representation of Lombards and Venetians as *meridionali*, and lumped with those people to whom, in Messedaglia's view, this label was more appropriately applied, namely southern Slavs such as the inhabitants of Dalmatia.[53] Indeed Messedaglia himself ranked peoples according to ethnic traits and placed them in a hierarchy of civilization, but his own ranking did not match that of the Austrian authorities: in the lecture notes he wrote for his students at the University of Padua in 1861 he declared that "The Italians and the Germans ... are the most civilized peoples of the Empire."[54]

While rehabilitating the image of his fellow Italians, the Veronese statistician simultaneously pointed a finger at the responsibilities of the Austrian system of justice which not only did not guarantee its Italian subjects an equal protection from crime, but also penalized them in various ways. It is not our purpose here to assess whether the working of the Austrian judicial system was fair or unfair towards its Italian subjects. The issue is one of contests over representations, struggles waged through and about numbers at a time when numbers were increasingly employed to portray and evaluate the performance as well as the essence of states and of peoples.

Throughout his whole analysis Messedaglia deployed the armory of tools – from reclassification to emphasis on large numbers – developed by the new statistics, and even introduced new techniques – such as adjustments or forms of standardization of rates and percentages – which might serve to make the data more comparable. This emphasis on methodology greatly increased the persuasive power of his arguments, and represented a departure from the character of earlier debate on moral statistics. In his rejection of what he considered faulty conclusions derived from lack of methodological rigor, Messedaglia appealed to the light of true science. According to him, figures speak only one

51 Ibid., p. 385. 52 Ibid., pp. 267–268.

53 This is apparent in his comment on the higher percentage of crime motivated by lust among adolescents in Dalmatia: it was due to their "difetto di cultura e rozzezza del vivere" ("lack of culture and primitive lifestyle"), ibid., p. 1023.

54 See "Note di statistica dell'Impero Austriaco precedute da alcune nozioni teoretiche fondamentali per servire all'insegnamento del Prof. Angelo Messedaglia. Padova 1861 per cura di T. A.," lithographed manuscript conserved at the Biblioteca Civica of Verona, Fondo Messedaglia, no. 1249. The quote is from p. 27.

language, which must be learned. Their interpretation is not a matter of preference, but of knowing and applying the correct method; there is only one correct interpretation. As Messedaglia wittily observed, tortured figures cannot speak the truth; his own interpretive *tour de force*, however, shows that they can speak truth (whatever one's conception of it is) but only in the hands of very careful and sophisticated analysts. If left in the hands of anyone else the risk of their speaking very little truth or none at all is very high. But there is also another lesson to learn from Messedaglia's analysis: that even very careful analysts, indeed scientists, need a strong motivation to do their job carefully, and such motivation is often not purely technical or methodological, but ideological and political. Indeed, at least at first sight, the methods of Quételetian social physics seemed to "fit" well, to use a term employed by modern statisticians, the politics of national identity.

THE IRONIES OF NATIONAL UNIFICATION

Needless to say, "southern" and "northern" are relative terms: geographically a place is always located south or north of somewhere else depending on the location and perspective of the observer. The same is true when the terms are used to qualify and rank whole peoples: the position and power of whoever does the ranking determines the outcome. The Lombards and Venetians, *meridionali* in the Austrian Empire, ceased to be so in the new Kingdom of Italy, but together with their fellow Italians continued to be southerners in Europe. When the new Italian state began collecting and publishing figures concerning the administration of civil and penal justice, and the phenomenon of crime, Italians discovered that their only "primacy" was a very "unfortunate" one: they surpassed other European nations only with respect to the rates of violent crimes.[55] When disaggregated by geographical areas, these rates appeared to be especially high in the southern regions. One is inevitably drawn to ask whether the same rigor that informed Messedaglia's critical work on Austrian figures informed also the analyses of these other numbers or whether the official analysts resorted to the "southern" nature of the inhabitants of the southern regions to make sense of the differential.

Before trying to answer this question, however, we must avoid making a misleading assumption. The stereotype of the "southern

[55] For one of the first studies on Italy's "infausto primato" see G. Curcio, *Gli omicidi in Italia. Studio legislativo e statistico* n. p., n. d. [but Florence, 1871]. The theme, and related lamentations, were to persist for quite a long time: see Davis, *Conflict and Control*, ch. 12.

nature" has never been an exclusive monopoly of "northerners", whoever they are or were. "Southerners" too used it, and, for that matter, continue to use it today, in a variety of ways. Commenting on the official crime statistics of the Kingdom of the Two Sicilies in 1835, Pietro Calà Ulloa – the Neapolitan magistrate and chief prosecutor at the Gran Corte Criminale (the highest criminal court of a province) of Trapani in Sicily – explained the high number of threats, insults and assaults by resorting to those "sudden emotions [*impronti moti*] originating from a southern nature."[56] When speaking of the relatively higher number of homicides in Calabria he referred to the passions "which are more lively in those provinces," a veritable south of the South, to be associated with Corsica, Spain, and Greece.[57] This kind of explanation was not uncommon among southern supporters of the status quo; Ulloa was among those publicists who also claimed that the Kingdom of Naples fared better than England and France with respect to crime, an opinion which did not find much support among liberal commentators inside or outside the Kingdom. In the *Annali universali di statistica* the Piedmontese economist Giovanni Giovannetti replied skeptically to Ulloa that in some areas of the Two Sicilies crime rates were higher than in Corsica, notoriously the most criminal of the French provinces. Nor was there much satisfaction to be found in the very low percentage of crime among women in relation to men in the Kingdom, a figure more than three times lower than those for England and France:

> Certainly women are better, more moral than us [men], but this difference between Neapolitan, and French and English, women is to be related to the greater social role played by French and English women, to their easier inclusion in male tasks [*uffizi*], to their having acquired a position that perhaps makes them more inclined to break the law, but is bound to bring immense advantages, far greater than this inconvenience.[58]

Thus while Bourbon rule was a favorite object of criticism on the part of liberal opinion in Italy and abroad, those who wanted to protect the government from accusations of inefficiency and backwardness invoked the natural predisposition of the people, although few went as far as the reactionary Marquis di Foresta who in 1805 had proclaimed that the causes of the current unhappy state of affairs were to be found in the "sweet fatalism" of the inhabitants of the Kingdom.[59]

56 "Dell'esposizione dei reati in Inghilterra e in Francia," p. 255.
57 Ibid., pp. 252–255.
58 Review of *Dell'amministrazione della Giustizia Criminale nel Regno di Napoli. Esame e paragone* by P. C. Ulloa, *AUS* 48 (1836), p. 252.
59 See A. Mozzillo, *Viaggiatori stranieri nel Sud* (Milan, 1964), p. 70.

In the new Italian state, statistical observers employed analogous arguments to account for the figures on crime. In fact, in the early statistical publications of the Ministry of Justice of Italy, which was in charge of criminal statistics until 1882, the data were presented with little elaboration, accompanied by some very general remarks on Italy as a whole and in comparison with other countries (France in particular). This paucity of elaboration did not satisfy expert readers such as the future director of the Italian statistical services, Luigi Bodio, who in 1867 in a publication addressing the International Congress of Statistics held in Florence complained of how little could be learned from that mass of poorly ordered information (not even the totals appeared in the tables).[60] Official commentators generally resorted to the very kind of explanations which Austrian statisticians had earlier applied to make sense of the "anomalous" percentages of crime found in certain areas of the Empire: thus in his introductory comments to the crime statistics for 1869, Minister of Justice Giovanni De Falco maintained that the low rates of prosecutions *vis-à-vis* the number of denunciations were caused by the "subtle devices of the criminals ... deplorable intimidation of witnesses ... the dangerous habit of replacing the legitimate work of justice with private *vendette*."[61] In an official publication of 1873, Giorgio Curcio, vice-president of the Tribunal of Florence and head of the statistical office of the Ministry of Justice, in his brief remarks on the geographical pattern of crime, observed that violent crime was more widespread in the South "because of the impetuous and vivacious temperament [*carattere*] of those populations."[62] However, there was room for hope to be found in this: since people were inclined to anger and to the "passionate desire for personal vendettas" particularly in some provinces "in which the moral and material progresses of civil life [were] less widespread," Minister of Justice De Falco remarked in 1871, it was "legitimate to hope that time and changing conditions will gradually eliminate the causes and the effects of these deplorable habits."[63]

Messedaglia would have agreed. In his lecture on the statistics of crime that opened his course at the University of Rome in 1878 and while upholding – as befitted a child of his time – the idea of the exis-

[60] *Sui documenti statistici del Regno d'Italia. Cenni bibliografici presentati al VI Congresso Internazionale di Statistica* (Florence, 1867), pp. XXX–XXXV.

[61] Ministero di Grazia e Giustizia, *Statistica giudiziaria penale del Regno d'Italia per l'anno 1869 e ragguagli comparativi con alcuni anni anteriori* (Florence, 1871), p. vii.

[62] *Della statistica giudiziaria civile e criminale nel Regno d'Italia* (Rome, 1873), p. 23 (originally published in *L'Italia economica nel 1873* [Rome, 1873], pp. 367–415).

[63] Ministero de Grazia e Giustizia, *Statistica giudiziaria penale del Regno d'Italia per l'anno 1869*, p. viii.

tence of a kind of criminality which was typical of "civilization" and of one which was typical of "barbarism," he expressed strong warnings against the danger of the misuse and simplistic reading of crime statistics, against various forms of deterministic thought, and against hasty generalizations and the racial explanations that were becoming popular at the time.[64] It is likely that the painstaking analysis he had made of the crime statistics of the Austrian Empire taught him a lasting lesson on the need for discrimination and analytical rigor, but this was a lesson lost on whoever looked at the figures to find a confirmation of their own prejudices.

64 "La statistica della criminalità. Prelezione al corso di statistica presso la R. Università di Roma (15 gennaio 1879)," *Archivio di statistica* 3 (1878 [*recte* 1879]), pp. 438–495. On the differences between the culture of state statisticians and that of the school of criminal anthropology see also C. Pazzagli, "Statistica 'investigatrice e scienze 'positive' nell'Italia dei primi decenni unitari," *Quaderni storici* 15 (1980), pp. 779–822.

CHAPTER 7

A MAP OF THE NEW NATION

> The unification of Italy makes increasingly possible, though not easy, those [statistical] studies to which the past opposed so many obstacles. The government itself participates in these researches and is anxious to communicate the results to the public. Reciprocally, the economic revolution, which necessarily follows every political revolution, is awakening all national forces and energies, and thus opening a new space to the spirit of enterprise which has been asleep for too long.[1]

These words from the *Annuario di economia sociale e di statistica pel Regno d'Italia*, by two Piedmontese commentators, P. Duprat and A. Gicca, are indicative of the hopes shared by liberal sectors of public opinion in the early 1860s, after the proclamation of the Kingdom of Italy. They express a still optimistic view of the future that awaited the new nation, and of its capability to mobilize "national forces and energies," as well as of the possibilities offered by the new state institutions to the development of that most stately and national of all forms of knowledge, statistics.

In the newly proclaimed Kingdom of Italy, the process of construction of a liberal order was to receive many benefits from the "social institution" of statistics – to use an expression recurrent in the vocabulary of the official and unofficial publications in those years.[2] It is known that the newly established Direzione di Statistica (Directorate

1 P. Duprat and A. Gicca, *Annuario di economia sociale e di statistica pel Regno d'Italia*, vol. I (Turin, 1863), p. 9. Gicca was at the time an employee of the Ministry of Agriculture, Industry, and Commerce; Duprat was a lawyer.

2 See, for example, C. Correnti and P. Maestri, eds., *Annuario statistico italiano. Anno II* (Turin, 1864), p. xxii, and Ministero di Agricoltura, Industria e Commercio (henceforth MAIC), *Statistica del Regno d'Italia. Popolazione. Movimento dello stato civile nell'anno 1863* (Florence, 1864), p. v (the author of the text is Pietro Maestri, the head of the Direzione di Statistica). On this view of statistics in those years see also C. Pazzagli, "Statistica 'investigatrice' e scienze 'positive' nell'Italia dei primi

of Statistics) put a special effort into the immediate production of the basic statistics of the nation, namely the population census. The first census, defined by the Minister of Agriculture, Industry, and Commerce, G. Manno, as the "statistical double-check of the national plebiscite," was taken in December 1861 "with the impatience of an act of possession."[3] This "act of possession," in which significantly the nation was endowed with the position traditionally assigned to the female gender, occurred while another, much more violent, act of possession was taking place in the southern provinces. There the army was engaged in a war against "brigands," allegedly instruments of the past regime, but in fact often an expression of the profound malaise and dissatisfaction of impoverished peasant masses who had hoped the new regime would meet some of their basic demands for land.[4]

In the first decade of existence of the new state, the condition of these masses was not the object of specific social investigations, as it would be later, beginning in the 1870s. Now, however, the new state's elites had more urgent tasks in terms of intelligence gathering: reviewing the forces of the nation, counting its resources, mapping its territory and institutions. In the 1860s all these tasks, performed by the statistical service of the Kingdom, amounted to a work of symbolic appropriation of the new nation on the part of the state and served to legitimate the new power. As the statisticians who performed this work were not simply neutral bureaucrats or technocrats, but were themselves participants in the political battles of the day, official statistics became a tool, sometimes a weapon, in the debate over the issue of the administrative organization of the new state that opposed different segments of the political elites. Believing they were bringing the contribution of a "positive" knowledge to the building of the new state, Maestri and his collaborators used official statistics to voice their concerns and suggest reform proposals intended to facilitate

decenni unitari," *Quaderni storici* 15 (1980), pp. 779–822, which reconstructs the theoretical debate on statistics between the 1870s and the 1890s.

3 The first quote is from MAIC, *Statistica del regno d'Italia. Popolazione. Censimento generale (31 dicembre 1861)*, vol. 1 (Turin, 1864), p. vi. The latter comes from *Annuario statistico italiano. Anno II*, p. 23. Romanelli has drawn attention to the meaning of this rushed operation and in general to the politics of statistical investigation in post-unification Italy in "La nuova Italia e la misurazione dei fatti sociali. Una premessa," introduction to a special issue of *Quaderni storici* devoted to "L'indagine sociale nell'unificazione italiana" (15 [1980], pp. 765–768).

4 The classic account is F. Molfese, *Storia del brigantaggio dopo l'Unità* (Milan, 1964). For a recent discussion of the meaning of the word "brigandage" see J. Dickie, "A Word at War: The Italian Army and Brigandage 1860–1870," *History Workshop* 33 (1992), pp. 1–24.

the integration within the unitary state of areas which were vastly different socially, economically, and in their political traditions.

The interconnection between statistical science and the politics of state building profoundly shaped the way in which the data collected were organized and read. A systematic observation of the country began then which anxiously scrutinized its territorial variations. In the 1850s, statistical publications informed by patriotic sentiment had produced a representation of the country in which the differences among the various areas of the peninsula, although extolled by patriotic rhetoric (which insisted on the handsome "variety" of the country), were submerged in a superior unity. In the 1860s, in the conviction – which bears resemblances to the outlook of the French reformers of 1789[5] – that an effective exercise of power depended on a rational administrative partition of the national territory, the Italian statisticians devised what they called a "scientific" division of the nation's "body." Like their French predecessors, they "divided in order to unify."[6] In doing so they introduced and consolidated in all official publications a partition of the national territory destined to become very entrenched in the reading of Italy both among ordinary people and in the scientific community. They were successful in dividing, but unifying was an altogether different matter.

THE INSTITUTION OF STATISTICS

The Divisione di Statistica Generale of the new Kingdom of Italy (henceforth Direzione di Statistica), created in October 1861, was a subdivision of the Ministry of Agriculture, Industry, and Commerce.[7]

5 See M.-V. Ozouf-Marignier, *La Formation des départements. La représentation du territoire français à la fin du 18e siècle* (Paris, 1989).

6 Ibid., p. 101.

7 The exact translation of Divisione di Statistica Generale is "Section of General Statistics." I will refer to it as "Direzione di Statistica" since this term was frequently used in the 1860s and became universally used when the Divisione was renamed "Direzione Generale di Statistica" by royal decree of February 10, 1878 which reorganized the whole section. It should also be noted that since the archives of the Direzione have disappeared (together with a large part of the archives of the Ministry of Agriculture, Industry, and Commerce), the reconstruction of the history of this office has to be based almost entirely on secondary literature (mostly written by the statisticians themselves): ISTAT [Istituto Centrale di Statistica], *Decennale 1926 IV – 1936 XIV* (Rome, 1936); R. Fracassi, "L'ordinamento dei servizi statistici," *Annali di statistica*, 8th s., 5 (1957), pp. 83–138; ISTAT, *Dal censimento dell'Unità ai censimenti del centenario. Un secolo di vita della statistica italiana* (Rome, n. d. [1961]). See also C. Corsini, "L'acculturazione mancata. Il processo formativo dell'organizzazione delle fonti demografiche nel periodo post-unitario," in *La popolazione italiana nell'Ottocento. Continuità e mutamenti* (Bologna, 1985), pp. 115–145.

Its history is thus linked to the vicissitudes of this department of state administration. The Ministry of Agriculture had been experiencing a precarious existence since its first establishment in the Piedmontese Kingdom. Created in 1848, apparently for reasons which had little to do with its sphere of competence, it was abolished by Cavour in 1852 under financial pressures (agriculture passed then to the Ministry of Interior, and commerce was absorbed into the Ministry of Finances) and then reconstituted in 1860.[8] Its re-establishment was motivated by the awareness that the economy of the new state needed government intervention: as the Minister of Finances, F. S. Vegezzi, observed in 1860, government intervention "certainly becomes more important in a state such as ours, which is formed by provinces in which agriculture, industry, and commerce, for natural reasons, or because of the existence of different legislative provisions and economic institutions, were and still are today in very different conditions."[9] In spite of the fact that the Ministry of Agriculture's share of the state budget was very modest (about 0.66 percent on average in the 1860s) and consequently its capacity for intervention very limited, the most dogmatic liberals were opposed to its existence and often demanded its elimination in parliamentary sessions only to have their demands rejected, at least in the 1860s, by the majority in parliament.[10] The story of the Minister of Agriculture is emblematic of the limited, if not altogether ambivalent, support that the elites of the new Italy gave to the activity of statistical investigations promoted by this department, and helps explain why the government in 1891 could make the decision of "skipping" altogether the decennial census because of financial considerations.[11]

In 1862, the Direzione di Statistica acquired a director who had obtained a remarkable reputation for his patriotism and his statistical studies before unification, Pietro Maestri. The appointment of Maestri, whose federalistic and democratic views were certainly a

[8] The vicissitudes of this department are narrated by A. Caracciolo in his *Stato e società civile. Problemi dell'unificazione italiana* (Turin, 1960), pp. 34–53.

[9] The quote comes from a bill presented by the Minister of Finances on May 22, 1860: see *Atti del Parlamento italiano* (henceforth *AP*), *Documenti*, leg. VII, p. 168.

[10] The Ministry of Agriculture, Industry, and Commerce was actually abolished in 1877, only to be reconstituted the following year within the Ministry of Interior after several protests from various economic and political quarters (see Caracciolo, *Stato e società civile*, p. 34). On the budgets of the various ministries see L. Izzo, *La finanza pubblica nel primo decennio dell'Unità italiana* (Milan, 1962).

[11] On this episode see Corsini, "L'acculturazione mancata." The annual budget for the Direzione di Statistica in the 1860s amounted to about 100,000 Liras (corresponding to about 1–2 percent of the Ministry of Agriculture, Industry, and Commerce budget). The census was assigned a budget of its own (300,000 Liras in 1861, but actually spent 640,000).

matter of public knowledge, could at first seem surprising in the political context of the 1860s, when the "party" of the moderate liberals, the so-called Destra Storica (Historic Right), dominated the political scene. But from the mid-1850s Maestri, like many others, had become closer to the moderate milieu and supported a unitary position on the issue of unification. Given his contribution to the struggle for independence and his studies on Italy, his political as well as scientific merits were well established in the eyes of the political class in power. The royal decree which instituted the Direzione created also a Giunta Consultiva (Advisory Committee) di Statistica to supervise and give direction to the work of the Direzione, and to guarantee the "conceptual unity" of the statistical investigations carried out by other state departments. Among the members of this committee were Maestri's former collaborator in the compilation of patriotic statistics, Cesare Correnti, now a member of the Chamber of Deputies and state councilor, the Sicilian economist Antonio Scialoia, professor of economics at the University of Turin, Attilio Zuccagni Orlandini, former director of the statistical office of Tuscany and at that time professor of statistics in the Regio Istituto di Studi Superiori in Florence, Antonio Rabbini, director of the land register, and Angelo Sismonda, professor of geology at the University of Turin.[12] Needless to say, the composition of the group reflected the prevalence of northerners in the political class in power.

The decree that instituted the Direzione di Statistica imposed at once the closing down of the statistical structures existing in the former capitals of the Italian states. Thus the personnel who used to staff those offices had to be transferred to other positions within the local or central administration, or made to retire within a few years. This was particularly the case of the employees of the Sicilian Direzione di Statistica, who did not arouse much sympathy and trust among the mostly northern new state elites, and were probably not eager to move to Turin, until 1865 the capital of the new state. The few who managed to find employment in the new central structure were people whose political record, intellectual background, and origins could be well assimilated to those of the new state elites: a close collaborator of Maestri was for example Giovanni Anziani, former employee of the Tuscan statistical office; another was Pietro Castiglioni, a Milanese physician

[12] The list of the nine members of this committee is in ISTAT, *Decennale*, p. 63. Royal decree no. 294, October 20, 1861, which instituted the statistical service, is in *Raccolta ufficiale delle leggi e dei decreti del Regno d'Italia anno 1861*, vol. 1 (Turin, 1861). A description of all the law provisions on the statistical service is in an appendix to Fracassi, "L'ordinamento," pp. 141ff.

who had migrated to Piedmont after 1848, found employment in the Ministry of Agriculture, Industry, and Commerce in 1862, as compiler of the *Annali del Ministero di Agricoltura, Industria e Commercio*, the official periodical of the ministry, and subsequently became part of the statistical section.[13]

The actual scientific autonomy of the head of the Direzione was limited by the inclusion of this office within the Ministry of Agriculture. Since the statistical service was associated with this ministry, it too became a target for attack from dogmatic liberals and was subjected to the budgetary limitations which, in the attempt to close the deficit, the governments of the Destra Storica imposed on many areas of public spending.[14] The royal decree of 1861, which instituted the Direzione, also provided for the establishment of special provincial offices with a personnel devoted entirely to the collection of data to be sent to the Direzione. In 1862, however, a new Minister of Agriculture, Gioacchino Napoleone Pepoli, decided to replace these paid employees with local notables, reviving an institution (the provincial statistical committees) which had already existed in the Kingdom of Sardinia and had been criticized for its ineffectiveness.[15] The relative independence and effectiveness of the Direzione di Statistica was thus altogether limited at the peripheral levels of the state by the lack of any specialized personnel exclusively assigned to the collection and elaboration of data.

In spite of its limited budget, in less than ten years the Direzione di Statistica produced a considerable number of works: about sixty volumes of statistical investigations, as its director proudly observed in a report he made to the Minister of Agriculture in 1869.[16] It was probably Maestri's and his collaborators' activism, originating in their faith in the politically crucial role of statistical science, that accounted for this remarkable (at least from a quantitative standpoint) accomplishment. Of these works, eleven volumes were devoted to population, three to meteorology and mineral waters, one to elections and the electoral body, one to a list of communes, two to the finances of

[13] The personal dossiers of the employees of the pre-unification statistical offices as well as of the new Direzione di Statistica can be found at the Archivio Centrale dello Stato in Rome, Ministero di Agricoltura, Industria e Commercio, Gabinetto, nos. 1–6. These appear to be the only documents concerning the Direzione di Statistica which have been preserved.

[14] On the budgetary problems of the new state see Izzo, *La finanza pubblica*.

[15] See ISTAT, *Dal censimento dell'Unità*, pp. 52–54.

[16] See P. Maestri, *Le pubblicazioni della Direzione di Statistica. Relazione a S. E. il Ministro di Agricoltura, Industria e Commercio con note bibliografiche e sommari statistici e parere della Giunta consultiva di statistica* (Florence, 1869).

communes and provinces, eight to industry (silk production, mining industry, industrial societies), twelve to commerce and related matters, one to a summary of data of various kind on Italy, two to savings banks and friendly societies, two to public health, four to charities, nine to public and private schools, and six to the International Statistical Congresses (the sixth took place in Florence). It was indeed a wide range of topics, typically covering the principal concerns of a ruling class which not only wanted to know the main resources of the new state in terms of population and economic activity, but also intended to extend its hold over the life of local administration and on those sectors traditionally in the hands of the Church, such as private charities and schools, whose control was strategic in the building of a national society.[17]

Remarkable for its total absence from this list is the country's principal economic resource and the source of occupation for the overwhelming majority of its population – agriculture. This striking absence turns out not to be a mere outcome of the difficult conditions in which the statisticians had to work in the 1860s, but a stable feature of Italian statistics in the next four decades.[18] Only in the early 1870s did the prefects begin to collect regularly data on the amount of land devoted to agriculture and on agricultural production, but they were never able to assemble and publish anything other than highly questionable estimates. Maestri, and in general all the statisticians after him, maintained that the absence of a general and uniform land register for the whole peninsula was the fundamental obstacle to the compilation of a statistics of agriculture. The absence of a uniform land register was in turn due to the resistance of landowners, reluctant to collaborate in an operation that could eventually increase their tax burden.[19] It was not until 1886 that a state law introducing uniform

[17] For an analysis of the political genesis and implications of the investigations on schools see P. Macry, "La questione scolastica: controllo, conoscenza, consenso (1860–1872)," *Quaderni storici* 15 (1980), pp. 894–917; on the statistics of charities see M. Piccialuti-Caprioli, "Il patrimonio del povero. L'inchiesta sulle opere pie del 1861," Ibid., pp. 918–935.

[18] On the vicissitudes of agricultural statistics and their poor reliability see G. Federico, "Per una valutazione critica delle statistiche della produzione agricola italiana dopo l'unità (1860–1913)," *Società e storia* 5 (1982), pp. 87–130.

[19] In his report on Italian statistics at the International Statistical Congress of The Hague in 1869, Maestri referred to the fear of many that a statistics of agriculture would be used to increase their taxes. This generated strong suspicions and recriminations (not necessarily unjustified, according to Maestri); thus any "scientific" inquiry appeared to be intrinsically tainted in the eyes of most people (see *Congrès international de statistique à la Haye. Septième session du 6 au 11 septembre 1869*, vol. III [The Hague, 1871], p. 66).

land taxes across the state was passed, and it took approximately seventy years before the land survey was fully completed.[20] Any investigation that could lead to increased taxation was of course looked on with suspicion by those who had to release the information, no matter what the statisticians said about confidentiality and the absence of any relation between their investigations and tax policies. The same applies, although to a lesser degree, to the statistics of industrial production, another sore spot in the enterprise. The census on manufacturing industry launched by Maestri soon after his appointment in 1862 led to dismal results because of both the very heterogeneity of the social and economic landscape of the country, and the informants' distrust.[21] Industrialists were willing to provide information only when this could lead to the implementation of measures designed to favor their interests: so they did collaborate in the inquest (not a statistical investigation, however) on the state of industry in the early 1870s, because they intended to pressure the government to introduce higher tariffs.[22]

Thus a mixture of reluctance on the part of the statisticians to alienate the propertied classes, and resistance on the part of the latter to reveal their income, resulted in a very poor performance in the field of the statistics of productive activities.[23] What the Direzione di Statistica was exclusively able to do was to collect information on aspects of the life of the new state which appeared *not* to threaten the power and position of property holders. Although in no country were statisticians endowed with the means of compelling the public to supply them with figures, and even in Britain (where statistics

[20] See G. Candeloro, *Storia dell'Italia moderna*, vol. VI, *Lo sviluppo del capitalismo e del movimento operaio* (Milan, 1970), pp. 217–218.

[21] Two volumes devoted to the industries of the provinces of Brescia and Bergamo constituted the only published outcome of this early attempt to make a statistics of industry: on this episode see A. Polsi, "La 'statistica dell'industria manifattrice' del 1862," *Quaderni storici* 15 (1980), pp. 894–917. See also V. Ellena, *La statistica di alcune industrie italiane* (Rome, 1879).

[22] I am referring to the inquest of 1870–1874 on which see G. Are, "Una fonte per lo studio della fondazione industriale in Italia: l'inchiesta del 1870–74," *Studi storici* 4 (1963), pp. 241–291, and M. Abrate *et al.*, *L'imprenditorialità italiana dopo l'unità. L'inchiesta industriale del 1870–1874* (Milan, 1970); see also S. Lanaro, *Nazione e lavoro. Saggio sulla cultura borghese in Italia 1870–1925* (Venice, 1979), pp. 183–184, which stresses the close linkage between the new protectionist orientations of industrialists and the launching of this inquest.

[23] In contrast, and perhaps not surprisingly, stands the French case. The French administration was able to generate fairly accurate statistics of agriculture during the whole nineteenth century: see G. Garrier, "Les Enquêtes agricoles décennales du XIXe siècle: essai d'analyse critique," in *Pour une histoire de la statistique* (Paris, 1977), pp. 269–279.

enjoyed great support in liberal circles) they were not always successful in obtaining solid results,[24] in Italy the statisticians' peculiar lack of success in the field of the economy reflected the wide gap between the culture of the men in charge of the new state apparatus and the social elites. To use the statisticians' own metaphor, "possession" of the nation on the part of the state was never as complete as the they had wished.

But let us not anticipate too much of what would become clear to the protagonists only later. The foregoing account serves to underline that, contrary to their expectations, in the 1860s (and even later) state statisticians found themselves still fighting a battle not too different from the one in which they thought they were engaged before 1861. They still had to convince those in power about the importance of what they did. They continued to publish their works with the purpose of influencing the formation of a national public opinion, of spreading a "scientific" knowledge of the country to those whom they perceived to be the backbone of the new state, the educated middle classes. In order to do so, they promoted the circulation of their publications by distributing about a thousand copies of each publication free of charge.[25] They also devised a "popular" edition of the works of the Direzione of which however only one volume, containing Maestri's own introductions to the census data, materialized in 1867.[26] The head of the Direzione and his collaborators never limited their task to a clinical presentation of the results of their investigations, but passionately interpreted and commented upon them. The publication in 1864 of the second volume of the *Annuario statistico italiano* (the first had appeared in 1857–1858) explicitly linked the enterprise of national statistics to that of patriotic statistics, and gave statistical science an all-important function in the building of the new nation.

[24] See L. Brown, *The Board of Trade and the Free-Trade Movement 1830–1842* (Oxford, 1958), p. 87.

[25] See ISTAT, *Decennale*, p. 38.

[26] MAIC, *Statistica del Regno d'Italia. Popolazione. Parte 1. Censimento generale (31 dicembre 1861)* (Florence, 1867). In the preface to this volume Maestri maintained that he had the intention of "making the factual information which reflects the Kingdom of Italy more accessible and more popular. [This series of publications] will be made up of several small volumes, each treating a particular subject matter, so that the series will comprise a general statistics of Italy." Although this project was not carried out, the size of later statistical publications became less monumental than that of the earlier large volumes. Several deputies had also demanded a smaller size for these publications during the discussion of the budget for statistics in 1869: see *AP*, *Discussioni*, leg. x, March 10, 1869.

IN SEARCH OF TRUE MEANS AND LAWS

Historians locate the advent of positivism in Italy in the mid-1860s when Pasquale Villari opened the debate with an essay on the historical method and positive philosophy.[27] More precisely, the period beginning in the 1860s was the start of a "second positivist era."[28] While the earlier phase had been characterized by the diffusion of theories of progress of Saint-Simonian origins and by an interest in the collection of "positive facts" and empirical investigations, the new phase was going to be characterized by the diffusion of the philosophies of Comte and Spencer and of Darwinian evolution among significant segments of the educated elites – especially in the northern regions and in Tuscany – and by an interest in the discovery of the "laws" of societal development. A heightened emphasis on the power and the benefits of science appeared then which fitted the ideological needs of the ruling elites of a state whose very existence was strongly opposed by the Catholic Church.

In the 1860s and early 1870s however, positivism was not yet the naturalistic and deterministic philosophy that it would become in later years. A critical and non-reductionistic perspective informed the hope that science could indicate solutions for the problems facing the new state. With equal enthusiasm Correnti and Maestri in the *Annuario statistico italiano* of 1864 greeted both the advent of a new Italy and the new perspectives brought about by the adoption of the scientific method in the study of society. Within statistics in particular – which the two patriots, with characteristic emphasis, defined as nothing less than the "discipline of democracy" and the "reflexive and experimental consciousness of mankind" – the introduction of mathematical methods allowed the observer to find light and certainty amidst the "infinite variety," "tumultuous confusion" of facts, and to identify the laws that ruled society and the life of mankind.[29]

What were the mathematical methods to which Correnti and Maestri referred? One should not imagine anything fancy: calculation and comparison of means, proportions, and rates were essentially the tools these authors referred to. The statistics of the 1860s did not differ

27 See E. Garin, "Il positivimo come metodo e come concezione del mondo," in *Tra due secoli. Socialismo e filosofia in Italia dopo l'Unità* (Bari, 1981), pp. 65–89; A. Asor Rosa, "La cultura," in *Storia d'Italia*, vol. IV, *Dall'Unità a oggi* (Turin, 1975), pp. 879ff.

28 See G. Sola, "Sviluppi e scenari della sociologia italiana: 1861–90. Percorsi di lettura da una bibliografia," in F. Barbano and G. Sola, *Sociologia e scienze sociali in Italia 1861–1890. Introduzioni critiche e repertorio biliografico* (Milan, 1985), p. 114.

29 *Annuario statistico italiano. Anno II*, p. xiii.

much from the empirical works of the 1850s. Nor did Italian practitioners abandon the idea that statistics was essentially a governmental science and had an eminently civil function. What changed more remarkably was, on the one hand, their approach to the use and meaning of average numbers, and, on the other, their conception of what properly constituted statistical science. Although examples of the purely descriptive conception were not lacking,[30] the Quételetian stress on mathematical methods and statistical laws permeated the most significant theoretical contributions of the time, as well as the works of the Direzione di Statistica. Nobody dared any longer compare absolute numbers across countries or over time as many used to do in the first half of the century. That a large number of observations was needed in order to build meaningful averages became a widely accepted notion; so did the distinction which Quételet had made since the 1840s between "true" means – calculated on the basis of homogeneous populations, namely populations with a "normal" distribution – and arithmetic means – expression of heterogeneous collectivities: he maintained that only the former revealed something about their collective character and pointed to underlying, constant causes, while the latter were artificial constructs, a pure fiction.[31] It is significant that, imported into the Italian literature of the 1860s, this kind of methodological notion was immediately grafted onto issues of governance and state building.

In 1861, in his *Dizionario della economia politica e del commercio*, the economist and geographer Gerolamo Boccardo observed that statistics was "an inventory, a classification, a more or less orderly, more or less complete, more or less systematic enumeration" of facts.[32] But this, he added, could not properly be described as science; for only when statistics inferred "a series of theorems and laws" from a "large number of

[30] See for example A. Zuccagni Orlandini's *Elementi di statistica* (Florence, 1869). The old conception persisted also through the institution of chairs of "geography and statistics" in the early 1860s following a Piedmontese law of 1859 which had created this kind of position at the University of Turin. Messedaglia criticized strongly this institution which, however, survived until the mid-1870s: see M. G. Ottaviani, "Note per una storia dell'insegnamento della statistica in Italia: la statistica nell'ordinamento didattico dell'istruzione superiore dalle origini al 1938," *Statistica* 47 (1987), pp. 621–630.

[31] T. M. Porter, *The Rise of Statistical Thinking 1820–1900* (Princeton, 1986), p. 107. On the history of the changing conceptions of the mean see also the essays in J. Feldman, G. Lagneau, and B. Matalon, eds., *Moyenne, milieu, centre. Histoires et usages* (Paris, 1991).

[32] G. Boccardo, *Dizionario della economia politica e del commercio così teorico come pratico utile non solo allo scienziato ed al pubblico amministratore, ma eziandio al commerciante, al banchiere, all'agricoltore ed al capitalista*, vol. IV (Turin, 1861), p. 428.

observations, of experiences, of facts collected not only in this or that country, but in several ones and amidst diverse circumstances," only then did it acquire the status of science.[33] As descriptive and classificatory practice, statistics was now confined to the status of a mere "art," while the prestigious label of "science" was assigned instead to the application of the numerical method to the search for laws, understood as those empirical regularities which characterized social phenomena when observed in large masses. The two-pronged definition offered by Boccardo's *Dizionario* exemplifies the merging of the descriptive conception of statistics with the conceptions of a new generation of statisticians who embraced with enthusiasm – Messedaglia is the best example in this respect – the methodology developed by Quételet. In the *Annali universali di statistica* Vincenzo Ferretti defined statistics as the "profound and complete knowledge of the manifestations of social facts, numerically determined in their quantity, in their relations, and movements." He declared with assurance that the study of the relations among the elements of a society would produce those statistical averages which "represent the invariable law of social movement."[34] At the same time he did not abandon the idea that statistics was a science of the state – the latter made into the "repository" of the forces of society.[35] It also continued to have its usual expository and classificatory functions, having to present in their "logical" order territory, population (defined as nothing less than "the soul of the country, its force, its power, its wealth, and its glory"),[36] production, and so on. The same two conceptions coexisted in the work of the future head of the Direzione di Statistica, Luigi Bodio who, in his inaugural lecture to the course on statistics at the Scuola Superiore di Commercio in Venice, underlined on the one hand the privileged relations of statistics with political economy, geography, law, and history, and on the other introduced the methodology developed by Quételet with its typology of causes and the worship of the mean, defined as nothing less than "a model on the basis of which men are forged by nature."[37]

In order to underline the scientific quality of official statistics and to instill a correct research methodology in the administrative personnel who were supposed to collect data at the local level, the Min-

[33] Ibid.

[34] "Il concetto di statistica ed il suo momento storico. Memoria," *Annali universali di statistica* (henceforth *AUS*), 4th s., 33 (1868), pp. 35–64 (the quotations are from p. 40).

[35] Ibid., p. 46.

[36] Ibid., p. 53.

[37] *Della statistica nei suoi rapporti coll'economia politica e colle altre scienze affini* (Milan, 1869).

ister of Agriculture, Industry, and Commerce, Filippo Cordova (who supervised the Direzione di Statistica in its very inception, before the appointment of Maestri) even announced a prize for the best "statistical manual to be used in the Kingdom of Italy."[38] The text had to cover the usual elements of this kind of publication: historical outline, bibliography, elementary postulates of the science, some practical instructions on general statistical procedures, and some applications to the statistics of the territory, population, agriculture, industry, and commerce. But it also had to include, it was recommended, a "summary of statistical laws, namely of the principal results – average, minimum and maximum – obtained in various countries with regard to every kind of information." We have no evidence that this prize was actually awarded, but contemporary textbooks of statistics show how this suggestion was followed. The author of a manual of statistics for technical schools, a professor of economic and legal sciences at a high school in Vercelli, gave the following examples of laws of the "economic and moral world": the positive influence of the increases in means of production on population growth (illustrated by the case of Lombardy); the positive relation between literacy and humane and civil government; the variation in illegitimacy rates according to good (the example was Piedmont) or bad (the example was Sicily) government.[39] This volume received the prize awarded by the Italian Pedagogical Society for the best work containing "statistical notions for the people."

In the examples of laws just cited the independent variable, to use the language of today's statistics, turns out to be, in two cases out of three, the action of government. Again, the new statistics was immediately put to political use in the construction of the national state. The same attitude informed reflection on the nature of averages. The heterogeneity of social conditions in the new state made national means a fictitious construct. In a booklet significantly entitled *Errori prodotti dalle medie nella statistica* and published in Milan in November 1861, Pietro Castiglioni[40] warned that averages summarizing very diverse situations should not be used by the government for policy-making purposes. He had in mind social services for the population, such as schools, hospitals, public relief institutions, and so on. In this respect it clearly made

38 See announcement in *AUS*, 4th s., 5 (1861), pp. 331–332. Filippo Cordova – a Sicilian émigré to Piedmont who was a member of Cavour's faction – had already been in charge of the census in the Kingdom of Sardinia: see ISTAT, *Dal censimento dell'Unità*, pp. 23–24.

39 L. Guala, *Elementi di statistica italiana teorica e pratica* (Biella, 1866).

40 On this author see ch. 4.

no sense to average places which had no such services and places which had some or plenty. But the same procedure applied also to measures of social behavior. As the author of the *Elementi di statistica italiana teorica e pratica* observed, one needed to disaggregate general averages and focus on local averages in order to obtain meaningful measures for comparative purposes.[41] What the boundaries of meaningful localities should be was decided by Maestri's division of Italy into *compartimenti* (compartments) which would provide the framework for the analysis of national statistics.

LOOKING FOR A "SCIENTIFIC" PARTITION OF THE COUNTRY

The genesis of the "compartments" shows the extent to which Maestri and his collaborators thought they could shape state administration through statistics, and conversely the extent to which their politics shaped the statistics of the new nation. We must recall that among the main institutional undertakings of the moderate liberals in power in the first decade after unification, there were the extension of the same laws to the whole of the national territory, the installation of the representatives and symbols of the new liberal order in every corner of the state, and a first attempt at "nationalizing" the population through a stimulus, albeit moderate, to schooling and the establishment of general conscription. But national laws were essentially Piedmontese laws, an extension of the legislation of the former Kingdom of Sardinia to the rest of Italy which its critics labeled *piemontesismo* (Piedmontization). This process aroused sharp criticisms everywhere, and especially in Lombardy, the area which in the pre-unification decades had expressed a strong current of federalistic thought, exemplified in the figure of Carlo Cattaneo. Now that federalism had been politically defeated, claims for some form of local autonomy within the new nation found strong support across the political spectrum, and also across regions, from Lombardy to Sicily.[42]

The establishment of an administrative structure informed by Piedmontese legislation and centralized practices of government on the national territory did not receive full legislative confirmation until 1865. In the preceding years, a long and complex debate over communal and

[41] Guala, *Elementi di statistica italiana*, p. 26.

[42] On this issue there is an extensive literature; probably the best account of the debate and its implications in the immediate period after unification is C. Pavone, *Amministrazione centrale e amministrazione periferica da Rattazzi a Ricasoli (1859–1866)* (Milan, 1964). See also R. Ruffilli, *La questione regionale dall'unificazione alla dittatura (1862–1942)* (Milan, 1971).

provincial legislation took place, in which Italian statesmen and politicians were split between supporters of centralization and supporters of greater local autonomy and a decentralized system of government. The former defended the policy of centralization as the only possible means of keeping together a new country threatened by centrifugal forces and internal opposition. The decentralizers' stances varied greatly: at the risk of over-simplification, the most moderate merely wanted a "bureaucratic" decentralization, namely a greater distribution of tasks to the organs of peripheral administration (such as the provinces), but always under the strict control of the center; Cattaneo and his followers insisted on local – municipal and regional – autonomy, which implied real subtraction of power from the center and the institution of autonomous organs of local government which would be more respectful of local traditions.[43] For decentralizers and autonomists, the methods of the centralizers were responsible for a purely "mechanistic" unity of the country; only a decentralized system, a system of local autonomies, could guarantee a real, "organic"[44] unity and a balance of interests between the different parts of the nation. The head of the Direzione di Statistica, Pietro Maestri, and his Lombard colleagues were themselves directly involved in these debates. In 1863 in a collection of essays which had originally been published in Cattaneo's *Politecnico*, Maestri denounced the policies of *piemontesismo* for having created only an illusory kind of uniformity.[45] The following year, Tullo Massarani, one of the contributors to the *Annuario statistico italiano* edited by Correnti and Maestri, proclaimed his preference for decentralization while granting the difficult circumstances in which political decisions had to be made in 1860–1861.[46] In the same work Maestri and Correnti maintained that the difficult issue of the state's administrative partition could profit from the enlightenment offered by science. One of

43 See Ruffilli, *La questione regionale*, chs. 1–2.

44 "Organic" was an expression used by Enrico Cenni in *Delle presenti condizioni d'Italia e del suo riordinamento civile* (Naples, 1862), p. 177, a work which demanded more autonomy for the provinces and the transfer of the capital to Naples.

45 See Maestri's *La Francia contemporanea. Studi economici ed amministrativi* (Milan, 1863). The critique of *piemontesismo* is in ch. 12, "Il principio di libertà e il decentramento amministrativo." While writing these essays, Maestri kept up a correspondence with Cattaneo in which he strongly criticized centralization and the projects of the moderates (see Museo del Risorgimento in Milan, Carte Cattaneo, no. 9).

46 T. Massarani (Mantua 1826 – Milan 1905) sat in the Chamber of Deputies until 1867, when he retired to devote himself to literary activity and to local politics in Milan; in 1876 he was appointed to the Senate. See *Dizionario del Risorgimento nazionale* (Milan, 1933), s. v., and A. Calani, *Il Parlamento del Regno d'Italia* (Milan, n. d.).

the functions of statistical studies was indeed that of offering a "*scientific* solution [emphasis added]" to the problem of "determining the administrative and civil topography" of the country, which "politics has obscured."[47] The statistical description of the country was thus to contribute in a concrete way to the pursuit of a rational and homogeneous administrative partition by identifying the real characteristics of the country; statisticians would thus bring to the attention of the government solutions based on the authority of science.

The claim clearly reflected a belief in a transcendent science of state and society, above factions, one able to provide solutions to political questions. To be sure, this was no Italian peculiarity, but a belief cultivated by social scientists throughout Europe in the nineteenth century.[48] In the case of Italian statisticians this faith – one that made Correnti proclaim that even military confrontations could be avoided once states knew the condition of their respective forces[49] – was joined to a widespread conviction, or should we say delusion, that political issues could find a solution at an administrative level, in "good government."

The analysis to which Maestri subjected the numerical data collected and published by the Direzione in the 1860s was informed by these objectives and convictions. This is evident not only in the comments that accompanied the publication of the results of the census and of other investigations, but also in the attention Maestri paid to the question of the size and nature of the various administrative subdivisions of the state: above all the communes, which supporters of a more decentralized administrative structure in particular considered as basic units in the social and political fabric of the new state.[50]

The commentary on the census opened with a classification of communes by demographic size and with the observation that there existed *profonde disformità* (profound differences) among municipalities. Modes of settlement varied on the national territory: small communes prevailed in the more populous northern areas of the country; larger com-

47 *Annuario statistico italiano. Anno II*, p. 47.

48 For a British example, see V. L. Hilts, "*Aliis exterendum*, or, the Origins of the Statistical Society of London," *Isis* 69 (1978), pp. 21–43.

49 See *Annuario statistico italiano. Anno II*, p. xxxv.

50 On the importance of the idea of the commune see E. Ragionieri, *Politica e amministrazione nella storia dell'Italia unita* (Bari, 1967), pp. 75–76, and Ruffilli, *La questione regionale*, pp. 19, 48–49, 105. For the debate on the role of communes in the new state see R. Romanelli, "La nazione e il campanile. Il dibattito intorno alle dimensioni dello scambio politico nell'Italia liberale," in F. Andreucci and A. Pescarolo, eds., *Gli spazi del potere. Aree, regioni, stati: le coordinate territoriali della storia contemporanea* (Florence, 1991), pp. 184–191.

munes prevailed in the less populous southern areas.[51] Were the small communes to be considered rural and the larger ones to be considered urban? Many large communes in southern Italy, which on the basis of their population could be classified as "urban," were on the contrary, as Maestri observed, no more than rural villages if considered from the standpoint of the actual social composition and function of their population.[52] The ostensible equivalences[53] created by conventional statistical classifications ran the risk of hiding more than they revealed of the characteristics of Italian society.

In order to come to terms with this problem, Maestri adopted a higher threshold of population than the one used in other countries to define an urban commune (6,000 inhabitants instead of the 2,000 used in French statistics at the time). He also added to the conventional system of classification by population size another system of classification – already in use in the Kingdom of Sardinia – which took into consideration the actual mode of residence of the inhabitants of a commune, namely whether they lived in *centri* (agglomerations) or in *casali e case sparse* (scattered houses).[54] The prevalence of a concentrated or a scattered mode of residence corresponded to distinct structures of property relations which characterized whole geographical areas. *Centri* were far more numerous in the South of the country where latifundia dominated the agrarian landscape and roads were scarce; there peasants lived concentrated in very large villages, far from the land they tilled. *Centri* also prevailed in a part of Lombardy where large properties, cultivated by wage laborers, dominated the rural landscape; there, however, the average size of communes was much smaller. By contrast, in areas of central Italy such as Tuscany, where most peasants worked as sharecroppers, scattered settlements were more common.[55]

By pointing out the differences in the demographic and territorial

[51] See MAIC, *Censimento generale (31 dicembre 1861)*, vol. 1, pp. xxii–xxv.

[52] Ibid., p. xxiii.

[53] For a discussion on the notion of "equivalence" on the basis of which statistical categories are created, see A. Desrosières, "How to Make Things Which Hold Together: Social Science, Statistics and the State," in P. Wagner, B. Wittrock, and R. Whitley, eds., *Discourses on Society. The Shaping of the Social Science Disciplines* (Dordrecht, 1991), pp. 195–218.

[54] MAIC, *Censimento generale (31 dicembre 1861)*, vol. 1, p. xxiii. Maestri also observed that "this scientific innovation, which mainly differentiates [the Italian] census from others, and gives it an honorable place among analogous operations undertaken by the other peoples of Europe," had already been devised by the makers of the Piedmontese census of 1857.

[55] Ibid., p. xxv. Contemporary historians continue to rely on this perceptive analysis: see P. Villani, "Gruppi sociali e classe dirigente all'indomani dell'Unità," in *Storia d'Italia. Annali 1. Dal feudalesimo al capitalismo* (Turin, 1978), pp. 884–893.

size of communes, and – with lesser emphasis – of districts and provinces, the state statisticians meant to attract the attention of "the government and representatives of the Nation," in support of the reform of the law on municipalities and provinces which was being discussed in parliament that same year (1864).[56] The emphasis they placed on the heterogeneity of the country represented a strong criticism of the administrative partition adopted by the new state, which merely reproduced the subdivisions existing in the former states. Their analysis translated into a demand for a more homogeneous partition – in which the average sizes and populations of communes and provinces would be more even – and gave implicit support to the arguments for administrative decentralization. It must also be added that the issue of the municipalities' size was politically relevant because of its linkage with the composition of the electoral body and thus with the overall working of the new representative regime. The minimum tax required for the franchise in local elections varied according to the demographic size of communes, being lower in the smaller ones. Yet very small municipalities still ran the risk of having too tiny a pool of eligible candidates; therefore a modification of the existing boundaries of villages and small towns appeared to be necessary for the correct functioning of liberal government. As Tullo Massarani observed in the *Annuario statistico italiano*, the "theory of majorities" (namely representative government) was particularly sensitive to numbers and the "probabilities" behind them.[57]

But not only did Maestri, by use of census figures, criticize the uneven administrative partition of the nation at the communal level, he also went so far as to create an altogether new partition of the national territory. Figures collected at the municipal level by the *giunte municipali di statistica* (municipal committees on statistics) were aggregated not only into reporting units that reflected the existing adminis-

56 MAIC, *Censimento generale (31 dicembre 1861)*, vol. I, p. xxiii.

57 See T. Massarani, "Italia politica," in *Annuario statistico italiano. Anno II*, p. 192. The electoral law is that of 17 December 1860. See also the comments on the data on elections in MAIC, *Statistica del Regno d'Italia. Elezioni politiche e amministrative. Anni 1865–66* (Florence, 1867), pp. xxxiii *et seq.*. This kind of argument found its way into parliamentary debates: see for example the intervention of a deputy of the opposition, G. Brunetti, on June 23, 1864 (*AP, Discussioni*, leg. VIII, p. 3097); he supported his requests for an extension of the franchise by making explicit reference to the *Annuario statistico italiano. Anno II*, and to the figures given to him by Pietro Castiglioni; see also the speech by a deputy of the Right, Stefano Castagnola who, on the same day, demanded a new, more uniform partition of the national territory and referred to the data he got from the "very beautiful work" on the census (pp. 3083–3086). No substantial change along these lines was actually implemented.

trative partition, such as *circondari* (districts) and provinces; in addition to these, Maestri also created for reporting purposes larger statistical units which had no administrative counterpart. These were the *compartimenti*, which some also referred to as "regions" (and they actually were called "regions" in official statistics from 1912 to 1918 and again from 1948 on). Through the creation of the *compartimento* as the basic unit of statistical reporting and analysis, a politically defeated but never extinct perspective of regional autonomy found its ultimate compensation in the rearrangement of national numbers.

The *compartimento* was not exclusively the product of Maestri's own mind. Indeed, the origin of this unit was in the idea of region elaborated by other Milanese intellectuals, Cattaneo and Correnti in particular, whose views we discussed in Chapter 5. Their definition of a region, we should recall, was based on geographical and historical criteria. They maintained that the Italian peninsula was in some degree "naturally" partitioned into regions, that is, it was formed of areas with distinct geographical and ethnographic characteristics; in the course of history these regions had often come to experience economic integration and in some cases even constitute distinct states, or districts within larger political formations.

The idea of the *compartimento* borrowed several elements from this conception, although at the same time it also partly modified it to meet current purposes. Consider the way in which Maestri presented it:

> our *compartimenti* are topographical or, in a word, statistical; they only reproduce the territorial divisions based on the nature of the soil and on the *laws of economic convenience*, territorial divisions which from ancient times and the Middle Ages *did not reflect political partitions*, nor the more inconstant administrative districts, but had a different and *solid basis in the laws of distance, movement, and work* [emphases added].[58]

Maestri's emphasis – evident in the repeated use of the term "law" – on the role of geography and economic relations in carving out distinct areas within the territory of the Italian peninsula shows how he was looking for criteria of partition which could represent a radical alternative to the present ones (the provinces) which were too dependent on the pre-existing political order.

There was, however, some wishful thinking, but also a degree of political astuteness, in the claim that nature, and the "naturalness" of

[58] MAIC, *Movimento dello stato civile nell'anno 1863*, p. vi. There were fourteen *compartimenti*: Piedmont, Liguria, Lombardy, Emilia, Umbria, the Marches, Tuscany, Abruzzi e Molise, Campania, Apulia, Basilicata, Calabria, Sicily, and Sardinia.

the economy – not politics – were the guiding principles of the new partition. In fact Maestri himself had to recognize that the *compartimenti* could not yet constitute the desired "scientific" partition of the nation. Only a better knowledge of the country could eventually allow for this creation. They represented, however, a first basic step in that direction, since they were constituted by the grouping of provinces linked together by "natural relations." Indeed Maestri went so far as to suggest that those purely statistical units were to be contemplated "from the standpoint of political and administrative convenience," since they could provide, better than did the existing ones (the provinces), a "homogeneous and proportionate partition of population and territory."[59] As this language shows, the statistician's aim went beyond mere description and analysis to prefigure the very transformation of the administrative structure of the new state according to an ideal of an almost geometric rationalization.[60] The example set by the French revolutionaries who had created the *départements* in the attempt to defeat old regional loyalties stood clearly in the background of this thinking, as was evident in the reference to the need to study the making of French *départements* (significantly called "topographic compartments") in the *Annuario statistico italiano*.[61]

But the *compartimenti* proposed by Maestri were far from reflecting this stated ideal; in fact they were simply aggregations of provinces, which, moreover, exhibited disturbing similarities with the former states, or with their former administrative subdivisions. Maestri had to concede that the *compartimenti* could "accidentally" recall the "image of the ancient states in which our country was divided, or of the regions in which some thought of conveniently dividing the national territory." However, he insisted that a careful observer would be convinced that the *compartimento* was "not modeled on the painful traces of a past which we hope is irrevocable, nor on the by now discredited prejudices of administrative federalism."[62]

This eagerness on the part of an advocate of regional devolution to emphasize the unity of the country should not be surprising. The

[59] Ibid., p. viii.

[60] See also Duprat and Gicca, *Annuario di economia sociale e di statistica*, pp. 15ff.

[61] Page 40. The fascination with *the* model for a "rational" partition of the nation, namely the French *département*, found a limit in the critical attitude towards Piedmontese institutions, whose French-like centralization was resented: see for example C. Norsa, *Considerazioni statistico-economiche sull'Italia* (Milan, 1864). On the making of *départements* in France see Ozouf-Marignier, *La Formation des départements*.

[62] MAIC, *Movimento dello stato civile nell'anno 1863*, p. vi.

Map 2 The Italian regions today. Maestri's *compartimenti* were the following: Piemonte, Liguria, Lombardia, Emilia, Umbria, Marche, Toscana, Abruzzo e Molise, Campania, Puglia, Basilicata, Calabria, Sicilia, and Sardegna (to which were added Veneto in 1866 and Lazio in 1870). The boundaries of these entities have changed very slightly over the years, following the addition or loss of territories in the north-east after the world wars, the creation of two new regions (Valle d'Aosta in 1945 and Molise in 1963), and some modifications to provincial boundaries.

recently attained and yet incomplete unification appeared as a fragile achievement, while the new nation was still under attack from the papacy and involved in a virtual war with so-called "brigandage" in the South. Unity and uniformity were the "obsession" of a time when even a very cautious and moderate project of bureaucratic decentralization providing for the constitution of regions – presented by the Minister of Interior, Marco Minghetti, on March 13, 1861 – was rejected by a parliamentary commission.[63] The head of the Direzione di Statistica clearly thought it necessary to protect himself from possible charges of undermining the "unity of the native soil" when introducing reporting units (those very *compartimenti*) which dangerously evoked previous divisions.[64] If the *compartimenti* showed similarities with the former states, Maestri further claimed, they were in fact the result of an adaptation on the part of former political subdivisions to natural constraints and cohesions: actually "the several groups of the Neapolitan provinces *predated* [emphasis added] the creation of the Southern Kingdom [of Naples]."[65]

Maestri's attempted reversal of the causal relation between state formation and geographical divisions could not hide the fact that several *compartimenti* (Piedmont, Tuscany, Liguria among others) continued to recall the political division of pre-unification Italy. The weight of history was particularly evident in the subdivision called "Provincie Napoletane" corresponding to the continental part of the former Kingdom of the Two Sicilies. This subdivision appeared in the volumes on the first census and in some other publications before being dropped and replaced by five *compartimenti* (Abruzzi and Molise, Campania, Apulia, Basilicata, and Calabria).[66] Ironically, the very tools that Maestri devised in order to promote a greater national integration turned out to reflect the country's past division, perhaps an inevitable, but certainly not a desired outcome.[67]

[63] A. Petracchi, *Le origini dell'ordinamento comunale e provinciale italiano. Storia della legislazione piemontese sugli enti locali dalla fine dell'antico regime al chiudersi dell'età cavouriana (1770–1861)*, vol. I (Venice, 1962), pp. 332–354. The text of the project is published in vol. III, pp. 335–343. Minghetti's "regions" followed Correnti's partition. E. Ragionieri speaks of "obsession" in "Politica e amministrazione nello stato unitario," in his *Politica e amministrazione*, p. 82.

[64] Even the use of the term "compartment" instead of "region" has to be read in this context: as a matter of fact, it was suggested that *compartimento* meant *parte unita* (united part) (see *Annuario statistico italiano. Anno II*, p. 41, footnote).

[65] MAIC, *Movimento dello stato civile nell'anno 1863*, p. vii. Maestri also recognized the utility of keeping the old boundaries for the sake of historical comparison (ibid.).

[66] This occurred in the volume on population movement in 1863 published in 1864.

[67] Let us recall the difficulty that Correnti had already encountered in the 1850s when he tried to propose a regional partition of the peninsula on purely "geogra-

THE MEANING OF REGIONAL AVERAGES

As we said earlier, the Direzione di Statistica published about sixty volumes of data in the 1860s. Each volume was introduced by detailed analyses (entitled "general considerations") aiming to define the patterns, extract the meanings and often provide an explanation of the numerical results. These analyses constructed and proposed a specific reading of the national territory; they made the numbers "speak." In contrast with the more technical and neutral tone that, as Pasquale Villani has observed, characterized the comments on the data in later years,[68] the commentaries of this period were passionate, and altogether candid about the beliefs and convictions of their authors. Indeed the wide gap between the expectations of the interpreter and what he "discovered" through his statistical lens is all too manifest. Less manifest, however, is the limiting logic of the language of numbers, which therefore requires a closer or, better, a "defamiliarizing" look.

The most important statistics of the first decade of existence of the new state were undoubtedly those on population. Demographic data played a fundamental role in the symptomatology of the national "body." The census formed the "basis of any statistical and economic inquiry," and the study of population movements was supposed to reveal the "laws according to which the nation preserves or renovates itself," as we read in the introduction to the first volume of the series.[69] Together these investigations constituted the review of the "vital forces" of the nation. Moreover, as state statisticians faced severe difficulties in collecting quantitative data on the economy, demographic data served as the main indirect indicator of the economic condition of the country.[70]

phical" or "geographical-dialectological" criteria and how he eventually gave up and settled for a compromise solution (see ch. 5).

68 P. Villani, "Gruppi sociali e classe dirigente," pp. 885–886. Villani links this characteristic to the separation between the bureaucratic and political class after the 1860s–1870s. By 1872 Luigi Bodio, who became the new head of the central statistical office in that year, emphasized that the task of official statistics was "to present the material according to the most open and transparent classifications, leaving to private researchers and the administration the task of combining the material according to their special criteria and purposes and of deriving their conclusions" – thus implicitly disavowing the kind of commentaries with which his predecessor had introduced the data collected (see MAIC, *Statistica del Regno d'Italia. Popolazione. Movimento dello stato civile nell'anno 1871* [Florence, 1874], p. xlvii).

69 MAIC, *Statistica del Regno d'Italia. Popolazione. Movimento dello stato civile nell'anno 1862* (Florence, 1864), p. viii.

70 On the reading of demographic statistics as indicators of the economic conditions of the population see, for example, A. Messedaglia's authoritative analysis of the

The three volumes on the census of 1861, published in 1864–1866, provided an overall picture of the population, what Maestri called with democratic emphasis "a description of the people made by the people themselves."[71] Volume I classified the population on the basis of its territorial distribution by communes, districts, provinces, and *compartimenti*. Volume II presented the age and sex composition of the population, its distribution by domestic status, rates of literacy; it also gave information on the electorate. Volume III showed the distribution of the population by occupation, language, origins, and religion.[72] In the introductory comments the figures were aggregated and compared by *compartimenti*, by provinces, and by rural and urban communes. The deployment of these various units of observation and reporting was very uneven. The reporting of data by *comuni rurali* or *comuni urbani*, although not absent, was rare, possibly owing to the many reservations, previously discussed, about inferring the status of a commune from its population. Data aggregated by provinces episodically accompanied the aggregation by *compartimenti*. Indeed it was the grid of the *compartimenti* that dominated the analysis of the data as a whole.

In the tables accompanying the comments, *compartimenti* were always arranged "geographically" – from the northern to the southern ones. Through the frame of the *compartimenti* the diversity of the peninsula was fully displayed and highlighted: modes of settlement, population densities, literacy, composition of the electorate followed distinct patterns across the national territory. In a few cases the differences seemed to cut the country into two parts, North and South. A line between North and South was drawn with respect to the stages in the life cycle of individuals, where a distinction was made between "sturdy boreal stock" and "precocious southern natures"; and with respect to the sex ratio, about which it was observed that "the smaller number of men, compared to women" in southern Italy was "a normal condition" accentuated by emigration and the scourge of brigandage."[73]

economic meaning of the "mean length of life" (*vita media*) in an essay published in 1866, "La vita media. Suo concetto, metodi di determinazione, criteri di applicazione," later reprinted in *Biblioteca dell'economista*, 5th s., 19 (Turin, 1908), pp. 193–270. On the importance of demographic statistics in the statistical production of post-unification Italy see also C. Ipsen, "The Statistics of Population in Liberal Italy," *Bollettino di demografia storica* 16 (1992), pp. 7–33.

71 MAIC, *Statistica del Regno d'Italia. Popolazione. Censimento generale (31 dicembre 1861)*, vol. II (Turin, 1865), p. xxii.

72 The second volume was published in Turin in 1865 and the third in Florence in 1866.

73 Ibid., vol. I, p. xxviii; the comments on life cycles are in vol. II, p. vii.

The census was nothing more than a static photograph of the nation, and thus satisfied only partially the need for a thorough knowledge of the population. In addition, it could only provide the portrait of the nation at the end of an old era. Necessarily this portrait was bound to reflect in every respect a situation inherited from the past: for example the fact that about 75 percent of the population was illiterate was clearly one of the heaviest legacies of past regimes. In fact, only the study of the evolution of demographic social indicators over time could show the effects of the new political and economic order. Figures on population movements therefore attracted even more anxious scrutiny than those of the census whose dismal percentages, it was hoped, would improve in the near future. In the belief that the positive effects of the new political situation should very soon make their appearance, the principal interpretive effort in the volumes on population movements aimed at determining whether, and where, the population increased or decreased.

The same units of reporting and comparison used in the volumes on the census, namely urban and rural communes on the one hand and, to a much a larger extent, the *compartimenti* on the other, were employed here to organize the analysis of the data; the provinces had virtually disappeared, with a couple of exceptions.[74] In the context of the urban–rural distinction, the statistician tried to determine whether it was the urban environment that favored the "vitality" (i.e. reproduction rates) of the population, or whether population growth was more substantial in the countryside. Even though Maestri maintained that the "heart of the nation"[75] was located in the numerous towns and cities with over twenty thousand inhabitants (and others even claimed that "from the fact of the greater agglomeration of the Italian populations [there derived] a higher degree of civilization"),[76] he did not cease to believe in the "normality" and health – both physical and social – of country life. The rural populations did not participate in the making of culture, yet they were a basic pillar of and stabilizing force in the nation – that is, when they owned their land or worked as share-

74 Provinces were used as reporting units in the commentary only in the case of literacy rates (measured by signatures in marriage registers) and for the relation between variations in grain prices and variations in the number of marriages and deaths. Several demographic measures reported by provinces can be found also in tables at the end of the volume on population movements for 1864.

75 MAIC, *Censimento generale (31 dicembre 1861)*, vol. 1, p. xxv.

76 The author of this remark was a Milanese writer close to Maestri, C. Norsa; it comes from his *Sul compartimento territoriale e sull'amministrazione del nuovo Regno d'Italia. Considerazioni statistiche ed economiche* (Milan, 1863), p. 13.

croppers.[77] Praise of urban culture and the idealization of small-scale landownership and sharecropping agriculture were actually the two sides of the same coin, of that paternalistically progressive ideology to which many Risorgimento figures such as Maestri subscribed.[78]

The first data on population movements, collected for 1862, seemed to point to a lesser "vitality" of the urban populations. However, as was observed, this was in part an "artificial" result, an outcome of the higher mortality rate to be found among the countryside poor who went to find relief in city hospitals and other institutions.[79] In fact, Italy did not seem to exhibit the pattern found in other European nations where peasant life appeared as the "normal" condition, namely healthier than urban life. In the peninsula, on the contrary, living in the countryside was not much healthier than living in the cities. When considered as a whole, moreover, the official commentator regretfully concluded, Italy seemed unable to preserve its "vital forces." Both in the countryside and in the cities, high levels of natality were accompanied by high levels of mortality, and especially by high infant mortality. Thus the head of the Direzione di Statistica had to admit

> that in Italy among both the populations of urban communes and those of rural municipalities, mortality reaches a proportion that has few examples elsewhere; to the point that the very increase of inhabitants is to be considered not as the effect of long life, but rather as the result of a larger number of births over deaths.[80]

In the following years, the comparison of the annual natural increase (measured by the excess of births over deaths) in rural and urban communes showed the former consistently above the latter, a result which was at the same time cause for surprise and some satisfaction. It was a cause for surprise because it ran against the overall developments observed at the time in other countries, such as France, where several cities grew rapidly at the expense of the countryside; but it was also a cause for satisfaction since the rural communes remained "a rich reservoir of forces for themselves and for the larger urban agglomerations."[81] Yet in the absence of data on emigration from the countryside to the cities these results were not really reliable and the

[77] For the praise of these categories see MAIC, *Censimento generale (31 dicembre 1861)*, vol. III, pp. ix–x.

[78] P. Villani, "Gruppi sociali e classe dirigente," pp. 891–893. On the so-called *modello mezzadrile* (sharecropping model) as a central component of the ideology of the Italian moderates see U. Carpi, "Egemonia moderata e intellettuali nel Risorgimento," in *Storia d'Italia. Annali IV. Intellettuali e potere* (Turin, 1981), p. 460.

[79] MAIC, *Movimento dello stato civile nell'anno 1862*, p. xxv.

[80] Ibid., p. xxvi.

[81] MAIC, *Movimento dello stato civile nell'anno 1864*, p. lxix.

overall comparison between the vitality of rural versus urban communes was abandoned after 1869.

In contrast, it was through the grid of the *compartimenti* that most demographic indicators were increasingly organized. These were the favored units for the assessment and comparison of the overall levels of "vitality" of the population. As summary results and indicators were all presented by *compartimenti*, these clearly offered the most important clue for making sense of the overall story that the figures could tell. Six demographic indicators (reduced to five in 1865) were chosen in order to determine synthetically the rank of the *compartimenti* on a kind of "ladder of vitality." They were the ratio of births to deaths, the fertility of marriages, population growth (replaced in 1863 by *male* population growth), mean and probable length of life (replaced by "longevity"[82] in 1865), and "premature" deaths (of people less than fifteen years old). It was a method of ranking that recalled the work of the French statisticians of the 1830s and 1840s, such as Adolphe D'Angeville, who had compared the *départements* of France according to a great variety of numerical indicators.[83]

With some exceptions, the southern *compartimenti* appeared consistently in the lowest positions. Initially, the commentator related the differences observed among *compartimenti* to the political conditions existing in the corresponding areas prior to unification and, in the rare case that a positive change occurred in a *compartimento*, to the revitalizing impact of unification. The well-marked territorial diversity revealed by these figures was considered a relic of the past, a relic which – it was the repeated wish – would disappear through the provisions of the new liberal institutions and the consequent revitalization of the economy.[84] The historical explanation, however, grew less and less adequate in view of the stability of the pattern which emerged through the years, and it was finally abandoned after 1866. Maestri avoided proposing alternative explanations. Indeed the comments on the results of ranking, which discredited optimistic expectations of the impact of liberal institutions, became increasingly laconic, in marked contrast

[82] "Longevity" corresponded to the percentage of people who died at eighty years of age or over. This was a rather crude indicator, probably chosen after criticisms had been aimed at the mode of calculating the mean and the probable length of life.

[83] *Essai sur la statistique de la population française considérée sous quelques-uns de ses rapports physiques et moraux* (Paris, 1836).

[84] For example, commenting upon the illiteracy data, Maestri observed how they "were not the fruit of the new civil institutions, but on the contrary the result of a state of things which fortunately does not exist any longer," adding that "the second Italian census will find the state of public education improved": MAIC, *Censimento generale (31 dicembre 1861)*, vol. II, p. xxvi.

with other issues on which the statistician expressed his definite convictions;[85] owing also to intrinsic flaws, ranking was abandoned in 1870.

Besides presenting and analyzing demographic data, the volumes on population movements reported also other kinds of data. They summarized the results of the analysis of the anthropometric data on conscripts collected by army doctors.[86] Following a practice which had first been established in France in the 1820s,[87] through these data the statisticians constructed indicators on the physical appearance and health of the population. Starting with the volume on vital statistics in 1863, they drew a picture of the physical characteristics of the population based on the percentages of rejections among conscripts on account of insufficient height, or illness and deformity. The results of these investigations showed that while in the regions of the North, and Tuscany, tall bodies were predominant, the southern population was characterized by shorter bodies.[88] In Italy heights followed a "topographical law, so to speak, which can be represented through a curve. This curve, reaching its peak in central Italy and precisely in Tuscany, slopes down on both sides in such a way that the slope is slightest northwards and maximum southwards."[89] While these results would later be interpreted by positivist criminal anthropology as evidence that two "races" inhabited the Italian nation,[90] in the 1860s and early 1870s the statisticians did not speculate on how the data should be interpreted. Moreover the data on the rejections of conscripts due to health problems showed a somewhat compensating pattern: people in bad health were relatively more numerous among the taller inhabitants of the northern and central regions (Lombardy surpassed all other areas in this respect), while, with the exception of Sicily, healthier bodies were more numerous among the shorter inhabitants of the southern *compartimenti*.

Other data seemed to offer more startling evidence on the existence of a cleavage among two distinctive parts of Italy. These were

85 For an example of the latter see the positive interpretation of the inconclusive statistics of natality and mortality in rice-growing communes in the volumes on population movements for 1867, 1868, and 1869.

86 On these studies see B. Farolfi, "Dall'antropometria militare alla storia del corpo," *Quaderni storici* 14 (1979), pp. 1056–1091.

87 See J.-P. Aron, P. Dumont, and E. Le Roy Ladurie, *Anthropologie du conscrit français d'après les contes numériques et sommaires du recrutement de l'armée 1819–26*, (Paris, 1972).

88 MAIC, *Movimento dello stato civile nell'anno 1863*, p. xl.

89 Ibid., p. xli.

90 See B. Farolfi, "Antropometria militare e antropologia della devianza 1876–1908," in *Storia d'Italia. Annali* VII. *Malattia e medicina* (Turin, 1984), pp. 1181–1219.

the data on "violent deaths," namely deaths by accident, suicides, and homicides. These numbers began to be published in the volume on the vital statistics of 1864. Inspired by the "desire to know through which unnatural exits people in our country see the end of their mortal career,"[91] this investigation was a first step towards the making of a much desired medical statistics. But the research also had direct relevance for measuring levels of "civilization" within the body of the national community. "Some of these deaths," Maestri maintained, referring to those brought about by certain crimes of passion, were "the consequence of little advanced civilization, while others, unfortunately, must be considered the result of passions and dangers to which modern progress exposes us."[92] This was the grim view of the fate of mankind that informed, as we have seen in the previous chapter, the contemporary discourse of "moral statistics." As Angelo Messedaglia observed, moral statistics ironically made crime (here appearing in its effects, as number of victims) the measure of morality in the same way as, in another branch of statistics, death was made the measure of life.[93]

Maestri viewed the investigation of violent deaths as a means to assess also the "different attitudes and moral predispositions of [the Italian] people."[94] The quantification of homicides exhibited a clear geographical pattern, according to which the Kingdom appeared divided into two distinct parts; in one of them the homicide rate was almost five times higher than in the other:

> With regard to the number of homicides Italy *can be divided topographically into two distinctive parts*, of which the first includes the northern and western provinces (Piedmont, Lombardy, Emilia, Liguria, and Tuscany) and the second the eastern and southern ones (the Marches, Umbria, Neapolitan provinces, Sicily, and Sardinia). In the former, homicides are relatively less numerous (on average 5.23 per 100,000 people); in the latter, this ratio reaches 23.64 per 100,000 [emphasis added].[95]

Three years later, the boundary dividing the areas with the different homicide rates had slightly changed. The Marches and Umbria were

[91] MAIC, *Statistica del Regno d'Italia. Popolazione. Movimento dello stato civile nell'anno 1864* (Florence, 1866), p. lvi.

[92] Ibid.

[93] A. Messedaglia, "Esposizione critica delle statistiche criminali dell'Impero austriaco, con particolare riguardo al Lombardo-Veneto, secondo i resoconti uffiziali del quadriennio 1856–1859, e col confronto dei dati posteriori," *Atti dell'I. R. Istituto Veneto di scienze, lettere ed arti*, 3rd s., 11 (1865–1866), p. 158.

[94] MAIC, *Movimento dello stato civile nell'anno 1864*, p. lvi.

[95] MAIC, *Statistica del Regno d'Italia. Morti violente. Anno 1866* (Florence, 1868), p. xxviii.

now included within the area defined by a lower homicide rate.[96] We should observe that in fact not all the "Neapolitan provinces" exhibited the same high rates of victims of homicides. For example Apulia and Basilicata often had lower rates than Umbria,[97] a fact that the construction of a general average for the whole South tended to obscure.

How was this polarity to be explained? The difference was immediately traced back to diverse and long-lasting local habits determining that in "certain Italian regions a given kind of violent, accidental or voluntary death is either very frequent, or it is altogether absent."[98] Yet it was not only a matter of local peculiarities originating in generic local habits. In the comment on the data for 1867 the two parts had become "very distinct." This was a cause of some worry for the commentator, who added emphatically (after quoting from a patriotic poem emphasizing the cultural unity of Italians) that "it is three centuries of bad foreign and local domination, of religious prejudice and baneful civil and economic theories that reflect and explain these strange figures."[99] Political history again, more than generic cultural traditions, explained the regional differential of crime, in the same way that it also explained the huge percentages of illiteracy in the southern regions.

Significantly, the urban–rural comparison, which made an appearance in the analysis of demographic data, did not appear at all in the comments on this kind of "moral statistics." In spite of the fairly frequent episodes of urban unrest, the social condition of cities was not perceived as a particular threat to the social order (nor were they growing enough at the time to be perceived as such).[100] In contrast, since 1860 the problem of "public order" and crime appeared primarily as a territorial–political issue to be "solved" through the adoption of harsh repressive measures in the areas concerned. The territorial grid of official statistics reflected and reinforced existing perceptions.

The territorial frame organized the reading of all kinds of statistics. We can mention here those on elections and the electorate, scrutinized in order to measure the greater or lesser "political vitality" of the various parts of Italy. In the effort to consider the question "in general and in large masses," two areas were distinguished: in the first, *Italia superiore* (including Tuscany but excluding Emilia), electors *per contribuzioni* (those who met the necessary tax requirement) were dominant,

96 Ibid., *Anno 1870* (Milan, 1871), p. xvi.

97 See for example the data for the period 1871–1876.

98 MAIC, *Statistica del Regno d'Italia. Morti violente. Anno 1866*, p. xxviii.

99 Ibid., *Anno 1867* (Florence, 1869), p. xxiv.

100 Only Turin and Florence experienced some demographic increase due to their position as capitals.

while in *Italia meridionale e media* electors *per capacità* (those who did not meet the tax requirements, but were electors by virtue of their belonging to a particular occupational category – state employees or people exercising a liberal profession) were more numerous. The commentator linked this differentiation to the "special economic conditions of the two parts of the Kingdom, namely to the more developed industrial activity and the more diffused and distributed wealth of the *settentrione* and to the opposite conditions in the *mezzodì*," as well as to other factors such as the level of literacy.[101] In contrast, if one examined the percentage of the electorate who turned out to vote, the Veneto and southern Italy exhibited much higher averages than the center and the North thus showing the greater "diligence" of their electors, but also reflecting the advantages of a more concentrated pattern of residence.

The tripartite (*Italia settentrionale, centrale e meridionale*), and the dualistic (*alta* and *bassa Italia*) schemes employed by Maestri and his collaborators were fairly conventional partitions of the peninsula used by geographers and statisticians in earlier times. The boundaries of these larger units, however, were quite mobile and opinions about them varied. For example Luigi Serristori in his *Saggio statistico dell'Italia*[102] had observed that Italy could be divided into two parts, one including the continental part of the Kingdom of Sardinia–Piedmont, Lombardy–Venetia, the Duchies of Parma and Modena, and the northern part of the Papal States north of the Apennines, and the other including all the rest of the peninsula and the islands. Distinctive geographies, climate, agricultural production, and people's mores, he claimed, characterized the two areas. In his *Corografia fisica, storica e statistica dell'Italia e delle sue isole*[103] Attilio Zuccagni Orlandini had divided Italy in four parts (northern, central, southern, and insular) claiming that the Italian peninsula let itself easily be divided in such a fashion along "natural" lines. These subdivisions began to acquire new meanings and eventually stable boundaries in the context of a politically unified Italy and in connection with the institutionalization of an official apparatus of data gathering and observation.

While the statistics collected by the Direzione during the first decade of its activity reflected the image of a divided Italy, this was an image that the statisticians appeared to acknowledge with a certain reluctance and, as we have seen, always with a profusion of references to the effects of the bad government of previous regimes, and thus to its

[101] MAIC, *Statistica del Regno d'Italia. Elezioni politiche e amministrative. Anni 1865–66*, p. xxxiii.

[102] (Vienna, 1833). [103] Vol. 1 (Florence, 1845), p. 123.

temporal and temporary dimension. What is striking in Maestri's comments is the constant attempt to exorcize the unpleasant reality shown by the figures, as if the very representation of this reality were a threat to national unity – without realizing that it was the very aggregative logic of the statistical gaze which tended inevitably to accentuate contrasts and oppositions. Thanks to the solid historicism of Maestri and his collaborators, the use of a language which spoke of trends and correlations as laws did not yet naturalize society . But there was room for less historically minded and less cautious social observers in the years to come to read the differential distribution of averages over the national territory as a sign of the existence of inherently different collectivities within the national "body."

A LASTING LEGACY

The units of observation and reporting created by Maestri were there to last. Although nobody can deny that their success was rooted in the antiquity of regional denominations in Italy, it is the enshrinement of regions in official statistics that gave them a stability and a life of their own in the new state. Indeed, as a result of the regular production of official statistics and of their diffusion and utilization by social investigators of all kinds, the description of the country through the lenses of the *compartimenti* – placed by Maestri at the core of official statistics – became a well-established mode of representing Italy in statistical studies and social investigations throughout the history of the country. From his first medical and anthropometric inquiries of the mid- and late 1860s, Cesare Lombroso adopted the regional frame to study the internal differentiation among the people of the country.[104] A regional partition was also used in preference to other possible ones in inquiries into the condition of agriculture. In the first investigation on Italian agriculture, whose results were published in 1876–1877,[105] the unit of analysis was the "agrarian region": there were twelve of these, amounting to groups of provinces, eight of which corresponded to Maestri's *compartimenti*. After a learned discussion on several possible modes of dividing the country into "agrarian regions" suggested by the studies of botanists and geographers, the compiler justified the choice of the more "conventional" aggregations of administrative entities, by

[104] See for example "Studi per una geografia medica d'Italia," *Gazzetta medica italiana. Lombardia*, (1865), and "Del peso del corpo nell'uomo sano e alienato," *AUS*, 4th s., 32 (1867), pp. 265–271.

[105] MAIC, Divisione di Agricoltura, *Relazione intorno alle condizioni dell'agricoltura nel quinquennio 1870–1874*, 4 vols. (Rome, 1876–1877).

observing – with a statement that only a few years earlier would have looked like heresy – that the regional partition

> [is] shaped to a large extent *on the basis of the former political division of Italy* [emphasis added], and since agriculture is such as people make it, it will not seem an exaggeration to say that each of our regions constitutes a distinct agrarian organism.[106]

A similar partition was adopted by the large parliamentary inquiry into the state of agriculture, the so-called Inchiesta Iacini (from the name of its main promoter), which began in 1877 and whose results were published in the early 1880s.[107]

The story told in this chapter points to the power of official numbers to create "things which hold," to use Alain Desrosières's expression,[108] namely categories and modes of representation which acquire the status of objective realities. Once the *compartimenti* were given official existence by the instrument of statistics, they entered the inertial life of administrative practices. Contrary to the wishes of Maestri, who insisted that these units must be periodically redrawn to allow them to better represent changing relations among the various parts of the country, they remained virtually unmodified for decades (new regions were added as new territories were acquired after World War I), gaining a solid existence. They even became the basis for the regional administrative partition of the Italian republic in 1948 – as some geographers then complained bitterly claiming that the *compartimenti* failed to reflect living, organic territorial entities.[109] Institutionalized in the volumes of official statistics and reproduced in countless works of social and scientific analysis, *compartimenti* became endowed with an aura of naturalness which obscured the circumstance of their actual creation and their status of artifacts. The same could be said in the case of the partition between North and South which we have seen in its incipient state in the official statistics of the 1860s, but which was to become a

106 Ibid., vol. 1, p. 30.

107 MAIC, *Atti della Giunta per l'Inchiesta agraria*, 15 vols. (Rome, 1882–1885). On the political context and the makers of this inquiry see A. Caracciolo, *L'inchiesta agraria Iacini* (Turin, 1958; 1973).

108 "How to Make Things Which Hold Together."

109 One of these was A. Sestini, "Le regioni italiane come base geografica della struttura dello stato," in *Atti del XIV Congresso geografico italiano* (Bologna, 1949). See also L. Gambi, "L'equivoco tra compartimenti statistici e regioni costituzionali," in *Avviamento allo studio geografico della regione* (Florence, 1967), pp. 119–145. For some of the arguments of this chapter see also S. Patriarca, "Statistical Nation Building and the Consolidation of Regions in Italy," *Social Science History* 18 (1994), pp. 359–376.

paradigm for the representation of the country in later years. As we shall see, in the new political and intellectual climate of the 1870s–1890s the "dualism" of Italian society which statistics seemed merely to reveal, and in fact contributed to shape by giving it the support of "scientific facts," acquired a dominant status in both scholarly and political discourse. Statistics in contrast never became the tool for the promotion of the country's greater homogeneity that its early makers had envisaged. To that end, other means were necessary, means that the elites in power, however, were not able, nor willing, to see.

CHAPTER 8

CENTER AND PERIPHERY

The previous chapter has shown how statisticians tended to downplay the grid of the urban and rural communes in favor of those all-powerful territorial units of reporting, the *compartimenti*. The role and performance of communes was however the center of attention of both politicians and state statisticians in the early phase of construction of the new nation. At the root of the discussion on the morphology of communes stood the larger issue of the difficult articulation between national and local elites, central state, and local government. The making of a liberal order required the spread of its institutions across the national territory. It was not only a matter of establishing representatives of state authority in all its peripheral offices, but also of favoring the establishment of the structures of a modern society by insuring the proper working of local government. A great responsibility fell on the local organs of government, as the legislation concerning the administrative organization of the new state assigned to communes more functions and resources than they had usually had in the former states. Communes came to enjoy a considerable autonomy in the imposition of taxes and the employment of their revenues. Distrusting the loyalty and skills of locally elected bodies, members of the state apparatus looked on this autonomy with concern and resisted demands to extend it.[1]

To a greater or lesser degree several of the statistical investigations carried out in the 1860s – the counting of schools, poor relief institutions, saving banks, post offices, manufacturing industry, the moni-

[1] On the power of local administrations see A. Polsi, "Le amministrazioni locali postunitarie fra accentramento e autonomia: il caso del Comune di Pisa (1860–1885)," *Società e storia* 6 (1983), pp. 828–867. For a recent overview of the historiography on the center–periphery relation see L. Riall, "Elite Resistance to State Formation: The Case of Italy," in M. Fulbrook, ed., *National Histories and European History* (London, 1993), pp. 46–68.

toring of the finances of communes and provinces, the collection of data on the results of elections at all levels – reflected a similar preoccupation on the part of the central state elites with the construction of the new liberal order and controlling the exercise of power held by local elites. This concern informed also a project that Maestri and Correnti presented at the meeting of the International Statistical Congress held in Florence in 1867: this project called for the making of a "statistics of communes," an item that appeared to be an Italian invention and that served to articulate a specifically Italian agenda. The examination of the discussion that took place at the congress on this subject will serve to point out some qualifying aspects of the use of statistics in the context of the process of nation building in liberal Italy.

It is necessary to mention that from the mid-nineteenth century on, statisticians across Europe began to strengthen their relations and form an internationally oriented community. In their work, Maestri and his collaborators could, and did, rely on examples of analogous works regularly published by the statistical services of other countries. In particular they referred to the resolutions of the International Congresses of Statistics, which met every two to four years in major European capitals between 1853 and 1876.[2] Indeed, in choosing the objects of investigation in national censuses and the categories in which they should be organized and classified, statisticians took into consideration the decisions made by their colleagues from other nations who, along with government delegates, met at the congresses.[3] The purpose of these meetings, as intended by their principal promoter, Adolphe Quételet, was precisely that of furthering the construction of a uniform system of categories and classifications which would make cross-national comparisons reliable, of developing exchanges and collaboration between savants and administrators, and of imposing the conceptions and practice of "science" on a growing number of amateur practitioners.

In the international congresses, however, the universalist ambitions of statistical science met with the actual contexts of production of social numbers, which were in fact national. The meetings were thus

2 Some information on the congresses can be found in H. Westergaard, *Contributions to the History of Statistics* (London, 1932), pp. 172–190, and J. and M. Dupâquier, *Histoire de la démographie. La statistique de la population des origines à 1914* (Paris, 1985), pp. 299–320. The congresses took place in Brussels (1853), Paris (1855), Vienna (1857), London (1860), Berlin (1863), Florence (1867), The Hague (1869), St. Petersburg (1872), and Budapest (1876).

3 On the decision to abide by the congresses' recommendations on census taking see Ministero di Agricoltura, Industria e Commercio (henceforth MAIC), *Statistica del Regno d'Italia. Popolazione. Censimento generale (31 dicembre 1861)*, vol. 1 (Turin, 1864), p. xlv.

bound to reflect the particular issues and preoccupations that characterized each host country. This was particularly the case with the Congress of Florence. The interests and language of the Italians were not necessarily shared by others; sometimes they appeared too abstract, philosophical, or, in contrast, too political; the "father" of statistics, Quételet, complained at the Italian Congress about the total absence of any discussion on the mathematical aspects of the science.[4] Indeed the Italian organizers made the International Congress into a mainly national event. They used the congress to talk about some of the issues that were particularly relevant to them, as appears clearly in the case of the "statistics of communes." The issue of what was to be considered an important and legitimate object of representation in the statistical description of a commune constituted a subject for ideological and political confrontation, and mirrored the conflicts that divided Italian society. At the same time the program of investigations envisioned by the Italian statisticians represented an attempt to lead conflicting local forces towards the achievement of similar objects.

A SHOWCASE FOR THE NEW STATE ELITES

If in their very role as international meeting points statistical congresses could be seen as fostering supra-national perspectives, at the same time they served to enhance the identity of member nations and provided the governments and ruling elites of the participating states with a showcase for the display of their involvement in scientific administration and the advancement of social science. For the new governing class of Italy the congress of 1867 represented a good occasion to obtain international recognition and to have the legitimacy of the new national state reaffirmed. This recognition was particularly necessary after a rather humiliating first performance on the international scene; the previous year Italy had suffered military defeat at the hands of the Austrians and was able to get hold of Venetia only through the mediation of France. In addition, the congress would also be an occasion to honor Florence, which had become the new capital of Italy in 1865. Thus a vast mobilization of manpower and energy accompanied the preparations for the Sixth International Congress of Statistics, which met in the Tuscan city in September 1867, and an exceptional

[4] In one of his very rare interventions at the congress: see *Compte-rendu des travaux de la VIe session du congrès international de statistique* (Florence, 1868), p. 309. He further stressed his position by asking that the next congress deal with "the great statistical questions in the language that is the most suitable for them, that is by using the probability calculus."

number of people showed up. Officially, the congress witnessed the participation of 741 members, of whom about 90 percent were Italian. The total number of members was the highest in any of the congresses, as well as the percentage of participants from the host nation.[5]

Unfortunately for the organizers, 1867 was not a good year for an international event such as this to take place in Italy. A cholera epidemic, which since 1865 had been ravaging several areas of the country, made the trip to Italy more adventurous than one would have liked it to be – although it hit the countryside and the small towns more than the largest cities, and the southern regions more than the northern ones.[6] The epidemic probably helped keep the foreign delegation to the congress quite small, as some French participants observed.[7] The foreign delegations at Florence included for the most part leaders or representatives of official statistical bureaus, such as the director of the British Registrar General's Office, William Farr, the director of the French Bureau de statistique générale, Alfred Legoyt, the president of the Belgian Commission centrale de statistique, Adolphe Quételet, the director of the Preussische statistische Bureau, Ernst Engel, and a number of delegates of statistical societies. The *Annali universali di statistica* observed how only two states (needless to say, European) had not sent representatives, Spain and the "government of the Pope," and took the occasion to remark polemically that nobody should be surprised that the two "worst governed states of the world" did not care to be present where "statistical science aimed to shed light and to enlighten [*far luce e dar luce su*] every social institution."[8]

In contrast, the unfavorable public health conditions did not deter the Italians, who went *en masse* to the meeting. All components of the various layers of the state apparatus along with local notables and the professional middle classes were represented in the crowd which gathered in the halls of the Uffizi, then the seat of the Italian Senate, between September 27 and October 5, 1867: from the Minister of

5 For these data see *Compte-rendu des travaux de la VIe session*, p. 642. At the Congress of Brussels the members were 236 (of whom 47 percent nationals), in Paris 264 (50 percent), at Vienna 541 (86 percent), in London 595 (85 percent), in Berlin 477 (73 percent).

6 On cholera epidemics in nineteenth-century Italy see A. L. Forti Messina, "L'Italia dell'Ottocento di fronte al colera," in *Storia d'Italia. Annali VII. Malattia e medicina* (Turin, 1984), pp. 459–462.

7 See "L'ultimo Congresso internazionale di statistica giudicato dai francesi," *Annali universali di statistica* (henceforth *AUS*), 4th s., 33 (1868), pp. 315–329. This is the report presented to the Société de statistique of Paris by the director of the French statistical bureau, A. Legoyt.

8 *AUS*, 4th s., 32 (1867), pp. 7–23 (the quote is from p. 7).

Agriculture, Industry, and Commerce, F. De Blasiis, to small-town mayors, from councilors of state to representatives of the provincial chambers of commerce, from university professors to members of local scientific societies, from countryside physicians and *medici condotti* (public doctors) to chief physicians of city hospitals. The overwhelming majority of the participants came from the northern regions and, unsurprisingly, from the host region, Tuscany, which alone accounted for about 37 percent of the members. Participants from the South and the rest of central Italy amounted to less than 20 percent of the overall membership. Leaving aside the figure for Tuscany, to be discounted by the fact that the meeting took place in Florence, the other percentages significantly reflect the markedly different degrees to which, during the rule of the Destra Storica (1861–1876), the regional elites participated in the public life of the new state, and espoused the ideology of its apparatuses. Not only did the holders of state power at the time originate mostly in the northern regions of the country and Tuscany, but also the institutions of "civil society," such as influential journals, academies, societies for the promotion of agrarian and industrial interests and so on, were far more numerous and influential in these areas than in the rest of Italy.[9] These people made up an overwhelmingly urban crowd (more than half of them came from the eight largest cities in Italy) where members of the liberal professions and the state bureaucracy were predominant. What appears especially clear is the large extent to which the event was able to mobilize the members of the several layers of the new state apparatus and many of the institutions which flanked it.

Besides exchanging ideas and taking resolutions on statistical matters proper, during the week or so they spent in Florence the members of the congress also participated in the social, as well as educational, activities, which it had become customary for the hosts to organize to honor congress members. The "program of the meetings and festivities" included a variety of activities, from a special visit to the National Museum to a walk in the Cascine, where the Florentine bourgeoisie loved to stroll, from a concert at the royal theater to a visit to an exhibition on gardening and horticulture; moreover, congress members had the opportunity to visit, without charge, all the "Galleries, the Museums, the Libraries, the relief Institutions, the Penitentiary, the Royal Printing-house" and other such edifying places (monuments dedicated to highly valued middle-class activities, such as education

[9] The complete list of the participants is in *Compte-rendu des travaux de la VIe session*, pp. 335–343. On the composition of the ruling elites see A. Caracciolo, *Stato e società civile. Problemi dell'unificazione italiana* (Turin, 1960), pp. 79–80.

and philanthropy) as were recommended by the guidebook specially prepared for the event.[10]

It is difficult to imagine these people earnestly involved in highly technical discussions on statistical methods. In any event, even the members of the organizing committee, with few exceptions, were essentially "practical men," mostly statesmen with little or no knowledge of mathematics or the calculus of probability of a kind that, for example, Quételet possessed.[11] The debates in which the delegates of governments, university professors, and amateur statisticians got involved in Florence, as well as in the previous and subsequent congresses, revolved closely around that cluster of middle-class concerns which was reflected in the broader program of activities prepared for their entertainment. Some of these concerns were common to the increasingly bourgeois elites of most European states, others were more localized.

CHOOSING THE CONGRESS'S AGENDA

By the time the participants of the Sixth Congress met in Florence, a remarkable number and variety of subjects had been examined and discussed in international meetings. The previous year in a preparatory publication Pietro Maestri – adopting a comprehensive classification established in 1863 by Ernst Engel[12] – listed the following topics which had been discussed in earlier congresses: statistical theory and techniques, organization of the statistical service, organization of statistical congresses, territory, *lieux d'habitation* (statistics of large towns), population, landed property, agriculture, fishing, mines and factories, manufactures, arts and crafts, public works, currencies, weights and measures, commerce, transport, banks and credit institutions, insurance, production and consumption, prices and salaries, working classes, providence institutions, charity and public relief, hygiene and sanitary service, education, press and literature, police and prisons, civil and

10 *Compte-rendu des travaux de la VIe session*, pp. 162–163.

11 One of the few exceptions was Angelo Messedaglia. A list of the members of the organizing committee is in *Compte-rendu des travaux de la VIe session*, pp. 4–5.

12 *Compte-rendu général des travaux du Congrès international de statistique dans ses séances tenues à Bruxelles, 1853, Paris, 1855, Vienne, 1857, et Londres, 1860* (Berlin, 1863). This classification of categories formed the basis for the works of the Prussian Statistical Bureau. According to its author, no social phenomenon escaped this classification (pp. v–vi). On E. Engel's work see I. Hacking, "Prussian Numbers 1860–1882," in L. Krüger, L. J. Daston, and M. Heidelberger, eds., *The Probabilistic Revolution*, vol. 1, *Ideas in History* (Cambridge, Mass., and London, 1987), pp. 377–394.

criminal justice, army, navy, finances.[13] Leaving aside the categories that referred to the congresses' procedural matters, and those that concerned subjects brought to public discussion to publicize a cause (such as that of the introduction of a uniform system of weights and measures), these were the objects on which statisticians of all countries had been producing data over the years.

Some of these topics of discussion (congress organization, territory, population) had been a constant matter of debate in all five previous congresses. The organization of statistical services, landed property, agriculture, hygiene and the sanitary service, civil and criminal justice had been discussed in four congresses out of five. Three congresses had tackled the questions of currencies, weights and measures, transport, finances; two those of insurance, commerce, working classes, providence institutions, charity and public relief, education, press and literature, the army. The other topics in Maestri's list had been on the agenda of only one preceding congress.

The recurrence of certain subjects in the congressional debates was a sign both of their importance to the statisticians and of the difficulty in finding a satisfactory agreement on the definition of the methods and procedures for investigating them; but it could also result from lack of success in the attempts to implement the resolutions taken in previous meetings. On the other hand, the fact that some questions did not appear at more than one session probably indicates that they had found a successful "solution" (agreed-upon procedures and categories had been adopted by governments), or that they had rapidly become obsolete and disappeared from the agenda of the meetings' organizers.

Since the congresses were organized each time by the statisticians and government of a different country, their agendas were bound to reflect, besides the state of the debate inherited from the previous meetings, the concerns peculiar to each host country. In fact every successive organizing commission made a point of introducing some new themes of discussion. The Austrian commission, for example, had proposed four new themes to the congress to be held in Vienna in 1857: a statistics of the ethnographic differences existing within a state, one of hospitals and sanitary structures, one of landed property, and one of banks and the credit system.[14] All reflected matters of special interest to Austria, but in particular the functionaries and statisticians of the multi-

[13] P. Maestri, *Compte-rendu général des travaux du Congrès international de statistique dans ses sessions de Bruxelles, 1853; Paris, 1855; Vienne, 1857; Londres, 1860; et Berlin, 1863* (Florence, 1866), pp. iv–ix.

[14] See "Projet de programme pour la troisième session du Congrès international de statistique," in *Compte-rendu de la troisième session du Congrès international de statistique*

national Austrian Empire could not but be the best equipped to deal with the categorization of ethnic groups.

Maestri's list of possible topics for the sixth congress included a number of matters which had not yet been discussed in a congress. These made up a heterogeneous group: administrative division of the country, forestry and hunting, fishing, religion, arts and sciences, constitutional and legislative power, fundamental laws of a country, constitution of departments and districts, communal system. Among these, some were matters of special interest to the Italian governing elites, matters to which Maestri and his collaborators were naturally very sensitive, in view of the importance they gave to statistics as a means of defining and solving what they considered to be the vital administrative and institutional issues of the new state. Not accidentally, the choice of the organizing committee fell on the one concerning the communal system, which they proposed to the debate in Florence under the name of "statistics of communes." The remaining themes brought to Florence (statistical theory and organization of congresses, topography, agriculture, monetary circulation, moral and judicial statistics, the army, and education) had all been examined in previous congresses; although they also received a particular imprint in the discussions held in Italy, reflecting the current and local preoccupations of Italian statisticians,[15] it was the debate on communal statistics that not only attracted many participants but was also particularly heated, clearly showing that the topic was of major relevance for the Italians.

ARTICULATING CENTRAL AND LOCAL POWER: THE "STATISTICS OF COMMUNES"

As Ernst Engel did not fail to point out during the Florence meeting, the statistics of communes at the Florence Congress bore some resemblances to the so-called "statistics of large towns," which had been discussed at the Congress of Paris in 1855.[16] Yet its meanings and the resolutions that it generated in the Italian context were entirely different. At the Paris Congress, the quantification of the phenomenon of

réuni à Vienne les 31 août, 1, 2, 3, 4, 5 septembre 1857 publiée sous la direction de M. Ch. baron de Czoernig (Vienna, 1858), p. 6.

15 See especially the sections of the Congress, as reported in the *Compte-rendu des travaux de la VIe session*, on the statistics of agriculture, which dealt with the study of land registers and the assessment of land revenue, urgent issues at the time in Italy, and on moral statistics, in which a classification of the poor was proposed (although no agreement was reached except on the proposal to establish a committee of patronage of the poor in each country with the aim of coordinating assistance).

16 *Compte-rendu des travaux de la VIe session*, p. 385.

large towns stemmed out of a time-honored French tradition of statistical studies on the capital city of France, and was strongly informed by the industrial and commercial concerns of the rapporteur of this project to that congress, the economist and statistician Charles Dupin.[17] Large towns – defined as cities with more than 50,000 inhabitants – were presented as a peculiar social phenomenon that required a specific study; their populations behaved differently from those scattered in the countryside; from an economic standpoint, in cities were concentrated the most refined and well paid arts and crafts. Dupin did not deny that smaller towns could exhibit interesting characteristics – as a representative from Switzerland had insisted – and were worth studying for their own sake; undoubtedly, however, they were of little interest to him.[18] Cities, in the French context, represented a very special set of social and moral problems. The Paris Congress produced a long questionnaire for the compilation of the statistics of large towns, a questionnaire which aimed to cover "methodically," that is through a comprehensive classification, all the categories that would generate a thorough description of a town, from its topographical position to the religion professed by its population. To a large extent the items followed those that could be found in the general statistics of a state.[19]

Although large towns could certainly be classified as a particular kind of commune, at Florence they were given only slight attention within the framework of the statistics of communes. To be sure, Correnti conceded that the study of "grandes villes" constituted the peak of this branch of statistics (and the Florence Congress especially recommended that communal authorities publish statistical yearbooks on large towns);[20] however, in Correnti's report, the ambivalence towards large towns (an ambivalence which easily turned into straightforward dislike

[17] This report was published in full in *AUS*, 3rd s., 8 (1855), pp. 117–129, with a few critical comments by G. Sacchi (pp. 129–131). The tradition of studies to which I refer (also directly mentioned by Dupin) included the works of Count G.-J.-G. Chabrol de Volvic (*Recherches statistiques sur la ville de Paris et le département de la Seine* [Paris, 1821–1829]) continued by his successors, Rambuteau and Haussmann, and large numbers of socio-medical investigations, on which see L. Chevalier, *Laboring Classes and Dangerous Classes in Paris during the First Half of the Nineteenth Century* (Princeton, 1973).

[18] In his critical comments to Dupin's report in *AUS* (see previous note), G. Sacchi supported the observations of the Swiss statisticians on the need to investigate also towns with fewer than 50,000 inhabitants since these, he observed polemically, "could exhibit results far more interesting than those memorable capitals that swallow – within a very limited space – millions of people who appear to be camping more than living together in a civil fashion" (p. 131).

[19] The questionnaire is reproduced in Engel, *Compte-rendu général*, pp. 76–83.

[20] *Compte-rendu des travaux de la* VI*e session*, p. 469.

among most members of the Italian elites) did not fail to emerge: "Large towns are in part the product of an artificial and violent concentration, in part the consequence of a free agglomeration."[21]

In contrast to the statistics of large towns, the Italian proposal for a statistics of communes was predicated upon a quite different view of urban landscape, and upon preoccupations concerning the relations between central and local power, a subject of ongoing debate and inquiry among Italian statesmen and officials towards the end of the 1860s. In 1869 the state of local government was the topic of an inquiry by the Minister of the Interior, Gerolamo Cantelli, who asked the prefects of the Kingdom about the extent to which they thought the autonomy of communes and provinces could extend itself "without weakening the *ingerenza* [action/intervention] of government authorities which is necessary to guarantee a state of lawfulness and maintain harmony among general interests."[22] The issues addressed by Correnti's report and the questionnaire produced by the congress clearly show the way in which the statistics of communes actually confronted the current state of relations between center and periphery in the new Italian state, and the centrifugal forces that threatened its existence.

To be sure, the genre of local statistics had always been cherished by Italian statisticians. Municipal statistics had made up a good number of those local descriptions whose classificatory logic we analyzed in Chapter 3.[23] Even when, before unification, Italian statisticians wrote their tracts on how to make a general statistics of a state, they included essays on how to do this at a local level. The Tuscan Attilio Zuccagni Orlandini had offered an example of communal statistics in his several volumes of *Ricerche statistiche* on Tuscany as a whole;[24] David Rabbeno, a patriot from Reggio Emilia, had done the same just before the proclamation of the Kingdom, in a work that also contained a strong appeal for the making of a statistics of the new nation.[25] These authors did not feel the need to explain the ra-

21 Ibid., p. 123. See also Sacchi's comments reported in note 18.

22 R. Romanelli, "Tra autonomia e ingerenza: un'indagine del 1869," in *Il comando impossibile. Stato e società nell'Italia liberale* (Bologna, 1988), p. 77. See also by this author in the same collection of essays "Il problema del potere locale dopo il 1865," pp. 31–75.

23 L. Bodio's bibliographical essay lists twenty-two of them published in the forty-year period preceding unification (*Annali di statistica. Saggio di bibliografia statistica italiana*, 2nd edn. [Rome, 1885], pp. 19–54).

24 See A. Zuccagni Orlandini, *Ricerche statistiche sul granducato di Toscana*, vol. v (Florence, 1854).

25 See D. Rabbeno, *Delle condizioni della statistica nell'Italia centrale e delle commissioni di*

tionale for these local descriptions. Others, who followed the lesson of Melchiorre Gioia, justified local statistical descriptions as the ideal space and the perfect scale (in contrast with descriptions of larger political–legal units) for the exercise of a totalizing, all-encompassing gaze.

As we saw in Chapter 5, many authors imagined the statistical description of Italy as emerging from the collection of all the local statistics of communes, districts, provinces. In this perspective communes were seen as akin to molecules, or, to be more precise, atoms, the simplest component of a larger organism to which they were supposed to bring their contribution harmoniously, without friction. After all, was not Italy, in the rhetoric of patriotic discourse, the classic land of communal liberties, the country of the "hundred cities," the blessed land of "unity in variety"?

Pietro Maestri and Cesare Correnti participated fully in the rhetoric of the "hundred cities." In this rhetoric the exaltation of the urban character of Italian civilization and of the "civilizing" role of cities combined with a view of a whole model of development, of the relations between town and countryside, a model which rejected the violent ruptures brought about by intense industrialism and capitalist transformations in agriculture. The "hundred cities" were in fact mostly middle-sized towns, with populations of between ten thousand and thirty thousand people, scattered in a territory where sharecropping agriculture was dominant and landless peasants were few; this was actually the landscape of northern–central Italy now erected into a model for the whole country.[26]

When Maestri and Correnti "invented" the statistics of communes they certainly had in mind this image of Italy and of its ideal development. But it was also Correnti's and Maestri's understanding of the role of communes in the new polity that gave the statistics of communes its peculiar characteristics. In the perspective of these two supporters of decentralization, the tasks and condition of communes within the national entity acquired a privileged status as indicators of general civilization and the nature of government. In Correnti's report to the congress the communes, defined as the "natural *rudimenti* [rudi-

statistica nell'Emilia instituite con legge 28 gennaio 1860 operazioni ed organizzazione di esse con un saggio di statistica comunitativa del Comune Parmense di Salsomaggiore (Parma, 1861). The commune of Salsomaggiore was the birthplace of G. D. Romagnosi.

26 For this interpretation of the model of the "hundred cities" see P. Villani, "Gruppi sociali e classe dirigente all'indomani dell'Unità," in *Storia d'Italia. Annali* 1. Dal feudalesimo al capitalismo (Turin, 1978), pp. 890–892. On the characteristics of the Italian urban structure and the "hundred cities" see also A. Gramsci's observations in *Quaderno 19. Risorgimento italiano*, ed. C. Vivanti (Turin, 1977), pp. 159–160.

ments] of human society," were endowed with a fundamental role in the making of the liberal polity; indeed, on their constitution, on their relative autonomy depended the implementation of the liberal ideal of "self-government." Only if their finances were adequate to their tasks, could they successfully perform their function.[27] Thus statistics must investigate the exact institutional position of communes, their finances and their fields of intervention in any given country so as to compare their role in respective polities and to evaluate the overall political institutions of a country.[28]

Yet Maestri and Correnti were also aware of the dangers involved in pushing the recognition of the role of communes too far, the dangers of legitimating the claims of supporters of the pre-1860 dynasties and Catholics alike who stirred the muddy waters of municipalism. As they had observed in the *Annuario statistico* of 1864:

> Some very famous writers, by denying any substance to the natural rights of regions, and even of provinces, incline to recognize the divine right of communes. Communes, they say, therefore are a product of nature; provinces are created by law ... If the question were to be posed in this way, we could demonstrate how the real commune has to be identified with the province, since usually the small rural commune is nothing more than a hamlet with neither a political nor an economic life. But neither the commune, nor the province has rights against [national] right and the law. National unity means the sovereignty of national reason.[29]

These qualifying concerns clearly appear in Maestri's and Correnti's project for a statistics of communes. Actually, this project seemed to respond also to a recent Italian phenomenon, namely the growing number of local statistics that had begun to appear right after the proclamation of the Kingdom.[30] A resurgence of local patriotism accom-

27 During the discussion Correnti underlined the urgency of this issue in Italy because of the tendency on the part of central government to lay upon the organs of local government an increasing number of tasks without adequate financial cover: see *Compte-rendu des travaux de la VIe session,* p. 238. There are similar arguments in Maestri's presentation of the statistics of communal and provincial budgets: *Le pubblicazioni della Direzione di Statistica. Relazione a S. E. il Ministro di Agricoltura, Industria e Commercio con note bibliografiche e sommari statistici e parere della Giunta consultiva di statistica* (Florence, 1869), pp. 45–51. The works to which Maestri referred were: MAIC, *Statistica del Regno d'Italia. Finanze. Bilanci comunali e provinciali. Anno 1863* (Florence, 1865), and *Bilanci comunali. Anno 1866. Bilanci provinciali. Anni 1866–67–68* (Florence, 1868).

28 "Statistique communale. Constitution démographique et économique des communes," in *Compte-rendu de la* VIe session, pp. 111–124. The same text appeared also in *Congrès International de Statistique à Florence. Programme de la sixième session du 29 septembre au 5 octobre 1867* (Florence, 1867).

29 *Annuario Statistico Italiano. Anno* II (Turin, 1864), p. 48.

30 Examples of these works published in the 1860s are P. L. Bembo, *Il comune di*

panied the creation of the national state. Several of these works were the product of amateur statisticians, local erudites or notables; others were the result of the activity of both old and newly founded civic bureaus of statistics.[31] The proliferation of local statistics after 1861 matched an analogous phenomenon in the field of historiography. Municipal histories too increasingly appeared at the time when the national question found a unitary solution. The sometimes publicly funded establishment, after 1870, of regional historical societies, which would contribute to the collection and publication of documents for the making of a general history of Italy, did not prevent students of local history from creating their own societies and from publishing their own histories.[32] The two genres, namely historical narratives and statistical descriptions, were often conflated in works which exalted both the antiquity and past glories, or remarkable events, of a town or small village, and its present institutions and prospects.[33]

Although communal statistics had been collected over many

Venezia nel triennio 1863–65 (Venice, 1866); G. Cotta Ramusino, *Cenni statistici e notizie patrie sulla città e provincia di Cremona* (Milan, 1861); L. Ferrario, *Busto Arsizio. Notizie storico-statistiche* (Busto Arsizio, 1864); R. Maini, *Quadri statistici sul comune di Cremona nel biennio 1866–67* (Cremona, 1869); *Notizie statistiche della città di Palermo* (Palermo, 1866); D. Pesci, *Statistica del comune di Ferrara* (Ferrara, 1869); A. Tassani, *Cenni topografici e statistici sulla città di Como* (Como, 1861); V. Zecca, *Monografia del comune di Chieti* (Chieti, 1866). In the following decades this phenomenon became even more remarkable, as is apparent from the numerous titles collected by L. Bodio in *Annali di Statistica. Saggio di bibliografia statistica italiana*, pp. 18–54.

31 See for example *Censimento della popolazione della città di Palermo al 31 dicembre 1861 pubblicato dall'ufficio comunale di economia e statistica* (Palermo 1862); *Censimento della popolazione della città di Torino al 31 dicembre 1861. Riassunto statistico sul movimento professionale ed industriale avvenuto in Torino nel quadriennio 1858–61* (Turin, 1863); *Statistica della città di Palermo* (Palermo, 1869). The creation of civic statistical offices had a parallel in Germany in those years, and also, but to a lesser extent, in other European countries: see I. Hacking, *The Taming of Chance* (Cambridge, 1990), p. 33.

32 The phenomenon has been noticed by A. Monti, *L'idea federalistica nel Risorgimento italiano. Saggio storico* (Bari, 1922), p. 77, and underlined by E. Sestan in "Origini delle società di storia patria e loro posizione nel campo della cultura e degli studi storici," *Annali dell'Istituto storico italo-germanico in Trento* 7 (1981), pp. 105–141. See also N. Raponi, "La Società Storica Lombarda e i suoi soci (1873–1899)," in *Milano fin-de-siècle e il caso Bagatti Valsecchi. Memoria e progetto per la metropoli italiana* (Milan, 1991), pp. 33–46.

33 See for example Ferrario, *Busto Arsizio. Notizie storico-statistiche.* This work was written to celebrate the status of "city" bestowed on this Lombard town of about 12,000 people in 1864. Historical narratives were included also in C. Bullo, *Delle condizioni statistiche e commerciali di Chioggia, aggiuntivi alcuni brevi cenni storici di quella città* (Padua, 1866); G. Cotta Ramusino, *Cenni statistici e notizie patrie sulla città e provincia di Cremona*; C. Dell'Acqua, *Villanterio. Cenni storici e statistici con documenti editi ed inediti* (Pavia, 1874); P. Guaitoli, *Della città e comune di Carpi. Cenni statistici e storici* (Carpi, 1874).

decades, their meaning was bound to change within the new context of the national state. Once the nation had been formed, the writing of local statistics would inevitably exhibit some ambiguous aspects in that it could work as the expression of purely local loyalties, if not anti-unitary stances, on the part of the notables.[34] Significantly, in his report Correnti observed:

> So far the commune has been considered by statisticians *as if it were a small State* [emphasis added]; it has been studied with the same methods of observation and according to the descriptive forms which apply to other political associations, with regard to the diverse elements which constitute it, such as territory, population, production, road conditions, crime, education.[35]

This mode of representation, Correnti went on, missed the "very essence of communal life, its constitutive principle, its *proportionality with other associations* [emphasis added], spontaneous or necessary, which surround or dominate it."[36] By making the "proportional" or balanced relation of the commune with other organizations a defining trait of the commune, Correnti pointed to an important aim of its statistics. This aim consisted in directing the compilation of local statistics to an acceptable, that is national, purpose, to the achievement of a full integration of local communities into the national "body" over and above the municipal sentiment that was still so vigorous in the new state and threatened the implementation of national policies. The production of local statistics certainly had to be encouraged, since it helped the central government in its effort to know the state of the country. At the same time it had to be supervised and directed to fit into the construction of a picture of the conditions of the nation as a whole in all of its local articulations. This applied not only to the activity of private or semi-private researchers, who in this perspective would have to collect data of relevance for the makers of policy at the central level (and not just for the exaltation of local glories or the praise of local notables), but also to the work of the civic bureaus of statistics. The question was

34 For an example of this use of statistics to defend municipalist claims see G. Caporale, *Dell'agro acerrano e della sua condizione sanitaria; ricerche fisiche, statistiche, topografiche e storiche* (Naples, 1859). The municipal authorities of the commune of Acerra, near Naples, had appointed the author to write this work (which claimed that the town enjoyed a healthy sanitary condition) in order to oppose the relocation of a local institution to another town.

35 "Statistique communale. Constitution démographique et économique des communes," in *Compte-rendu de la VIe session*, p. 111. The first sentence in the original reads: "La commune jusqu'ici a été considérée par les statisticiens comme un petit Etat."

36 Ibid.

well summarized by the Minister of Agriculture, Industry, and Commerce, F. De Blasiis, in his opening speech at the congressional general assembly:

> Our provincial and communal statistics are equally valuable [compared to general ones] and are getting better. The special condition of our Italy, where there is no center which has acquired a superiority over the others ... is the cause that each province, and even some communes, cultivate the legitimate ambition of showing, by means of statistical figures, their demographic and moral entity ... It would really be desirable that these partial works were better coordinated, with regard to both form and content, in order to draw more easily comparative evaluations and general summaries from them. The Direzione di Statistica has given suggestions on how to proceed in this respect: but this advice would be more authoritative and effective if it were supported by a vote of this congress requiring a more uniform direction and conception in all official statistics.[37]

The idea underlying the statistics of communes was thus an attempt to curb the dark side of the land of "hundred cities," that is local, municipal patriotism, what the Italians call *campanilismo*, a phenomenon in which the feeling of attachment and identification with one's place of origin combined with the strenuous defense of purely local interests, and stood up as an irreducible antagonist to the "higher" interests of the nation in its making. A long-standing characteristic of political life in the Italian peninsula, its disruptive potential had manifested itself clearly even in the politically critical decades before unification, when the struggle of the liberal movement had been hampered by tensions between capital cities and provincial towns.[38] After 1861, fiercely municipalist traditions, behind which stood often the defense of established privileges, impeded the very attempts to build the institutional structures of the new nation on more rational bases, as the new state elites had to realize quite early in many instances.[39] It is to this harmful

37 *Compte-rendu des travaux de la VIe session*, p. 348. The provincial statistics to which the minister refers here were mostly those produced by a few prefects of the Kingdom, zealous and loyal functionaries. These works have been studied by L. Gambi, "Le 'statistiche' di un prefetto del Regno," *Quaderni storici* 15 (1980), pp. 823–866.

38 See J. A. Davis, *Conflict and Control: Law and Order in Nineteenth-Century Italy* (London and Atlantic Highlands, N. J., 1988), ch. 4.

39 Caracciolo, in *Stato e società civile*, pp. 95–100, points out the victorious opposition by municipalist interests in the 1860s to the government's plans for a reform of the university system which would concentrate and strengthen higher education in a few centers and abolish several minor universities. Of course the argument can be made that the national elites themselves were prey to their own kind of municipalism when they imposed Piedmontese legislation on the country as a whole.

defense of local interests that Correnti certainly referred when he maintained that in the new political conditions those cities that – in the former Italian states – had enjoyed, as capitals, the privileges which derived from conditions of "monopoly and political centralization" needed to acquire now a new role as "productive cities."[40] Correnti probably well remembered the violent rioting into which Turin had burst only three years earlier at the announcement that Florence would replace it as capital of Italy.[41]

How did the participants in the congress react to Correnti's proposals? Interest in the topic was certainly remarkable. Both the Fourth Section of the congress (which had to discuss Correnti's report) and the general meeting (where the resolutions of every section were presented and discussed further for a final decision by the congress) witnessed a particularly lively debate, which contrasted with the more subdued tone of other discussions. The Fourth Section, with its 124 members, was one of the best attended; and several of the members took part in the discussion. The debate concentrated on the mode of making the proposal more attuned to a statistical congress, by translating this highly political issue into a set of questions that could lead to quantification. To some of the participants, Correnti's proposal appeared to be too "philosophical";[42] it did not make clear enough the kind of questions that a statistics of communes had to ask and what objects it had to investigate.[43] Others seemed to defend the way in which things had always been done, and insisted that all that existed in a commune had to be included in its statistics,[44] or that all that belonged to general statistics, that is to the statistics of the nation, had to appear in communal statistics as well.[45]

Some of the statisticians simply did not grasp what preoccupied Correnti, and continued to think about the relation between the whole and its parts, between general and local statistics, as if it were a problem of degrees of complexity as in the great chain of being. We have a

40 *Compte-rendu des travaux de la VIe session*, p. 123.

41 During these riots twenty-three people were killed: see D. Mack Smith, *Italy: A Modern History* (Ann Arbor, Mich., 1959), p. 66.

42 See the speech of A. Marescotti, a professor from the university of Bologna in the proceedings of the Fourth Section in *Compte-rendu des travaux de la VIe session*, p. 234.

43 Ibid., speeches by M. Block, a French statistician, p. 240, and by P. Mantegazza, a professor of medicine, p. 234.

44 See the speech by G. Piantanida, a statistician from Milan, at the general assembly (ibid., pp. 382ff.).

45 See the speech by R. Pareto, a high-ranking official of the Ministry of Agriculture, Industry, and Commerce, at the general assembly (ibid., p. 380).

perhaps excessively candid example of this attitude in the following statements from one of the participants to the congress, a physician and professor of statistics in a technical school in Naples:

> In the social hierarchy, facts of every kind and first-order needs appear almost individually, and therefore in a better condition to be analyzed, in communes; they become complex and less clear at the provincial level, and arrive complicated and almost muddled at the central government. And if in the polity it is necessary that analyses of facts and needs constantly proceed from the simple to the complex and from unities to additions, it clearly follows that the researches of communal statistics have naturally to precede those of provincial statistics, and these have to precede those of governmental statistics.[46]

To the various criticisms, Correnti responded by further emphasizing that his purpose was not so much that of positively indicating objects and categories of investigation, or of drawing up a list of questions,[47] as that of posing the question of the articulation between general and local statistics, that is of the harmonious articulation between the state and its peripheral organs. A compromise solution was eventually achieved between Correnti and his critics so that the final resolution contained both Correnti's original proposal and a detailed questionnaire compiled by a special subcommittee formed by some members of the Fourth Section.[48]

In general all the questions aimed at investigating the way in which communes performed their tasks as organs of local power in a dynamic interaction with central power. The first nineteen sets of issues of the questionnaire concerned basic information about the population and the number of agglomerations composing a commune, its topographical location, public services, voluntary associations, political constitution (what kind of power is in the hands of the mayor and of the municipal representatives, how they are elected, and so on), finances (modes of taxation and expenses, relationships with central finances). The subsequent fifteen sets of questions (actually fifteen specifications of the twentieth question) concerned the issue of communal hygiene and sanitary structures. The questions on the sanitary conditions of communes in particular touched upon the articulation between center and periphery in an especially delicate site of intervention, that con-

46 G. Caporale, *Sunti storici del VI Congresso di Statistica tenuto a Firenze* (Naples, 1868), p. 48. In the general assembly the author also asked that his own questionnaire for communal statistics (which he was unable to present because of lack of time) be printed in the congress's proceedings; his request was satisfied: see *Compte-rendu des travaux de la VIe session*, pp. 315–317.

47 Ibid., p. 234.

48 The text of the questionnaire is in ibid., pp. 469–470.

cerning the health of the population, the main resource of the national state. It is on this issue in particular that we will focus our attention in the rest of this chapter.

STATISTICAL INVESTIGATIONS AND THE LIBERAL ORDER

The will to give a specific direction to local statistics which was embodied in the project for the statistics of communes was linked to the wider aims assigned by Italian statisticians to their quantitative investigations. We saw in the previous chapter how statistics was to contribute to the building of the new state by promoting a reduction in the heterogeneity of the country. This enterprise implied the integration of local communities within the national entity, which meant in its turn an effort at transforming them to meet the requirements of a modern society as imagined by the national elites. Now, local communities exhibited extremely diverse conditions from a demographic, economic, and civic standpoint. Divergence implied incompleteness; by the standards of "civilized norms," rural settlements were not wholes. Dramatic differences placed "uncultured" rural communes apart from "civilized" and populous ones; only the latter were able to produce what Correnti called "raison impersonnelle," that is, an informed, bourgeois "public opinion";[49] only the latter took effective care of the complex needs and extended interests of the populations.

It is particularly significant that, in Correnti's view, the separation between the social life of cities and towns on the one hand, and of rural villages on the other, could best be measured through their widely different endowments in terms of relief and hygiene structures, of bodily and social healing. Cities were replete with relief institutions, such as hospitals, medical schools, pharmacies, asylums, health regulations concerning building, streets, and so on. In contrast, none of all this could be found in the countryside. The countryside completely lacked the discipline, the "domination over nature" that was embodied by towns:

> The agglomeration of inhabitants contributes to discipline and enlightens them about their hygienic interests, while *agriculteurs* [peasants], scattered in isolated houses or small villages, yield themselves to nature, which unfortunately shows very little of that idyllic character that poets assign to it.[50]

This representation of the virtues of city life as opposed to the countryside not only mirrored an actual process of growth and concentra-

[49] Ibid., p. 116. [50] Ibid., p. 121.

tion of welfare functions and institutions in towns, but also constituted a "self celebration" of the reforming middle classes, who saw in assistance to the poor and sanitation primary fields of intervention for the "improvement" of society.[51] The lack of adequate sanitation structures in the countryside was necessarily a sign of the weakness of this intervention, but, more importantly, of the want of initiative on the part of the local authorities who were supposed to intervene. Correnti seemed to echo concerns diffused among the national elites towards the end of the 1860s. As Raffaele Romanelli has shown, peripheral state officials often complained that in small, rural communes, especially in the South of the country, the modernizing plans of the national elites – such as the building of roads, the diffusion of elementary education, the improvement of the sanitary system, all tasks which by law had been assigned to municipalities – were more often than not completely neglected by the local elites. On the basis of these considerations officials even denounced existing legislation for allowing too much autonomy to municipal authorities, who often, and, again – especially in small communes – were not up to their tasks.[52] Self-government in harmony with the national interest did not arise spontaneously in the rural communes.

The consequences of this situation seemed all too evident. The far from idyllic character of the countryside could be further observed, according to Correnti, in the strained relationships between peasants (whom our author called, perhaps too idyllically but perhaps with mild irony as well, "nature's workers") and the "higher classes, landowners, capitalists and merchants" – a national class, let us say, mainly located in the truly national "hundred cities." Therefore, it seems likely that the attention given by Correnti to the question of hygiene in the countryside aimed at finding means of appeasing social tensions and guaranteeing the implementation of the modernizing programs of the new liberal governing class in what turned out to be a particularly weak link in the chain of power in the new state.

As vehicles of a program of reform, the bearers of hygiene, namely the *medici condotti* and especially the countryside medical doctors, were identified as the ideal protagonists of this investigative effort. The

51 On the rise and organization of relief institutions in nineteenth-century urban Italy see the essays collected in E. Sori, ed., *Città e controllo sociale in Italia tra* XVIII e XIX secolo (Milan, 1982).

52 See R. Romanelli, "Il problema del potere locale dopo il 1865," and "Tra autonomia e ingerenza: un'indagine del 1869," in *Il comando impossibile*. Even though the law defined some local expenses as mandatory, there were no legal provisions obliging local administrations to comply with it (ibid., p. 21).

medici condotti were paid by the communes to take care of the health problems of the poor or of the general population. Being half public officials and half scientists, they had always been principal sources of statistical intelligence, for they among all other members of the new state elite or national class were best situated to "penetrate" rural society: "In this respect" – Correnti further observed – "countryside physicians have to exercise a real civil ministry [*un véritable sacerdoce civil*]. They could give us a complete and, so to speak, photographical representation of the moral and economic conditions of the least known and least important section of the people."[53] The other main candidates for the role of emissary to the rural classes might have been the clergy; but in the new Italy they were not eligible. In Correnti's perspective, informed by the anti-clericalism of many protagonists of the Risorgimento and yet imbued with a Catholic imagery, physicians, and in particular *medici condotti*, were the equivalent of lay priests, but equipped with scientific training, skills of observation, and healing powers. Like priests, *medici condotti* had access to the houses of the poor and could have a first-hand knowledge of their conditions. Like priests, they often spoke of their work as a "mission."[54]

Correnti's identification of the role of doctors in contributing to an official knowledge of the conditions of rural communes expressed a desire that was not to receive institutional translation into practice until fairly late, in the mid-1880s, with the publication of the *Risultati dell'inchiesta sulle condizioni igieniche e sanitarie nei comuni del Regno*,[55] an investigation which, however, had no direct links with Correnti's plans. As a matter of fact, until the sanitary reform of the late 1880s, the institutional position of *medici condotti* was not one of power in the local context, that is *vis-à-vis* their employers, the communal administrations; the early sanitary legislation of the new state (which was essentially that of Piedmont, partially revised in 1865) did not enhance their local status; it was up to communal authorities to appoint the *medici condotti* and to decide whether to give tenure to the physicians or not: they could easily dismiss them and did not provide pensions for their retirement. In fact this legislation constituted a setback for physicians in those areas of the country, such as Lombardy and the Kingdom of the

[53] *Compte-rendu des travaux de la VIe session*, p. 122.

[54] See G. Cosmacini, *Storia della medicina e della sanità in Italia. Dalla peste europea alla guerra mondiale, 1348–1918* (Bari, 1987), pp. 344–345.

[55] 3 vols. (Rome, 1886). On this inquiry see P. Frascani, "Medicina e statistica nella formazione del sistema sanitario italiano: l'inchiesta del 1885," *Quaderni storici* 15 (1980), pp. 942–965. The primary responsibility in the collection of data was, however, assigned to the mayors.

Two Sicilies, which, before 1861, had enjoyed relatively more advanced legislation on the matter of public health provisions.[56]

Doctors, and especially Lombard ones, were therefore particularly vocal in their criticism of the laws of the Italian state on the matter of hygiene and public health, and they constantly demanded statistical investigations, or produced them themselves by using whatever data they could assemble in order to call attention to issues on which they wanted to exercise their professional jurisdiction.[57] While in countries such as Great Britain and France the role of physicians in the production of knowledge about society in the nineteenth century paralleled a growing "medicalization" of society, in Italy doctors' investigative activism was more linked to the attempts made by this professional group to organize itself and further its position *vis-à-vis* the political authorities.[58] Just as some representatives of the national elites engaged in the building of the new Italy, such as Correnti, saw medical doctors as the potential means for penetrating resilient local realities, so too the latter relied on liberal reformers for support in their struggle to establish the power of their profession. In some cases the two roles overlapped, as for example with Pietro Castiglioni, the Milanese doctor and reformer who founded the Associazione nazionale dei medici condotti and from the early 1860s actively promoted the compilation of sanitary and medical statistics and the reform of sanitary legislation. In the periodical he directed, the *Annali di medicina pubblica, igiene e idrologia medica*, the performance of institutions of local government was judged on the basis of, among other things, their involvement in the production of "positive" information. And here again, it was noted how great was the gap between urban and rural realities, and how great was the difference between the performance of communes in various areas of the country. Consequently the doctors demanded more trenchant action from the state to overcome the omissions and limitations of municipal administrations.[59] The involvement of physicians in the activity of

[56] On the condition of *medici condotti* as a social category see A. L. Forti Messina, "I medici condotti all'indomani dell'Unità," in M. L. Betri and A. Gigli, eds., *Salute e classi lavoratrici in Italia dall'Unità al fascismo* (Milan, 1982), pp. 663–697. On the sanitary legislation of the Kingdom of Italy see F. Della Peruta, "Sanità pubblica e legislazione sanitaria dall'Unità a Crispi," *Studi storici* 21 (1980), pp. 713–760.

[57] For some examples of this investigative activity see Frascani, "Medicina e statistica," and Forti Messina, "I medici condotti."

[58] On the less advanced process of medicalization in Italy see the observations of P. Frascani, "Il medico nell'Ottocento," *Studi storici* 23 (1982), pp. 617–637.

[59] See in particular E. Fazio, "Sulla importanza di una statistica medica italiana," *Annali di medicina pubblica, igiene e idrologia medica* 10 (1875), pp. 141–160.

social investigation recalled the role that many of them had had in the development of statistics in the pre-unification period.

The politics of investigations into the conditions of communes planned at the Congress of Florence paralleled those of other statistical inquiries carried out in this first decade of life of the new state. They were part of an attempt to impose a liberal order on a society which appeared in many of its parts dangerously removed from any order at all, but especially from the idea of order held by the bourgeois national elites, who, although in control at the center, feared the uncertainty of their rule in the "periphery."[60] That statistics appeared to some of them as an instrument particularly apt for bringing order out of chaos should not be surprising after all, since as an investigative practice it strives to bring order to reality by making its components measurable and comparable, symmetrical and complete.

If we consider the Congress of Florence from the perspective of the development of statistical science in the second half of the nineteenth century – or even from the perspective of many statisticians from other European countries who were annoyed by the sometimes obscure talk of the Italians – the event will not appear as a bright episode in the path towards modern statistics. Yet it epitomizes well the way in which a component of the new state elite of Italy looked at statistics, and at science in general, as the midwife of a healthy "national body" and a privileged instrument of liberal government. In their view it was a form of intervention which was not in opposition to the tenets of liberalism and in fact should favor the establishment and working of a liberal order. Maestri and Correnti even imagined that the information provided by statistics could serve to smooth the rough movements the market economy:

> free competition, accused – not always unjustly – of being the cause of the disorders of production, cannot free itself from these accusations and avoid the dangers that it generates unless it increases the light of publicity, unless it multiplies information, and unless it impedes, through the help of industrial and commercial statistics, the confusion produced by the improvident crowding of consumers and producers [onto the market].[61]

60 Romanelli too stresses the function of statistics in liberal Italy as a particular form of government intervention that was to help produce a "behavior compatible with a modern liberal regime" (introduction to *Il comando impossibile*, p. 21). In order to understand why statistics was so important for the national administrative elites it is, however, essential to emphasize the context of uncertainty in which they had to operate.

61 *Annuario statistico italiano. Anno* II, p. xviii.

This statement expresses what in the eyes of some of their contemporaries, not to mention those of late twentieth-century observers, will certainly appear as a naive trust in the virtues of numbers and in the possibility of bringing some order into the anarchy of capitalistic production. But besides displaying the great faith in scientific knowledge of these Italian liberals (a faith which of course they shared with many other members of the European educated classes in the mid-nineteenth century), it also points to the keen understanding that a segment, however small, of the state elites had of the new forms of "government" required for the "constitution of liberty" within the boundaries of a new national state.[62]

[62] The expression "constitution of liberty" comes from a speech given by the deputy N. Nisco, a supporter of the Ministry of Agriculture, Industry, and Commerce during a debate on this Ministry's possible abolition on June 6, 1863: see *Atti del Parlamento italiano*, Camera, leg. VIII. Sessione Seconda.

CHAPTER 9

EPILOGUE

Rapid change took place in Italian politics and in official Italian statistics in the 1870s. After Pietro Maestri, one of the main promoters of patriotic statistics and the principal architect of official statistics in the 1860s, died in 1871, he was replaced in 1872 by Luigi Bodio (Milan 1840–1910). Bodio was a protégé of the then general secretary of the Ministry of Agriculture, Industry, and Commerce, Luigi Luzzatti, himself a student of Messedaglia and representative of the so-called "Lombard–Venetian school" of political economy which traced its roots back to the teaching of Romagnosi.[1] Bodio was also strongly convinced of the high political and scientific mission of statistics, but clearly faced new tasks and a different intellectual climate. He introduced new methodological concerns and a more technical language into the official statistics of Italy, and gained a solid international reputation as head of the International Statistical Institute. Necessarily statistics in the new Italy could no longer be the charged symbol of liberalism, of representative institutions, and of rational government that it had been in the previous period in the midst of "autocratic" regimes. It had become an established institution in the state apparatus, a discipline taught in most universities and high schools.

With the annexation of Rome in 1870, the "dispersed limbs" of the national "body" had all (or almost all) been assembled together; the body was finally a whole, although its surface was still marked by wounds. In the view of Risorgimento politicians and scholars such as Maestri, statistics – the "anatomy" of nations – was supposed to heal those wounds, to help generate a healthy whole. We know that this is not what happened. Although statistics continued to occupy a respectable place in the ideology and practice of sectors of the state apparatus throughout the 1880s – before entering a period of intense decline in

[1] On Bodio see *Dizionario Biografico degli Italiani*, s. v. (by F. Bonelli).

the 1890s – it was not invested again with the hopeful expectations that characterized the period of Italian history we have examined in this study. This change was partly related to the transformations that overtook the country as a whole. Already in the 1870s patriotic enthusiasm was replaced by a mood of disenchantment *vis-à-vis* the new political entity that had resulted from the process of unification – a mood well expressed in the coining of the derogatory word *Italietta.* Great hopes gave way to an appreciation of the mediocrity of the outcome. Official numbers provided ample evidence of mediocrity, revealing the persistently poor performance of Italy compared to other European nations with respect to what were considered the indicators of "civilization" *par excellence*: rates of violent crime first of all, and illiteracy. The mood of disenchantment was to be found also among the state statisticians. Representatives of both the old (such as Cesare Correnti) and the new generation (such as Luigi Bodio and Aristide Gabelli) began to complain about the relative isolation in which they were left to operate, about the lack of means, the skepticism and absence of collaboration on the part of the public and the very elites of the country.[2]

In the 1870s, in the aftermath of the Paris Commune, the governing classes began also to face what by then was called the "social question"; conservative reformers such as Pasquale Villari, Leopoldo Franchetti, and Sidney Sonnino began to articulate the social problems of the southern regions in terms of a "southern question" thus asserting the distinctive character of the South as a whole. In many respects the "southern question" came to be understood as the specific form that the "social question" assumed in Italy. The myth of the naturally wealthy South that Maestri's generation had cultivated faded away to be replaced by the negative image of southern society and politics which had also been present among the national elites at the time of the annexation of the Kingdom of the Two Sicilies – when Cavour referred to the South as the " weakest and most corrupt part of Italy."[3] Recent readings of the representation of the South among Cavour's collaborators, among military men and public officials – many of

[2] See for example C. Correnti, "Cesare Correnti a Luigi Bodio," *Archivio di statistica* 1 (1876), pp. v–xvii; A. Gabelli, "Gli scettici della statistica," *Archivio di statistica* 2 (1877), pp. 9–28. L. Bodio's critical views are reported by M. Lungonelli, "Sul servizio statistico del Regno d'Italia nel 1871 (Una lettera di Luigi Bodio a Luigi Luzzatti)," *Clio* 18 (1982), pp. 295–299. See also C. A. Corsini, "L'acculturazione mancata. Il processo formativo dell'organizzazione delle fonti demografiche nel periodo post-unitario," in *La popolazione italiana nell'Ottocento. Continuità e mutamenti* (Bologna, 1985), pp. 115–145.

[3] Reported in M. Salvadori, *Il mito del buongoverno. La questione meridionale da Cavour a Gramsci* (Turin, 1960), p. 24.

whom were southerners themselves – during the repression of "brigandage" have stressed how utterly Other the South appeared to them. Comparisons with Africa, the very epitome of the "primitive" in their eyes, were common. From the beginning of its history as a unitary state, "Italy" was constructed by the national elites as the opposite of its South.[4]

Whether these images constituted a continuous and distinctive presence in the outlook of the governing elites of liberal Italy remains to be seen. We still lack a detailed reconstruction of the perceptions and representations of socio-economic and cultural differences within the national territory in post-unification Italy.[5] Our investigation of the history of statistics points to the need to pay attention to the role that this particular "technology of knowledge" and the representations it generated played in the shaping of the elites' perceptions. We saw earlier that in the 1860s and early 1870s the comparison of statistical indicators across the national territory singled out geographical variations in the distribution of averages. Although the statisticians pointed out from time to time the differences between the southern and northern regions, the national territory was far from being represented along the lines of a simple North–South opposition. Quantitative differences, moreover, did not imply a qualitative or essential opposition between North and South, but mostly a difference of degree – of quantity, precisely – which (it was hoped) would disappear with time. The averages for the northern areas, in any event, were not something the statisticians felt very proud of. Their faith in the virtue of good administration, their conviction that change was possible, and in fact inevitable, if only the principles of good, namely liberal, government were applied, made the results look like a temporary outcome.

In the later part of the century, in connection with the spread of biological evolutionism and racial determinism, the opposition between "barbarism" and "civilization" which informed the dominant

4 See J. Dickie, "A Word at War: The Italian Army and Brigandage 1860–1870," *History Workshop* 33 (1992), pp. 1–24, and N. Moe, "'Altro che Italia!' Il Sud dei piemontesi (1860–61)," *Meridiana. Rivista di storia e scienze sociali* 15 (1992), pp. 53–89. See also D. Pick, *Faces of Degeneration: A European Disorder, c. 1848—c. 1918* (Cambridge, 1989), ch. 5.

5 Some material can be found in old studies such as G. Talamo, "Il problema delle diversità e degli squilibri regionali nella cultura politica italiana dal periodo dell'unificazione alla caduta della Destra," in *Gli squilibri regionali e l'articolazione dell'intervento pubblico. Atti del Convegno di studio svoltosi a Torino e Saint Vincent dal 3 al 7 settembre 1961* (Milan, 1962); most studies, however, tend to focus on the North–South contrast. For a recent analysis of the beginning of this opposition see N. Moe, "Representing the South in Postunification Italy ca. 1860–1880," Ph.D. dissertation, Johns Hopkins University (Baltimore, 1994).

understanding of human societies in the nineteenth century was given new content and more exclusionary meaning. In this new intellectual climate, and in the context of the social unrest and the severe crisis of state institutions of the 1890s, the persistence of statistical differentials between North and South came to be read as the expression of an essential difference, a difference inscribed in the bodies and minds not only of individuals but of whole peoples, and characterizing whole societies. This reading of Italian society came from the followers of Lombroso, in particular from Alfredo Niceforo who was one of the first, along with Giustino Fortunato (who, however, claimed paternity for the expression),[6] to talk of the "two Italies." To look briefly at this later development is instructive. It will allow us to make some concluding remarks about the role that statistics had in the imagining and the construction of the nation in nineteenth-century Italy and the legacy of this enterprise today. This seems also to be the appropriate place to conclude our story because the strongest image of Italy that we possess today is that of a divided and heterogeneous country: a country of regions certainly, but even more fundamentally a country divided between a North and a South.

Perhaps the most vocal representative of the school of criminal anthropology at the turn of the century, Alfredo Niceforo proclaimed his faith in statistics at the very beginning of his *L'Italia barbara contemporanea* (1898):

> Statistics – which ignorant or malicious people consider a humble handmaid to this or that opinion ... is in fact a truly marvelous precision instrument for investigating social phenomena. These can be placed in two great categories: those that pertain to inferior civilizations, and those that characterize superior civilizations. Statistics has this magic power: it can – through the miraculous language of its figures – indicate if a given phenomenon, expressed by numbers, belongs to one or the other civilization."[7]

Accordingly he proceeded to show how North and South exhibited distinctive patterns of crimes (as epigraph to the chapter devoted to these data, he used a distinction made by Angelo Messedaglia – in a different context – in 1879 between a criminality distinctive of "civilization" and one distinctive of "barbarism"),[8] levels of literacy, natality rates, suicides and mortality rates, distribution of modern industry, modes of agriculture. In fact there was much else in the book besides

6 See his letter to Federico Severini (March 27, 1911) cited in S. F. Romano, *Storia della questione meridionale* (Palermo, 1945), p. 80.

7 *L'Italia barbara contemporanea. Studi e appunti* (Milan and Palermo, 1898), p. 15.

8 In his lecture of 1879 on the statistics of criminality on which see ch. 6.

numbers: claims about general features of social life in the southern regions and especially representations of the character of southerners that had little to do with any "positive" evidence, but seemed to be the repository of all sorts of stereotypes of both old and new coinage. There is no need here to recall here Niceforo's vast apparatus of oppositions which defined the two "races" ("Mediterranean" southerners, by nature individualist and not inclined to life in society, and "Aryan" northerners who had a more developed social sentiment and were thus better placed to deal with the requirements of modern society; colder and more thoughtful northerners versus more instinctive and easily aroused southerners, and so on). Some of them are still present in the collective imagination of contemporary Italians.

In a new work published three years later with the purpose of responding to his critics, *Italiani del Nord e Italiani del Sud*,[9] Niceforo restated more fully his theses about the two different "races" inhabiting the Italian peninsula and subjected his description more systematically to the "discipline" of statistics. In this tome he deployed a much larger armory of numerical evidence both in the text and in numerous tables and bar charts at the end of each chapter: 133 numerical and 31 graphs to be exact, as the subtitle specified. As a good positivist, Niceforo invited his critics to pay attention to the authority of the "long series of facts" that he had accumulated in support of his views. Now even more indicators pointed to the alleged radical difference between the two Italies: both fairly traditional ones such as population density and patterns of urbanization, among others, which had been used as indicators of "civilization" even by patriotic statisticians in the first half of the century, and new ones, such as the diffusion of democratic ideas as measured by the percentage of socialists in municipal councils. Niceforo was actually in trouble when attempting to show the different psychology of the two "races" with the help of numbers and had to infer it from the different climates of the places they inhabited. But consistency was apparently not of great concern in his discourse.

Ironically, while speaking of "two Italies," in his statistical tables Niceforo always reported data on three Italies: North, Center, and South – with Sicily and Sardinia often apart as extreme examples of "southernness." Conveniently, most of the time the Center exhibited values which were in between those of the North and those of the South thus appearing like a vast borderland in which the two races coexisted (Aryans in Tuscany and Mediterraneans in the rest) and gen-

9 The complete title was *Italiani del Nord e Italiani del Sud (con 133 tavole numeriche e 31 tavole grafiche)* (Turin, 1901).

erated median values. Since dividing the Center along the Apennines and allocating one half to the North and the other to the South would probably have broken the symmetry of the whole construction, Niceforo did not choose this option. However, he did not explain the rationale for the partition he used in his data presentation and it is very likely that he resorted to it in order to generate more contrasting averages between North and South and thus strengthen his thesis about the two civilizations.

Of course, as critics were quick to point out, Niceforo's "facts" were not enough to prove his racial explanation; competing explanations which pointed to the role of socio-historical factors in determining the conditions of the South made more sense.[10] It must also be said that the scholars and statisticians who were responsible for the making of official statistics did not approve the work of Lombroso and his followers.[11] Although they all shared a faith in "objective" science and "positive" knowledge, the simplifying biological determinism of the Lombrosians – not to mention their more than cursory treatment of numbers – never appealed to those researchers who embraced more complex explanatory paradigms. A political contrast also opposed state statisticians and Lombrosians as the former remained wedded to liberal and unitary views while the latter were staunch critics of the centralized state and called for regional devolution if not for a federalist reform of the state which would deal more adequately, in their view, with the profoundly different character of the northern and southern peoples.[12] But questionable new "sciences" such as criminal anthropology and the biological and racial readings of Italy and its people that these sciences proposed received unwitting support from the steady accumulation of statistical data on the physical characteristics of the population which was generated by state statisticians. In the mid-1860s,

10 Niceforo did not completely exclude the role of socio-economic and historical factors, but he always inscribed them within his anthropological and racial scheme of interpretation, particularly emphasized in *Italiani del Nord e Italiani del Sud*. For example when he stressed the more widespread poverty among the population of the South, he saw it as a factor that had contributed in the course of history to higher levels of "organic degeneration" and had slowed, if not paralyzed, the advancement of that population. On Niceforo's work see B. Farolfi, "Antropometria militare e antropologia della devianza 1876–1908," in *Storia d'Italia. Annali VII. Malattia e medicina* (Turin, 1984), p. 1209, and V. Teti, *La razza maledetta. Origini del pregiudizio antimeridionale* (Rome, 1993).

11 See C. Pazzagli, "Statistica 'investigatrice' e scienze 'positive' nell'Italia dei primi decenni unitari," *Quaderni storici* 15 (1980), pp. 807–808.

12 On the regionalist policies of the Lombroso school see A. Mulas, "Il regionalismo nell'opera di Cesare Lombroso e della sua scuola," *Archivio storico sardo* 32 (1981), pp. 311–347.

we will recall, Maestri and his collaborators, in an attempt to provide a detailed picture of the national population, added the anthropometric data collected by army doctors to the yearly publication of data on natality, mortality, and marriages. Colorful tables showing the differential distribution of the height of conscripts on the national territory began to be published by the Direzione di Statistica in the late 1870s and 1880s.[13] These collections were important sources of data for those private researchers such as Niceforo and the other Lombrosians who were looking for signs of essential difference on the surface of bodies.

Beyond making available a specific kind of quantitative information concerning human beings, official statistics established generally what constituted significant and authoritative evidence about the state of a country, or a region, and its level of "civilization." Far from merely reflecting a state of things inherited from the past and strengthened in the course of Italian unification, statistics translated complex differences between the conditions of social life in various parts of the country into "facts," and established what the facts to be considered should be: by definition only those phenomena that could be translated into figures, and actually only those that the state was willing and able to measure. Official statistics did something else too: by constantly presenting comparisons between the various Italian regions, and aggregates of regions, they reinforced the primacy of the regional grid in the reading and imagining of the country. When we face these objects it was and it continues to be easy to forget that what we see is the result of aggregative procedures; that the figures are averages, means of means, summary measures which hide a great diversity of individual cases; and that the regions and macro-regions, like the nation, are not natural entities, but historical artifacts.

I must make clear that by drawing attention to the contribution of statistics to the making of the "two Italies" I am not denying the existence of differences in the texture of economic, social, and political life within Italy – clearly an absurd position – but I simply intend to question the effects of identifying these differences primarily by means of statistics and by the facts constructed by statistics. Undoubtedly many of these facts are revealing, and indeed in many cases indispensable for the objectives of reformers and modern state administrations alike, but they are profoundly marked by their genealogy and cannot be appropriated without cost: the cost of a mode of representation based on a classificatory and oppositional way of thinking. In late nineteenth-

13 Ministero di Agricoltura, Industria e Commercio, *Carte e diagramma di demografia italiana* (Rome, 1878) and *Atlante statistico del Regno d'Italia. Diagrammi di demografia italiana* (Rome, 1882).

century Italy the use of statistics established the kind of "comparativist" approach that has characterized the debate on the "southern question" since its origins.[14]

It is ironic that the image of a profoundly divided Italy was built with the very material produced by an activity which was originally intended to contribute to the making of a more cohesive state, to the real unification of the country. Patriotic statisticians such as Maestri and Correnti attempted to gather a comprehensive statistical knowledge of the society enclosed in the new state, its mirror-like representation which they thought was needed in order to found enlightened administrative choices and eventually a rational "science of government." This aspiration did not amount to a technocratic form of social engineering and to the totalitarian "management of the population" which would later characterize the use of statistics by the fascist regime.[15] Although they operated within the limits of a definitely classist outlook, the protagonists of the epoch of statistical investigations examined in this book intended to educate the "public" and contribute to the building of liberal institutions, not to replace them.

We have seen throughout this book how these indefatigable collectors of figures, as the statisticians liked to see themselves, were well aware of the constitutive power of statistics. Before unification, they promoted statistics to give a spur to the "improvement" of society; they counted to find out what still needed to be done by wise legislators to achieve the level of the most advanced "civilization"; they produced new taxonomies for the ordering of the polity. Eventually patriots used statistics to evoke the "body" of a political entity which did not yet exist. After unification, they enrolled statistics in the very construction of the nation state; they provided a map of the country, monitored its movements and looked for its laws, counted its resources, its people above everything, and tried to link together more closely center and periphery. Although their procedures did not differ much from those employed by statisticians elsewhere, the strong belief in the constitutive power of numbers remains a remarkable peculiarity of Risorgimento statisticians. Yet they did not realize the extent to which the outcomes of statistical practices tend to acquire a life of their own and escape our control. But this is something that we could not reproach them for since we still live under the dominance of statistical representations.

14 On the predominance of the comparativist approach see also G. Galasso's introduction to D. Abulafia's *Le due Italie* (Naples, 1981; orig. English edn. 1977), p. 2.

15 On the fascist use of statistics see D. Horn, *Social Bodies. Science, Reproduction and Italian Modernity* (Princeton, 1994), ch. 3.

APPENDIX

NUMBERS OF STATISTICAL PUBLICATIONS

Table 1 *Statistical publications in the Italian states by decade, 1800–1859*

	1800–1809	1810–1819	1820–1829	1830–1839	1840–1849	1850–1859	Total
Lombardy–Venetia	13	10	12	25	27	44	131
Two Sicilies	5	9	8	28	20	37	107
Piedmont	1	3	6	13	18	36	77
Papal States	1	1	2	4	11	12	31
Tuscany	–	1	3	5	6	9	24
Total	20	24	31	75	82	138	370

Source: This table is based on the bibliography edited by Luigi Bodio, *Annali di statistica. Saggio di bibliografia statistica italiana*, 2nd edn.(Rome, 1885). In this bibliography Bodio included also those works of geography that, according to him, appeared to present statistical notions "to a greater extent than usual, or discussed [them] with real competence" (from the Introduction to the volume). On the other hand, Bodio excluded works that he considered too limited in scope, such as hospital reports. He also listed articles on statistics published in major periodicals which, however, I have excluded from my count. I have also excluded a very few works published in other smaller Italian states (Duchy of Modena, Duchy of Parma e Piacenza).

Table 2 *Statistical publications (both Italian and non-Italian) reviewed in the* Annali universali di statistica, *1824–1869*

Years	Number of works
1824–1829	33
1830–1839	97
1840–1849	72
1850–1859	74
1860–1869	79

Note: I have included in my count all works whose titles contain the word "statistics" or "statistical" (even in hyphenated form).

SELECT BIBLIOGRAPHY

PRIMARY SOURCES

Manuscript

Académie Royale de Belgique: Correspondence of Adolphe Quételet
Archivio Centrale dello Stato, Rome: Gabinetto
Archivio di Stato, Florence: Ministero dell'Interno; Segreteria di Gabinetto; Segreteria di Gabinetto Appendice
Archivio di Stato, Milan: Studi. Parte Moderna
Archivio di Stato, Palermo: Direzione Centrale di Statistica; Intendenza di Palermo
Archivio di Stato, Turin: Prima Sezione: Regno di Sardegna, Materie Economiche-Statistica; Sezioni Riunite: Intendenza di Susa
Biblioteca Civica, Verona: Fondo Messedaglia
Biblioteca Nazionale, Florence: Raccolta Tordi; Carte Vieusseux
Museo del Risorgimento, Milan: Carte Cattaneo
Museo Nazionale del Risorgimento, Turin: Archivio Petitti

Printed

Abbreviations

AUS *Annali universali di statistica*
GdS *Giornale di statistica*

Almanacco aretino, Arezzo, 1837.
Amari, E., "Difetti e riforme delle statistiche de' delitti e delle pene," *GdS* 3 (1838), pp. 270–334.
"Società statistica di Londra, e i suoi lavori," *GdS* 3 (1838), pp. 41–60.
Anastasio, G., review of *Statistica dell'Italia* by L. Serristori. In *GdS* 1 (1836), pp. 60–70.
Annali di statistica. Saggio di bibliografia statistica italiana, ed. L. Bodio, 2nd edn., Rome, 1885.
Annali universali di statistica (*AUS*), Milan, 1824–1871.

Annuario geografico italiano, Bologna, 1845.

Annuario statistico italiano. Anno I 1857–58, [ed. C. Correnti], Turin and Milan, n. d. [1858].

Annuario statistico italiano. Anno II, [ed. C. Correnti and P. Maestri], Turin, 1864.

[Anonymous] "Giuseppe Sacchi," *Bulletin de l'Institut International de Statistique*, Rome, 1890 [*recte* 1891], pp. 327–328.

[Anonymous but G. D. Romagnosi] "Necrologia. Melchiorre Gioia," *Biblioteca italiana* 52 (1828), pp. 392–407 (reprinted as "Elogio storico di Melchiorre Gioia," in *Florilegio di eloquenza italiana*, vol. II [Pistoia, 1839], pp. 55–86).

Atti del Parlamento Italiano [1861–1871].

Atti della Prima Riunione degli scienziati italiani tenuta in Pisa nell'ottobre del 1839, Pisa, 1840.

Atti della Quarta Riunione degli scienziati italiani tenuta in Padova nel settembre del 1842, Padua, 1843.

Atti della Quinta Unione degli scienziati italiani tenuta in Lucca nel settembre del 1843, Lucca, 1844.

Atti della Seconda Riunione degli scienziati italiani tenuta in Torino nel settembre del 1840, Turin, 1841.

Atti della Settima Adunanza degli scienziati italiani tenuta in Napoli dal 20 di settembre al 5 di ottobre del 1845, Naples, 1846.

Balbi, A., "Rapport du nombre de crimes à l'état de l'instruction publique en France," *Bulletin des sciences géographiques, économie publique, voyages. 6e section du Bulletin universel publié ... par la Société pour la propagation des connaissances scientifiques et industrielles* 20 (1829), pp. 252–264.

Compendio di geografia compilato su di un nuovo disegno conforme agli ultimi trattati di pace e alle più recenti scoperte, 2nd edn., Turin, 1840.

"Dell'Italia e del saggio statistico del conte Luigi Serristori," in *Scritti geografici, statistici e vari pubblicati in diversi giornali d'Italia, di Francia e di Germania da Adriano Balbi raccolti ed ordinati per la prima volta da Eugenio Balbi*, vol. I, Turin, 1841.

Miscellanea italiana. Ragionamenti di geografia e statistica patria. Raccolti e ordinati da Eugenio Balbi, Milan, 1845.

Basevi, E., *Della necessità d'un uffizio di statistica in Toscana sue immediate applicazioni nelle riforme municipali e daziarie*, Florence, 1847.

Bembo, P. L., *Il comune di Venezia nel triennio 1863–65*, Venice, 1866.

Bianchi, A., summary translation of *Sur l'homme et le développement de ses facultés, ou Essai de physique sociale* by A. Quételet, in *AUS* 53 (1837), pp. 9–33, 129–154, 241–258; 54 (1837), pp. 9–28, 157–169, 301–307; 55 (1838), pp. 191–203; 56 (1838), pp. 37–47; 57 (1838), pp. 156–166; 58 (1838), pp. 39–52.

Bianchi, M., *Geografia politica dell'Italia*, Florence, 1851.

Bicchierai, "Statistica delle comunità di Albiano, Bagnone, Filattiera, Groppoli e Terrarossa," *Calendario lunese per l'anno 1835*, Fivizzano, [1835].

[Bicchierai], "Statistica delle comunità di Fivizzano e di Casola," *Calendario lunese per l'anno 1834*, Fivizzano, [1834].

Boccardo, G., "Degli studi geografici e del loro stato presente in Italia," *Archivio storico italiano*, n. s., 5 (1857), pp. 60–87.

Dizionario della economia politica e del commercio così teorico come pratico utile non solo allo scienziato ed al pubblico amministratore, ma eziandio al commerciante, al banchiere, all'agricoltore ed al capitalista, 4 vols., Turin, 1857–1861.

Bodio, L., *Sui documenti statistici del Regno d'Italia. Cenni bibliografici presentati al VI Congresso Internazionale di Statistica*, Florence, 1867.

Della statistica nei suoi rapporti coll'economia politica e colle altre scienze affini, Milan, 1869.

Bonomi, review of *Stato agrario economico del Ferrarese* by A. Casazza, *AUS*, 2nd s., 5 (1845), pp. 328–330.

Bonstetten, Ch.-V. de, *L'Homme du Midi et l'homme du Nord ou l'influence du climat*, Geneva, 1824.

Bowring, J., *Statistica della Toscana, di Lucca, degli Stati Pontifici e Lombardo-Veneti e specialmente delle loro relazioni commerciali*, n. p., n. d. [London, 1838].

Bruno, G., "Difetti e riforme delle statistiche commerciali," *GdS*, 2nd s., 7 (1852), pp. 14–67.

Bullo, C., *Delle condizioni statistiche e commerciali di Chioggia, aggiuntivi alcuni brevi cenni storici di quella città*, Padua, 1866.

Busacca, R., review of *Archives statistiques du Ministère des Travaux publics de l'agriculture et du commerce, publiés par le ministre secrétaire d'état de ce département*. Paris imprimerie royale [sic], 1837. In *GdS* 3 (1838), pp. 146–169.

Cagnazzi, L. de Samuele, *Elementi dell'arte statistica*, Naples, 1808–1809.

Saggio sulla popolazione del Regno di Puglia ne' passati tempi e nel presente. Parte prima che contiene lo stato de' tempi passati, Naples, 1820; *Parte seconda che contiene lo stato presente*, Naples, 1839.

"Cenno sulla statistica," *Il progresso delle scienze, delle lettere e delle arti* 1 (1832), pp. 44–57.

La mia vita, ed. A. Cutolo, Milan, 1944.

Cantù, C., review of *Saggio sulla popolazione del Regno di Puglia ne' passati tempi e nel presente* by L. de Samuele Cagnazzi. In *AUS* 67 (1841), pp. 161–168.

Caporale, G., *Dell'agro acerrano e della sua condizione sanitaria; ricerche fisiche, statistiche, topografiche e storiche*, Naples, 1859.

Sunti storici del VI Congresso di Statistica tenuto a Firenze, Naples, 1868.

Castiglioni, P., *Errori prodotti dalle medie nella statistica*, Milan, 1861.

"Introduzione storica dei censimenti delle popolazioni italiane dai tempi antichi sino al 1860," in *Statistica del Regno d'Italia. Popolazione. Censimento degli antichi Stati Sardi (1 gennaio 1858) e censimenti di Lombardia, di Parma e Modena (1857–58)*, vol. I, Turin, 1862.

Cattaneo, C., "Sulla densità della popolazione in Lombardia e sulla sua relazione alle opere pubbliche," *Il politecnico* 1 (1839), pp. 29–52 (now in *Scritti economici*, ed. A. Bertolino, vol. II, Florence, 1956, pp. 131–165).

"Di varie opere sulla Sardegna," *Il politecnico* 4 (1841), pp. 219–273 (now in *Scritti storici*, pp. 189–254).

"Osservazioni sulle guide di Pisa, Torino, Firenze e Padova pubblicate in onore del congresso degli scienziati," *Il politecnico* 6 (1843), pp. 471–483 (now in *Opere edite ed inedite*, ed. A. Bertani, 2nd edn., Florence, 1948, pp. 106–121).

"Prospetto d'una raccolta di notizie naturali e civili sulla Lombardia, proposta da alcuni studiosi per l'occasione del Congresso scientifico di Milano," *Il politecnico* 7 (1844), pp. 212–222 (now in *Scritti storici*, pp. 309–324).

Notizie naturali e civili sulla Lombardia, Milan, 1844.

La città considerata come principio ideale delle istorie italiane, Milano, 1858.

review of *Annuario geografico italiano*. In *Rivista Europea. Giornale di scienze morali, letteratura ed arti*, (1845) (now in *Scritti storici*).

Epistolario, ed. R. Caddeo, Florence, 1949.

Scritti storici e geografici, vol. I, ed. G. Salvemini and E. Sestan, Florence, 1957.

Cavour, C., *Tutti gli scritti di Cavour*, ed. C. Pischedda and G. Talamo, vol. II, Turin, 1976.

Cenni, E., *Delle presenti condizioni d'Italia e del suo riordinamento civile*, Naples, 1862.

Censimento della popolazione della città di Palermo al 31 dicembre 1861 pubblicato dall'ufficio comunale di economia e statistica, Palermo, 1862.

Censimento della popolazione della città di Torino al 31 dicembre 1861. Riassunto statistico sul movimento professionale ed industriale avvenuto in Torino nel quadriennio 1858–61, Turin, 1863.

Cevasco, M., *Statistique de la ville de Gênes*, Genoa, 1838.

Chabrol de Volvic, G.-J.-G., *Statistique des provinces de Savone, d'Oneille, d'Acqui et de partie de la province de Mondovì formant l'ancien département de Montenotte*, Paris, 1824.

Recherches statistiques sur la ville de Paris et le département de la Seine, Paris, 1821–1829.

Chadwick, E., *Report on the Sanitary Condition of the Labouring Population of Great Britain 1842*, ed. M. W. Flinn, Edinburgh, 1965.

[Commissione di statistica giudiziaria] *Statistica giudiziaria degli Stati Sardi*, vol. I, *Statistica giudiziaria civile, commerciale, e del contenzioso amministrativo degli Stati Sardi per gli anni 1849–50*, Turin, 1852; vol. II, *Statistica giudiziaria penale degli Stati Sardi per l'anno 1853 e ragguagli comparativi con gli anni 1854 e 1855 e con gli altri anni anteriori*, Turin, 1857.

Compte-rendu de la troisième session du Congrès international de statistique réuni à Vienne les 31 août, 1, 2, 3, 4, 5, septembre 1857 publiée sous la direction de M. Ch. baron de Czoernig, Vienna, 1858.

Compte-rendu des travaux de la VIe session du congrès international de statistique, Florence, 1868.

Compte-rendu des travaux du congrès général de statistique réuni à Bruxelles les 19, 20, 21 et 22 septembre 1853, Brussels, 1853.

Congrès International de Statistique à Florence. Programme de la sixième session du 29 septembre au 5 octobre 1867, Florence, 1867.

Congrès international de statistique à la Haye. Septième session du 6 au 11 septembre 1869, 3 vols., The Hague, 1869–1871.

Cordova, F., *I discorsi parlamentari e gli scritti editi ed inediti preceduti dai ricordi della sua vita*, Rome, 1893.

Correnti, C., "Teoria della statistica. Memoria," *AUS* 71 (1842), pp. 13–24.

"Indicazioni storiche e statistiche sulla provincia di Bergamo," *AUS* 81 (1844), pp. 48–69; 82 (1844), pp. 138–165, 265–278; 2nd s., 3 (1845), pp. 45–56.

"Fisionomia delle regioni italiane," *Il Nipote del Vesta Verde* 5 (1852), pp. 42–61 (now in *Scritti scelti*, vol. II, pp. 372–399).

"Ancora delle nostre regioni," *Il Nipote del Vesta Verde* 8 (1855), pp. 144–163 (now in *Scritti scelti*, vol. II, pp. 446–447).

"Cesare Correnti a Luigi Bodio," *Archivio di statistica* 1 (1876) (partially reprinted in *Scritti scelti*, vol. III, pp. 213–215).

Scritti scelti di Cesare Correnti in parte inediti e rari, ed. T. Massarani, vol. II, Rome, 1891; vol. III, Rome, 1894.

[Correnti, C.,], *L'Austria e la Lombardia*, Italia [but Milan], 1847.

[Correnti, C., ed.], *Annuario statistico italiano. Anno I 1857–58*, Turin and Milan, n. d. [1858].

[Correnti, C., and P. Maestri, eds.], *Annuario Statistico Italiano. Anno II*, Turin, 1864.

Cotta Ramusino, G., *Cenni statistici e notizie patrie sulla città e provincia di Cremona*, Milan, 1861.

Cuoco, V., *Statistica della Repubblica Italiana. Scritti inediti*, ed. V. Gatto, Rome, 1991.

Curcio, G., *Gli omicidi in Italia. Studio legislativo e statistico*, n. p., n. d. [but Florence, 1871].

Della statistica giudiziaria civile e criminale nel Regno d'Italia, Rome, 1873.

D'Angeville, A., *Essai sur la statistique de la population française considérée sous quelques-uns de ses rapports physiques et moraux*, Paris, 1836.

De Bartolomeis, G. L., *Notizie topografiche e statistiche sugli stati sardi*, 5 vols., Turin, 1840–1847.

De Luca, P., *Principii elementari di statistica*, Naples, 1857.

De Renzi, S., *Topografia e statistica medica della città di Napoli con alcune considerazioni sul Regno intero, ossia guida medica per la città e pel Regno*, Naples, 1845.

Deliliers, F. M., "Discorso sull'insegnamento statistico," *AUS*, 2nd s., 25 (1850), pp. 166–214.

Dell'Acqua, C., *Villanterio. Cenni storici e statistici con documenti editi e inediti*, Pavia, 1874.

D'Emarese, F., *Cenni statistici sulla provincia di Mondovì*, Mondovì, 1842.

Destombes, D., *Annuaire statistique du Département de la Stura pour l'an 1809, faisant suite à celui de 1806*, Cuneo, n. d. [1809].

D'Ondes Reggio, V., review of *Notices statistiques sur les colonies françaises*, *GdS* 3 (1838), pp. 260–269.

Dufau, P.-A., *Traité de statistique ou théorie de l'étude des lois d'après lesquelles se dé-*

veloppent les faits sociaux; suivi d'un Essai de statistique physique et morale de la population française, Paris, 1840.

Dupin, C., *Forces productives et commerciales de la France*, Paris, 1827.

Duprat, P. and A. Gicca, *Annuario di economia sociale e di statistica pel Regno d'Italia*, vol. I, Turin, 1863.

Eandi, G., *Statistica della provincia di Saluzzo*, 2 vols., Saluzzo, 1833–1835.

Ellena, V., *La statistica di alcune industrie italiane*, Rome, 1879.

Engel, E., *Compte-rendu général des travaux du Congrès international de statistique dans ses séances tenues à Bruxelles, 1853, Paris, 1855, Vienne, 1857, et Londres, 1860*, Berlin, 1863.

Fabi, M., *Corografia d'Italia. Gran dizionario storico-geografico-statistico delle città, borghi, villaggi, castelli, fiumi, monti, ecc. della penisola*, Mantua, 1854.

Fazio, E., "Sulla importanza di una statistica medica italiana," *Annali di medicina pubblica, igiene e idrologia medica* 10 (1875), pp. 141–160.

Ferrara, F., *Opere complete edite e inedite*, vol. I, *Scritti di statistica*, ed. B. Rossi Ragazzi, Rome, 1955.

Ferrario, G., *Statistica medica di Milano dal secolo* XV *fino ai nostri giorni*, Milan, 1838.

Ferrario, L., *Busto Arsizio. Notizie storico-statistiche*, Busto Arsizio, 1864.

Ferretti, V., "Il concetto di statistica e il suo momento storico. Memoria," *AUS*, 4th s., 33 (1868), pp. 35–64.

Ferrussac, A.-E., *De la nécéssité de fixer et d'adapter un corps de doctrine pour la géographie et la statistique*, Paris, 1819.

Fiori, C., *Del metodo statistico considerato specialmente nei suoi rapporti co' fatti della medicina*, Milan, 1871.

Francolini, F., "Della utilità di una statistica che pubblicasse il prezzo degl'-Immobili, congiuntamente alle loro principali qualità. Memoria del socio ordinario Felice Francolini, letta nell'Adunanza ordinaria del dì 7 Gennaio 1844," *Atti dell'I. e R. Accademia dei Georgofili* 22 (1844), pp. 20–34.

Gabelli, A., "Gli scettici della statistica," *Archivio di statistica* 2 (1877), pp. 9–28.

Galanti, G. M., *Della descrizione geografica e politica delle Sicilie*, ed. D. Demarco, 2 vols., Naples, 1969.

Garnier, J., "Statistique des Etats Sardes. Population," *Journal des économistes* 5 (1846), pp. 381–387.

Gioia, M., *Logica statistica abbassata da M. G. alla capacità de' giovani agricoltori, artisti, commercianti, novizi in ogni altra professione privata o pubblica*, Milan, 1808.

Tavole statistiche ossia norme per descrivere, calcolare, classificare tutti gli oggetti d'amministrazione privata e pubblica, Milan, 1808.

"Indole, estensione e vantaggi della statistica" [1809], in Gioia, *Opere minori*, vol. VII, Lugano, 1834, p. 87.

review of *L'Homme du Midi et l'homme du Nord* by Ch.-V. de Bonstetten, *AUS* 5 (1825), pp. 245–301.

"Esame di un'opinione intorno all'indole, estensione e vantaggi delle statistiche," *AUS* 8 (1826), pp. 3–57.

Filosofia della statistica, Milan, 1826; 2nd edn. Milan, 1829.
"Nota sui suicidi avvenuti nelle Provincie Lombarde," *AUS* 17 (1828), pp. 67–68.
Giornale di statistica, Palermo, 1836–1848, 1852–1859.
Giovannetti, G., review of *Dell'amministrazione della Giustizia Criminale nel Regno di Napoli. Esame e paragone* by P. C. Ulloa, *AUS* 48 (1836), pp. 249–252.
Giulio, C. I., "Osservazioni sui fatti principali e sulle leggi del movimento della popolazione negli Stati di S. M. il Re di Sardegna," in *Informazioni statistiche raccolte dalla Regia Commissione Superiore per gli Stati di S. M. in Terraferma. Movimento della popolazione*, Turin, 1843, pp. 635–736.
Gråberg till Hemsö, J., *Della statistica e dei suoi progressi in Italia*, Tangier, 1818.
"Dell'attuale condizione della scienza statistica, e di alcune opere statistiche novellamente pubblicate," *Il progresso delle scienze, delle lettere e delle arti* 3, no. 8 (1834), pp. 235–262, no. 9 (1834), pp. 53–71.
review of *Informazioni statistiche raccolte dalla Regia Commissione Superiore per gli Stati di S. M. in Terraferma. Censimento della popolazione* and *Movimento della Popolazione*, *Rivista Europea* (1840).
Guaitoli, P., *Della città e comune di Carpi. Cenni statistici e storici*, Carpi, 1874.
Guala, L., *Elementi di statistica italiana teorica e pratica*, Biella, 1866.
Guerry, A.-M., *Essai de statistique morale de la France*, Paris, 1833.
Statistique morale de l'Angleterre comparée avec la statistique morale de la France, Paris, 1864.
Hassel, J. H., *Statistique de l'Europe*, Brussels, 1827.
Heuschling, X., *Manuel de statistique ethnographique universelle*, Brussels, 1847.
Informazioni statistiche raccolte dalla Regia Commissione Superiore per gli Stati di S. M. in Terraferma. Censimento della popolazione, Turin, 1839.
Informazioni statistiche raccolte dalla Regia Commissione Superiore per gli Stati di S. M. in Terraferma. Movimento della popolazione, Turin, 1843.
Informazioni statistiche raccolte dalla Regia Commissione Superiore per gli Stati di S. M. in Terraferma. Statistica medica, 2 vols., Turin, 1847–1852.
"Istruzione della Regia Segreteria di Finanze, ai signori Intendenti generali ... per la compilazione della Relazione statistica" (4 marzo 1820), in *Raccolta degli Atti del Governo di S. M. il Re di Sardegna dall'anno 1814 a tutto il 1832*, vol. x, Turin, 1845.
Jacquet, *Mémoire sur la statistique de l'arrondissement de Suse addressée au général Jourdan conseiller d'état ... par le citoyen Jacquet sous-préfet du même arrondissement*, Turin, Year x [1802].
La Marmora, A., *Voyage en Sardaigne, ou description statistique, physique et politique de cette île avec des recherches sur ses productions naturelles et ses antiquités*, Turin, 1839.
Laboulinière, R., *Plan d'une statistique générale d'arrondissement communal publié par ordre du général Jourdan conseiller d'Etat, administrateur général de la 27e division militaire*, Turin, Year IX [1802].
[Lampato, F.], "Su i progressi dell'industria in Inghilterra," *AUS* 7 (1826), pp. 234–235.

Lampertico, F., "Sulla statistica teorica in generale e su Melchiorre Gioia in particolare," *Annali di statistica*, 2nd s., 7 (1879).

Lichtenstern, J. M., *Saggio di una statistica dell'Impero d'Austria considerato nelle sue attuali circostanze*, Milan, 1819.

L'Italia economica nel 1873, Rome, 1874.

Lombroso, C., "Studi per una geografia medica d'Italia," *Gazzetta medica italiana. Lombardia*, 1865.

"Del peso del corpo nell'uomo sano e alienato," *AUS*, 4th s., 32 (1867), pp. 265–271.

"L'ultimo congresso internazionale di statistica giudicato dai francesi," *AUS*, 4th s., 33 (1868), pp. 315–329.

Maestri, P., *La Francia contemporanea. Studi economici ed amministrativi*, Milan, 1863.

Compte-rendu général des travaux du Congrès international de statistique dans ses sessions de Bruxelles, 1853; Paris, 1855; Vienne, 1857; Londres, 1860; et Berlin, 1863, Florence, 1866.

L'Italia economica nel 1868, Florence, 1868.

Le pubblicazioni della Direzione di Statistica. Relazione a S. E. il Ministro di Agricoltura, Industria e Commercio con note bibliografiche e sommari statistici e parere della Giunta consultiva di statistica, Florence, 1869.

L'Italia economica nel 1870, Florence, 1870.

[Maestri, P.], *Annuario economico-politico*, Turin, 1852.

Annuario economico e statistico dell'Italia per l'anno 1853, contenente studi e statistiche riguardanti la popolazione d'ogni stato e provincia, le istituzioni comunali, l'agraria, l'industria, le amministrazioni politiche, la marina militare d'Italia, Turin, n. d. [1853].

Maini, R., *Quadri statistici del comune di Cremona nel biennio 1866–67*, Cremona, 1869.

Mandarini, F., *Statistica della Provincia di Basilicata*, Potenza, 1839.

Marmocchi, F. C., *Prodromo della storia naturale generale e comparata d'Italia*, Florence, 1844.

Geografia d'Italia. Libri due. Nel primo de' quali si discute della geografia fisica e nel secondo della geografia politica e storica d'Italia da' più remoti tempi a' dì nostri, Italia [Bastia], 1850.

Geografia politica, storica, etnografica e statistica d'Italia da' più remoti tempi a' dì nostri, Italia [Bastia], 1851.

Massarani, T., "Italia politica," in [Correnti, C., and P. Maestri, eds.], *Annuario statistico italiano. Anno II*, Turin, 1864.

Messedaglia, A., *Della necessità di un insegnamento speciale politico-amministrativo e del suo ordinamento scientifico*, Milan, 1851 (now in A. Messedaglia, *Opere scelte di economia e altri scritti*, ed. L. Messedaglia, vol. 1, Verona, 1920, pp. 207–307).

Della teoria della popolazione principalmente sotto l'aspetto del metodo. 1. Malthus e dell'equilibrio della popolazione colle sussistenze, Verona, 1858.

"Relazione critica sull'opera di M. A. Guerry: Statistica morale dell'Inghilterra comparata alla statistica morale della Francia," *Atti dell'I. R. Istituto Veneto di scienze, lettere e arti*, 3rd s., 10 (1864–1865), pp. 1068–1085, 1135–1168.

"Esposizione critica delle statistiche dell'Impero austriaco, con particolare riguardo al Lombardo-Veneto, secondo i resoconti uffiziali del quadriennio 1856–1859, e col confronto dei dati posteriori," in *Atti dell'I. R. Istituto Veneto di scienze, lettere ed arti*, 3rd s., 11 (1865–1866), pp. 151–211, 331–409, 483–510, 599–652, 993–1051, 1237–1258; 12 (1866–1867), pp. 227–268.

"La vita media. Suo concetto, metodi di determinazione, criteri di applicazione" [1866], *Biblioteca dell'economista*, 5th s., 19, *Scritti di statistica teorica ed applicata*, ed. P. Iannacone, Turin, 1908, pp. 193–270.

Le statistiche criminali dell'Impero austriaco nel quadriennio 1856–59 con particolare riguardo al Lombardo-Veneto e col confronto dei dati posteriori fino al 1864 inclusivamente. Esposizione critica, Venice, 1866–1867.

"La statistica della criminalità. Prelezione al corso di statistica presso la R. Università di Roma (15 gennaio 1879)," *Archivio di statistica* 3 (1878), pp. 438–495.

Michelini, G. B., "Degli studi economici e statistici in Italia," *Rivista contemporanea* 16 (1859), pp. 142–164.

Milano e il suo territorio, 2 vols., Milan, 1844.

Ministero di Agricoltura, Industria e Commercio [MAIC], *Statistica del Regno d'Italia. Popolazione. Censimento generale (31 dicembre 1861)*, 3 vols., Turin and Florence, 1864–1866.

Statistica del Regno d'Italia. Popolazione. Movimento dello stato civile, 9 vols. [*1862–1870*], Florence and Milan, 1864–1872.

Statistica del Regno d'Italia. Finanze. Bilanci comunali e provinciali. Anno 1863, Florence, 1865.

Statistica del Regno d'Italia. Elezioni politiche e amministrative. Anni 1865–66, Florence, 1867.

Statistica del Regno d'Italia. Popolazione. Parte 1. Censimento generale (31 dicembre 1861), Florence, 1867.

Statistica del Regno d'Italia. Bilanci comunali. Anno 1866. Bilanci provinciali. Anni 1866–67–68, Florence, 1868.

Statistica del Regno d'Italia. Morti violente, 5 vols. [*1866–1870*], Florence and Milan, 1868–1871.

Carte e diagramma di demografia italiana, Rome, 1878.

Atlante statistico del Regno d'Italia. Diagrammi di demografia italiana, Rome, 1882.

Atti della Giunta per l'Inchiesta agraria, 15 vols., Rome, 1882–1885.

Risultati dell'inchiesta sulle condizioni igieniche e sanitarie dei comuni del Regno, 3 vols., Rome, 1886.

Divisione di Agricoltura, *Relazione intorno alle condizioni dell'agricoltura nel quinquennio 1870–1874*, 4 vols., Rome, 1876–1877.

Ministero di Grazia e Giustizia, *Statistica giudiziaria penale del Regno d'Italia per l'anno 1869 e ragguagli comparativi con alcuni anni anteriori*, Florence, 1871.

Molossi, L., "Cenni statistici intorno ai Ducati di Parma Piacenza e Guastalla," in *Vocabolario topografico dei Ducati di Parma Piacenza e Guastalla*, Parma, 1832–1834.

Monitore Toscano, 1847–1851.

Monnier, E., "Le popolazioni italiane: nuovi studi statistici," *AUS*, 4th s., 37 (1869), pp. 241–269.

Montezemolo, M., review of *Statistica dell'Italia* by L. Serristori. In *Giornale agrario toscano* 14 (1840), pp. 59–77.

Morozzo della Rocca, E., *Saggio statistico della Valsesia*, Varallo, 1856.

Niceforo, A., *L'Italia barbara contemporanea. Studi e appunti*, Milan and Palermo, 1898.

Italiani del Nord e Italiani del Sud (con 133 tavole numeriche e 31 tavole grafiche), Turin, 1901.

Norsa, C., *Sul compartimento territoriale e sull'amministrazione del nuovo Regno d'Italia. Considerazioni statistiche ed economiche*, Milan, 1863.

Considerazioni statistico-economiche sull'Italia, Milan, 1864.

Notizie economico-statistiche della provincia di Casale raccolte e pubblicate dal suo comizio nella solenne occasione del Quinto Congresso Generale dell'Associazione Agraria, Casale, 1847.

Notizie statistiche della città di Palermo, Palermo, 1866.

"Nuova statistica dell'Impero austriaco," *AUS*, 2nd s., 25 (1850), pp. 273–278; 30 (1852), pp. 297–315; 31 (1852), pp. 65–86, 295–321.

Orario delle Prelezioni dell'Imperial Regia Università di Padova per l'anno scolastico 1815–1816 (Padua, 1815).

Padovani, A., *Introduzione alla scienza della statistica*, Pavia, 1819.

Palluel, *Annuaire statistique du département du Mont-Blanc, rédigé par le secrétaire-général de la préfecture*, Chambery, 1803–1806.

Paolini, A., "Studio e progressi della statistica in Toscana," *AUS* 36 (1833), pp. 209–230.

Perez, F., "Idea del Perfetto Civile riguardata come norma della statistica," *GdS* 5 (1840), pp. 140–207.

Perini, A., *Statistica del Trentino*, Trento, 1850.

Pesci, D., *Statistica del comune di Ferrara*, Ferrara, 1869.

Petitti di Roreto, C. I., "Del lavoro dei fanciulli nelle manifatture. Dissertazione," *Memorie della reale Accademia delle scienze di Torino*, 2nd s., 3 (1841), pp. 209–306.

Opere scelte, 2 vols., ed. G. M. Bravo, Turin, 1969.

Petroni, R., *Censimento ossia Statistica de' Reali Dominii di Qua dal Faro del Regno delle Due Sicilie*, Naples, 1826.

Piccinetti, G., "Censimento nominativo della Popolazione del Granducato di Toscana eseguito nell'anno 1841. Memoria letta dal socio ordinario Giovanni Piccinetti nell'Adunanza del dì 5 Maggio 1844," *Atti dell'I. e R. Accademia economico-agraria dei Georgofili* 22 (1844), pp. 105–123.

Piola, A., *Statistica della provincia d'Alessandria*, Alessandria, 1831.

Plebano, T., *Statistica del Mandamento di Baldichieri provincia d'Asti*, Turin, 1832.

Statistica del Mandamento di Riva presso Chieri, Turin, 1836.

"Programma dell'Accademia Reale delle scienze in Torino," *AUS* 57 (1838), pp. 237–240.

Prospetto degli studi dell'Imperial R. Università degli studi di Padova per l'anno scolastico MDCCCXXIX–XXX, Padua, 1830.

Quadri, A., *Storia della statistica dalle sue origini sino alla fine del secolo XVIII per servire d'introduzione ad un prospetto statistico delle provincie venete*, Venice, 1824.

Prospetto statistico delle provincie venete, Venice, 1826.

Atlante statistico delle provincie venete, Venice, 1827.

Quadro statistico generale sull'amministrazione della giustizia penale ne' reali domini di qua del Faro per l'anno 1833 rassegnato a Sua Maestà il Re dal Ministro Segretario di Stato di Grazia e Giustizia, Naples, 1835.

Quételet, A., *Sur l'homme et le développement de ses facultés, ou Essai de physique sociale*, Paris, 1835.

Lettres à S. A. R. le Duc Régnant de Saxe-Coburg et Gotha sur la Théorie des probabilités appliquée aux sciences morales e politiques, Brussels, 1846.

Rabbeno, D., *Delle condizioni della statistica nell'Italia centrale e delle commissioni di statistica nell'Emilia istituite con legge 28 gennaio 1860 operazioni ed organizzazione di esse con un saggio di statistica comunitativa del Comune Parmense di Salsomaggiore*, Parma, 1861.

Racca, C., *Notizie statistiche e descrittive della Valsesia*, Vigevano, 1833.

Raccolta degli Atti del Governo di S. M. il Re di Sardegna dall'anno 1814 a tutto il 1832, vol. X, Turin, 1845.

Raccolta delle leggi del Granducato di Toscana, Florence, 1817.

Raccolta ufficiale delle leggi e dei decreti del Regno d'Italia, vol. I, Turin, 1861.

Racioppi, G., *Del principio e dei limiti della statistica. Saggio*, Naples, 1857.

Rampoldi, G. B., *Corografia dell'Italia*, 3 vols., Milan, 1832.

Raso, G. R., *Quadro statistico dei distretti di Palmi e Gerace nella Prima Calabria Ultra*, Naples, 1843.

Repetti, E., *Dizionario geografico fisico storico della Toscana*, 5 vols., Florence, 1833–1846.

Review (signed C. M.) of *Corografia fisica storica e statistica dell'Italia e delle sue isole* by A. Zuccagni Orlandini, *Biblioteca Italiana* (February 1837), pp. 153–172.

Review (signed D.) of "Influence de l'instruction et de la civilization en général sur la diminution des délits et des crimes (extrait de l'ouvrage ci-après cité; par C. Lucas)," *Bulletin des sciences géographiques, économie politique, voyages. 6e section du Bulletin universel publié . . . par la Société pour la propagation des connaissances scientifiques et industrielles* 14 (1828), pp. 106–116.

Review (signed M. C.) of *Statistica generale della Francia* by Schnitzler, *AUS*, 2nd s., 13 (1847), pp. 125–126.

Review (signed X. X.) of *Filosofia della statistica* by M. Gioia, *Antologia* 22 (1826), pp. 72–92.

Romagnosi, G. D., "Questioni sull'ordinamento delle statistiche," *AUS* 14 (1827), pp. 281–298; 15 (1828), pp. 113–131; 16 (1828), pp. 170–191; 17

(1828), pp. 3–15; 25 (1830), pp. 131–202 (also publ. as separate vol.: Milan, 1830).

"Necrologia. Melchiorre Gioia," *Biblioteca italiana* 52 (1828), pp. 392–407 (reprinted as "Elogio storico di Melchiorre Gioia," in *Florilegio di eloquenza italiana*, vol. II, Pistoia, 1839, pp. 55–86).

"Osservazioni statistiche al proposito del libro intitolato: *Conto generale dell'Amministrazione della giustizia criminale in Francia durante l'anno 1827 presentato al Re dal Guarda-Sigilli. Parigi 1828. Stamperia Reale*," *AUS* 19 (1829), pp. 1–25.

Dell'indole e dei fattori dell'incivilimento con esempio del suo Risorgimento in Italia, Milan, 1832; ed. A. De Giorgi, Palermo, 1859.

review of *Atlante geografico, fisico e storico del Granducato di Toscana* by A. Zuccagni Orlandini, *AUS* 37 (1833), pp. 238–242.

review of *Grundzüge einer Allgemeinen Statistik* by W. E. Schlieben, *AUS* 44 (1835), pp. 5–6.

Scritti filosofici, vol. I, *Gnoseologia, psicologia, morale*, ed. S. Moravia, Milan, 1974.

Roncaglia, C., *Statistica generale degli Stati Estensi*, Modena, 1849–50.

Sacchi, D., review of *Calendario lunese per l'anno 1835* and *Lo spigolatore novarese. Almanacco per l'anno 1835*, *AUS* 43 (1835), pp. 6–7.

Sacchi, G., review of *Filosofia della statistica* by M. Gioia, *AUS* 15 (1828), pp. 309–323.

"Melchiorre Gioia," *AUS* 19 (1829), pp. iii–xxviii (reprinted as "Notizie storiche intorno alla vita e alle opere di Melchiorre Gioia," in M. Gioia, *Filosofia della statistica*, 2nd edn., Milan, 1829, pp. iv–xxx).

review of *Commentari dell'Ateneo di Brescia per l'anno 1829*, *AUS* 30 (1831), pp. 130–132.

review of *Statistica del mandamento di Baldichieri provincia d'Asti*, by T. Plebano, and of *Statistica della provincia di Saluzzo* by G. Eandi, *AUS* 37 (1833), pp. 185–186.

"Istruzioni diramate dalla Commissione superiore di statistica residente a Torino alle Giunte Provinciali del Regno, per la compilazione della Statistica Generale degli Stati di S. M. il Re di Sardegna," *AUS* 58 (1838), pp. 129–142.

"Progetto di una statistica generale dell'industria italiana in relazione al miglior essere della popolazione operaia," *AUS* 71 (1842), pp. 186–204.

review of *Rendiconto sulla beneficenza dell'Ospitale Maggiore e del Pio Istituto di Santa Corona in Milano* by A. Verga, and of *Sulla beneficenza del pubblico manicomio La Senavra* by C. Castiglioni, *AUS*, 3rd s., 2 (1854), pp. 3–5.

Salari, G., *Statistica generale della regia città e provincia di Milano*, Milan, 1839.

Salvagnoli Marchetti, A., *Saggio illustrativo le tavole della statistica medica delle Maremme Toscane compilata per ordine del Granduca di Toscana*, Florence, 1844.

Memorie economico-statistiche sulle Maremme Toscane, Florence, 1846.

Saussay, *Statistique du département du Mont-Blanc par le citoyen Saussay, préfet, publié par ordre du Ministre de l'Intérieur*, Paris, Year IX [1801].

Say, J.-B., *Traité d'économie politique ou simple exposition de la manière dont se forment, se distribuent et se consomment les richesses*, Paris, 1802.

"De l'objet et de l'utilité des statistiques," *Revue encyclopédique* 9 (1827), pp. 529–553.

Schlözer, A. L. von, *Theorie der Statistik*, Göttingen, 1804.

Schnabel, J. N., *Statistica generale degli Stati Europei*, Pavia, 1835.

Serristori, L., *Saggio di un Atlante Statistico dell'Italia*, Vienna, 1833.

Saggio statistico dell'Italia, Vienna, 1833.

Primo supplemento al Saggio statistico dell'Italia, Vienna, 1834.

Statistica dell'Italia contenente le statistiche del Regno di Sardegna, Principato di Monaco, Isola di Corsica, Ducato di Parma, Modena e Lucca, Repubblica di S. Marino, Granducato di Toscana, Stati Pontifici, Regno delle Due Sicilie, commercio dell'Italia con alcune delle principali nazioni, Florence, 1835–1839; 2nd edn.: Florence, 1842.

"Progetto di una commissione di statistica in Toscana," *AUS* 64 (1840), pp. 314–315.

"Sulla possibilità di un'unione doganale tra gli stati italiani," *AUS* 75 (1843) pp. 292–300.

"Di una unione doganale tra gli stati italiani," *AUS* 78 (1843), pp. 197–202.

Sinclair, J., *Statistical Account of Scotland*, Edinburgh, 1792.

Springer, J., *Statistica dell'Impero d'Austria*, Pavia, 1840.

"Statistica criminale dell'Impero austriaco," *AUS*, 2nd s., 26 (1850), pp. 279–285.

Statistica della città di Palermo, Palermo, 1869.

"Statistica delle comunità di Albiano, Bagnone, Filattiera, Groppoli e Terrarossa," *Calendario lunese per l'anno 1835*, Fivizzano, [1835].

"Statistica delle comunità di Fivizzano e di Casola," *Calendario lunese per l'anno 1834*,Fivizzano, [1834].

Statistisches Jahrbuch der Oesterreichischen Monarchie für das Jahr 1863, Vienna, 1864.

Tafeln zur Statistik der Oesterreichischen Monarchie, Vienna, 1846–1863.

Tassani, A., *Cenni topografici e statistici sulla città di Como*, Como, 1861.

Übersichtstafeln zur Statistik der Oesterreichischen Monarchie, Vienna, 1863–1864.

Ulloa, P. C., "De' reati e della civiltà," *Il progresso delle scienze, delle lettere e delle arti* 10 (1835), pp. 161–173.

"Dell'esposizione dei reati in Inghilterra e in Francia, e del quadro statistico dell'amministrazione della giustizia penale nel Regno di Napoli," *Il progresso delle scienze, delle lettere e delle arti* 10 (1835), pp. 240–259.

Vanneschi, G., review of *Statistica del mandamento di Baldichieri* and *Statistica del mandamento di Riva presso Chieri* by T. Plebano, *GdS* 2 (1837), pp. 104–105.

Elementi di statistica, Palermo, 1859.

Vieusseux, G. P., "Società toscana di geografia, statistica e storia naturale patria," *Antologia* 24 (1826), pp. 184–193.

Zecca, V., *Monografia del comune di Chieti*, Chieti, 1866.

Zizius, J., *Teorie preliminari e introduttive alla statistica*, Pavia, 1822.

Zuccagni Orlandini, A., *Atlante geografico, fisico e storico del Granducato di Toscana*, Florence, 1832.

Corografia fisica, storica e statistica dell'Italia e delle sue isole, 12 vols., Florence, 1835–1845.
Ricerche statistiche sul granducato di Toscana, 5 vols., Florence, 1848–1854.
Annali statistici, Florence, 1856.
Elementi di statistica, Florence, 1869.
Zuradelli, G., *Saggio di una teoria della scienza statistica*, Pavia, 1822.
Preliminari alle teorie statistiche, 2nd edn., Pavia, 1844.

SECONDARY LITERATURE

Abrate, M., *et al.*, *L'imprenditorialità italiana dopo l'unità. L'inchiesta industriale del 1870–1874*, Milan, 1970.
Abulafia, D., *Le due Italie*, Naples, 1981; orig. English edn. 1977.
Albertoni, E. A., *La vita degli stati e l'incivilimento dei popoli nel pensiero politico di Gian Domenico Romagnosi*, Milan, 1979.
Ambrosoli, L., "Cattaneo e i problemi del territorio," *Nord e Sud* 21 (1974), pp. 80–94.
Anderson, B., *Imagined Communities. Reflections on the Origins and Spread of Nationalism*, 2nd edn., London and New York, 1991.
Anderson, M. J., *The American Census. A Social History*, New Haven and London, 1988.
Are, G., "Una fonte per lo studio della fondazione industriale in Italia: l'inchiesta del 1870–74," *Studi storici* 4 (1963), pp. 241–291.
Ariotti, R., "Primi apporti di Carlo Cattaneo all'introduzione delle tecniche statistiche nella ricerca economica e sociale. La collaborazione agli Annali di Statistica," *Statistica* 29 (1969), pp. 739–753.
Armatte, M., "Une discipline dans tous ses états: la statistique à travers ses traités," *Revue de synthèse*, n. s., 4 (1991), pp. 161–206.
Aron, J.-P., P. Dumont, and E. Le Roy Ladurie, *Anthropologie du conscrit français d'après les contes numériques et sommaires du recrutement de l'armée 1819–26*, Paris, 1972.
Asor Rosa, A., "La cultura," in *Storia d'Italia*, vol. IV, *Dall'Unità a oggi*, Turin, 1975, pp. 819–1664.
Augello, M. M., M. Bianchini, G. Gioli, and P. Roggi, eds., *Le cattedre di economia politica in Italia. La diffusione di una disciplina "sospetta" (1750–1900)*, Milan, 1988.
Baker, K. M., "The Early History of the Term 'Social Science,'" *Annals of Science* 20 (1964), pp. 21–226.
Condorcet: From Natural Philosophy to Social Mathematics, Chicago and London, 1975.
Balan, B., *L'Ordre et le temps. L'anatomie comparée et l'histoire des vivants au XIXe siècle*. Paris, 1979.
Bandettini, P., *L'evoluzione demografica della Toscana dal 1810 al 1889*, Turin, 1960.

Bandettini, P., ed., *La popolazione della Toscana alla metà dell'Ottocento*, Turin, 1956.

Barbano, F. and G. Sola, *Sociologia e scienze sociali in Italia 1861–1890. Introduzioni critiche e repertorio bibliografico*, Milan, 1985.

Bartoccini, F. and S. Verdini, *Sui congressi degli scienziati*, Rome, 1952.

Barucci, P., *Il pensiero economico di Melchiorre Gioia*, Milan, 1965.

Battistini, M., "Le relazioni di Adolfo Quételet con i dotti italiani," *Rivista delle scienze mediche e naturali di Firenze* 20 (1929), pp. 115–127.

"Documenti italiani nel Belgio. La corrispondenza del Conte Ilarione Petitti di Roreto con Adolfo Quételet," *Rassegna storica del Risorgimento* 23 (1936), pp. 1024–1080.

Beck, H.,"Geographie und Statistik. Die Lösung einer Polarität," in M. Rassem, ed., *Statistik und Staatsbeschreibung in der Neuzeit*, Paderborn, 1980, pp. 269–276.

Bellettini, A., "Contenuto e tecnica degli ultimi censimenti dello Stato Pontificio," in Comitato Italiano per lo Studio della Demografia Storica, *Le fonti della demografia storica in Italia. Atti del seminario di demografia storica, 1971–72*, Rome, 1974, pp. 463–478.

Berengo, M., "Intellettuali e centri di cultura nell'Ottocento italiano," *Rivista storica italiana* 87 (1975), pp. 132–166.

"Intellettuali e organizzazione della cultura nell'età della Restaurazione," in *La Restaurazione in Italia. Struttura e ideologie. Atti del XLVII Congresso di Storia del Risorgimento italiano*, Rome, 1976, pp. 297–307.

Intellettuali e librai nella Milano della Restaurazione, Turin, 1980.

"Antonio Quadri e le statistiche venete della Restaurazione," in *Studi veneti. Offerti a Gaetano Cozzi*, Vicenza, 1992, pp. 391–407.

Berti, G., *I democratici e l'iniziativa meridionale nel Risorgimento*, Milan, 1962.

Biagioli, G., *L'agricoltura e la popolazione in Toscana all'inizio dell'Ottocento. Un'indagine sul catasto particellare*, Pisa, 1975.

Bianchini, M., *Alle origini della scienza economica. Felicità pubblica e matematica sociale negli economisti italiani del Settecento*, Parma, 1982.

Bobbio, N., *Una filosofia militante. Studi su Carlo Cattaneo*, Turin, 1971.

Bollati, G., "L'Italiano," in *Storia d'Italia*, vol. 1, *I caratteri originali*, Turin, 1972, pp. 949–1022.

Bordone, R., "Spunti archeologici nelle descrizioni erudite fra Sette e Ottocento," in V. Fumagalli and G. Rossetti, eds., *Medioevo rurale. Sulle tracce della civiltà contadina*, Bologna, 1981, pp. 139–154.

Bourdieu, P., "The Social Space and the Genesis of Groups," *Theory and Society* 14 (1985), pp. 723–744.

Bourguet, M.-N., *Déchiffrer la France. La statistique départementale à l'époque napoléonienne*, Paris, 1988.

Bowen, M., *Empiricism and Geographical Thought. From Francis Bacon to Alexander von Humboldt*, Cambridge, 1981.

Bravo, G. M., *Torino operaia. Mondo del lavoro e idee sociali nell'età di Carlo Alberto*, Turin, 1969.

Brown, L., *The Board of Trade and the Free-Trade Movement 1830–42*, Oxford, 1958.

Brunello, P., *Banditi, questuanti, ribelli: proteste contadine in Veneto e in Friuli 1814–1866*, Venice, 1981.

Buck, P., "Seventeenth-Century Political Arithmetic: Civil Strife and Vital Statistics," *Isis* 68 (1977), pp. 67–84.

"People Who Counted: Political Arithmetic in the Eighteenth Century," *Isis* 73 (1982), pp. 28–45.

Burchell, G., C. Gordon and P. Miller, eds., *The Foucault Effect. Studies in Governmentality*, Chicago, 1991.

Burckhardt, J. *The Civilisation of the Renaissance in Italy*, 8th edn., London, 1921.

Cafagna, L., *Dualismo e sviluppo nella storia d'Italia*, Venice, 1989.

Calani, A., *Il Parlamento del Regno d'Italia*, Milan, n. d.

Camurri, R., ed., *La scienza moderata. Fedele Lampertico e l'Italia liberale*, Milan, 1992.

Candeloro, G., *Storia dell'Italia moderna*, vol. VI, *Lo sviluppo del capitalismo e del movimento operaio*, Milan, 1970.

Caracciolo, A., *L'inchiesta agraria Iacini*, Turin, 1958; 1973.

Stato e società civile. Problemi dell'unificazione italiana, Turin, 1960.

Carpi, U., *Letteratura e società nella Toscana del Risorgimento. Gli intellettuali dell'"Antologia"*, Bari, 1974.

"Egemonia moderata e intellettuali nel Risorgimento," in *Storia d'Italia. Annali IV. Intellettuali e potere*, Turin, 1981, pp. 429–471.

Chabod, F., *L'idea di nazione*, ed. A. Saitta and E. Sestan, Bari, 1961.

Chartier, R., "Les Deux Frances. Histoire d'une géographie," *Cahiers d'histoire* (1978), pp. 393–415.

"Intellectual History or Sociocultural History? The French Trajectories," in D. LaCapra and S. L. Kaplan, eds., *Modern European Intellectual History: Reappraisals and New Perspectives*, Ithaca, 1982, pp. 13–45.

"Science sociale et découpage régional. Notes sur deux débats 1820–1920," *Actes de la recherche en sciences sociales* 8 (1982), pp. 27–36.

Cultural History: between Practices and Representations, London, 1988.

Chevalier, L., *Laboring Classes and Dangerous Classes in Paris during the First Half of the Nineteenth Century*, Princeton, 1973.

Ciuffoletti, Z., "I moderati toscani e la tradizione leopoldina," in Rotondi, C., ed., *I Lorena in Toscana*, Florence, 1989, pp. 121–138.

Cohen, P. C., *A Calculating People: The Spread of Numeracy in Early America*, Chicago and London, 1982.

Cohn, B. I., "The Census, Social Structure and Objectification in South Asia," in Cohn, *An Anthropologist among the Historians and Other Essays*, Dehli and Oxford, 1987, pp. 224–254.

Cole, J., "The Chaos of Particular Facts: Statistics, Medicine, and the Social Body in Early Nineteenth-Century France," *History of the Human Sciences* 7 (1994), pp. 1–22.

Coleman, W., *Death is a Social Disease. Public Health and Political Economy in Early Industrial France*, Madison, 1982.

Coppino, R. P., "Restaurazione e ceti dirigenti in Toscana," in Rotondi, C., ed., *I Lorena in Toscana*, Florence, 1989, pp. 81–105.

Il Granducato di Toscana dagli anni "francesi" all'Unità, Turin, 1993.

Corsini, C., "L'acculturazione mancata. Il processo formativo dell'organizzazione delle fonti demografiche nel periodo post-unitario," in *La popolazione italiana nell'Ottocento. Continuità e mutamenti*, Bologna, 1985, pp. 115–145.

Corsini, C., ed., *Da osservazione sperimentale a spiegazione razionale. Per una storia della statistica in Italia*, Pisa, 1989.

Cosmacini, G., *Storia della medicina e della sanità in Italia. Dalla peste europea alla guerra mondiale, 1348–1918*, Bari, 1987.

Cova, A., "Osservazioni sulla origine delle statistiche del Regno italico," *Annali dell'Istituto Italo-Germanico in Trento* 5 (1979), pp. 117–141.

Croce, B., *Storia della storiografia italiana nel secolo decimonono*, 2nd edn., Bari, 1930.

Cullen, M. J., *The Statistical Movement in Early Victorian Britain. The Foundations of Empirical Social Research*, New York, 1975.

D'Alessandro, V. and G. Giarrizzo, *La Sicilia dal Vespro all'unità d'Italia*, Turin, 1989.

Daston, L., "The Domestication of Risk: Mathematical Probability and Insurance 1650–1830," in L. Krüger, L. J. Daston, and M. Heidelberger, eds., *The Probabilistic Revolution*, vol. 1, *Ideas in History*, Cambridge, Mass., and London, 1987.

Davis, J. A., *Conflict and Control: Law and Order in Nineteenth-Century Italy*, London and Atlantic Highlands, N. J., 1988.

De Certeau, M., *The Writing of History*, New York, 1988.

De Lorenzo, R., "Strategie del territorio e indagini statistiche nel Mezzogiorno fra Settecento e Ottocento," in R. De Lorenzo, ed., *L'organizzazione dello stato al tramonto dell'Antico Regime*, Naples, 1990.

De Luca, I., ed., *La vita letteraria in Piemonte e in Lombardia 1850–1859. Carteggio inedito Tenca–Camerini*, ed. I. De Luca, Milan and Naples, 1973.

De Seta, C., "Città e territorio in Carlo Cattaneo," *Studi storici* 16 (1975), pp. 439–460.

Della Peruta, F., *I democratici e la rivoluzione italiana. Dibattiti ideali e contrasti politici all'indomani del 1848*, Milan, 1974.

"Sanità pubblica e legislazione sanitaria dall'Unità a Crispi," *Studi storici* 21 (1980), pp. 713–760.

Conservatori, liberali e democratici nel Risorgimento, Milan, 1989.

Dematteis, G., *Le metafore della Terra. La geografia umana tra mito e scienza*, Milan, 1985.

Desrosières, A., "How to Make Things Which Hold Together: Social Science, Statistics and the State," in P. Wagner, B. Wittrock and R. Whitley, eds., *Discourses on Society. The Shaping of the Social Science Disciplines*, Dordrecht, 1991, pp. 195–218.

Di Carlo, E., *L'influsso del pensiero di Romagnosi in Sicilia*, Palermo, 1959.

Dickie, J., "A Word at War: The Italian Army and Brigandage 1860–1870," *History Workshop* 33 (1992), pp. 1–24.

Dizionario biografico degli italiani, s. v. "Balbi, Adriano" by M. Gliozzi; "Bodio, Luigi" by F. Bonelli; "Cagnazzi, Luca de Samuele" by C. P. Scavizzi; "Castiglioni, A." by C. Capra; "Cattaneo, Carlo" by E. Sestan; "Correnti, Cesare," by L. Ambrosoli.

Dizionario del Risorgimento nazionale, Milan, 1933.

Dockès, P., *L'espace dans la pensée économique du XVIe au XVIIIe siècle*, Paris, 1969.

Dotti, U., ed., *Il progresso delle scienze, delle lettere e delle arti (1832–34)*, Rome, 1970.

Dupâquier, J. and M., *Histoire de la démographie. La statistique de la population des origines à 1914*, Paris, 1985.

Eastwood, D., "Amplifying the Province of Legislature: the Flow of Information and the English State in the Early Nineteenth Century," *Historical Research* 62 (1989), pp. 276–294.

Eisenstein, E., *The Printing Press as an Agent of Change*, Cambridge, 1979.

Eley, G., "Nations, Publics, and Political Culture: Placing Habermas in the Nineteenth Century," in C. Calhoun, ed., *Habermas and the Public Sphere*, Cambridge, Mass., and London, 1992, pp. 289–339.

Elias, N., *The Civilizing Process*, vol. 1, *The History of Manners*, New York, 1978.

Faccini, L., "Carl Czoernig e la statistica agraria in Lombardia," *Società e storia* 21 (1980), pp. 931–950.

Farolfi, B., "Dall'antropometria militare alla storia del corpo," *Quaderni storici* 14 (1979), pp. 1056–1091.

"Antropometria militare e antropologia della devianza 1876–1908," in *Storia d'Italia. Annali VII. Malattia e medicina*, Turin, 1984, pp. 1181–1219.

Federico, G., "Per una valutazione critica delle statistiche della produzione agricola italiana dopo l'unità (1860–1913)," *Società e storia* 5 (1982), pp. 87–130.

Feldman, J., G. Lagneau, and B. Matalon, eds., *Moyenne, milieu, centre. Histoire et usages*, Paris, 1991.

Forti Messina, A. L., "I medici condotti all'indomani dell'Unità," in M. L. Betri and A. Gigli, eds., *Salute e classi lavoratrici in Italia dall'Unità al fascismo*, Milan, 1982, pp. 663–697.

"I medici e la professione del medico nell'Ottocento," *Società e storia* 8 (1984), pp. 101–162.

"L'Italia dell'Ottocento di fronte al colera," in *Storia d'Italia. Annali VII. Malattia e medicina*, Turin, 1984, pp. 431–494.

Foucault, M., *The Order of Things. An Archeology of the Human Sciences*, New York, 1970.

Discipline and Punish. The Birth of the Prison, New York, 1977.

The History of Sexuality, vol. 1, *An Introduction*, New York, 1980.

Fracassi, R., "L'ordinamento dei servizi statistici," *Annali di statistica*, 8th s., 5 (1957), pp. 83–138.

Frascani, P., "Medicina e statistica nella formazione del sistema sanitario italiano: l'inchiesta del 1885," *Quaderni storici* 15 (1980), pp. 942–965.

"Il medico nell'Ottocento," *Studi storici* 23 (1982), pp. 617–637.

"Les Professions bourgeoises en Italie à l'époque libérale (1860–1920)," *Mélanges de l'Ecole Française de Rome* 97 (1985), pp. 325–340.

Fraser, N., "Rethinking the Public Sphere: A Contribution to the Critique of Actually Existing Democracy," in C. Calhoun, ed., *Habermas and the Public Sphere*, Cambridge, Mass., and London, 1992, pp. 109–142.

Freeman, T. W., *A Hundred Years of Geography*, London, 1961.

Gabaglio, A., *Teoria generale della statistica*, 1st edn., Milan, 1880; 2nd edn., Milan, 1888.

Gabba, E., and D. Zanetti, eds., *Defendente Sacchi: filosofo, critico, narratore*, Milan, 1992.

Galante Garrone, A. and F. Della Peruta, *La stampa italiana del Risorgimento*, Bari, 1979.

Gambi, L., "L'equivoco tra compartimenti statistici e regioni costituzionali," in *Avviamento allo studio geografico della regione*, Florence, 1967, pp. 119–145.

"I valori storici dei quadri ambientali," in *Storia d'Italia*, vol. 1, Turin, 1972.

"Uno schizzo di storia della geografia in Italia," in *Una geografia per la storia*, Turin, 1973, pp. 3–37.

"Le 'statistiche' di un prefetto del Regno," *Quaderni storici* 15 (1980), pp. 823–866.

Garin, E., "Il positivismo come metodo e come concezione del mondo," in *Tra due secoli. Socialismo e filosofia in Italia dopo l'Unità*, Bari, 1981, pp. 65–89.

Garrier, G., "Les Enquêtes agricoles décennales du XIXe siècle: essai d'analyse critique," in *Pour une histoire de la statistique*, Paris, 1977.

Ghetti, M. C., "Struttura e organizzazione dell'Università di Padova dal 1798 al 1817," *Quaderni per la storia dell'Università di Padova* 17 (1984), pp. 135–185.

Ghiringhelli, R., "Gian Domenico Romagnosi e gli 'Annali universali di statistica' (1827–1835)," in Ghiringhelli and Invernici, eds., *Per conoscere Romagnosi.*

Ghiringhelli, R. and F. Invernici, eds., *Per conoscere Romagnosi*, Milan, 1982.

Gramsci, A., *Quaderno 19. Risorgimento italiano*, ed. C. Vivanti, Turin, 1977.

Greenfield, K. R., *Economics and Liberalism in the Risorgimento. A Study of Nationalism in Lombardy 1814–1848*, 2nd edn., Baltimore, 1965.

Guglielminetti, M., "'Decadenza' e 'progresso' dell'Italia nel dibattito fra classicisti e romantici," in *La Restaurazione in Italia. Strutture e ideologie. Atti del XLVII Congresso di storia del Risorgimento italiano (Cosenza, 15–19 settembre 1974)*, Rome, 1976, pp. 251–307.

Habermas, J., *The Structural Transformation of the Public Sphere. An Inquiry into a Category of Bourgeois Society*, Cambridge, Mass., 1989 (orig. German edn.: Neuwied, 1962).

Hacking, I., "Biopower and the Avalanche of Printed Numbers," *Humanities in Society* 5 (1982), pp. 279–295.

"Making Up People," in T. C. Heller, M. Sosna, and D. E. Wellbery, eds., *Reconstructing Individualism: Autonomy, Individuality, and the Self in Western Thought*, Stanford, Calif., 1986, pp. 222–236.

"Prussian Numbers 1860–1882," in L. Krüger, L. J. Daston, and M. Heidelberger, eds., *The Probabilistic Revolution*, vol. 1, *Ideas in History*, Cambridge, Mass., and London, 1987, pp. 377–394.

The Taming of Chance, Cambridge, 1990.

Hecht, J., "L'Idée de dénombrement jusqu'à la Révolution," in *Pour une histoire de la statistique*, Paris, 1977, pp. 21–82.

"Johan Peter Süssmilch point alpha ou omega de la science démographique naïve?," *Annales de démographie historique* (1979), pp. 101–144.

Hilts, V. L., "*Aliis exterendum*, or, the Origins of the Statistical Society of London," *Isis* 69 (1978), pp. 21–43.

Hobsbawm, E. J., *Nations and Nationalism since 1780. Programme, Myth, Reality*, Cambridge, 1990.

Hoock, J., "D'Aristote à Adam Smith: quelques étapes de la statistique allemande entre le XVIIe et le XIXe siècle," in *Pour une histoire de la statistique*, Paris, 1977, pp. 477–491.

Horn, D., *Social Bodies. Science, Reproduction and Italian Modernity*, Princeton, 1994.

Horvath, R., "Statistische Description und nominalistische Philosophie," in M. Rassem, ed., *Statistik und Staatbeschreibung in der Neuzeit*, Paderborn, 1980, pp. 37–52.

Hull, P., *Von Thünen's Isolated State*, Oxford and New York, 1966.

Hunt, L., ed., *The New Cultural History*, Berkeley and Los Angeles, 1989.

Hutchinson, E. P., *The Population Debate. The Development of Conflicting Theories up to 1900*, Boston, 1967.

Ipsen, C., "The Statistics of Population in Liberal Italy," *Bollettino di demografia storica* 16 (1992), pp. 7–33.

ISTAT [Istituto Centrale di Statistica], *Decennale 1926 IV – 1936 XIV*, Rome, 1936.

Dal censimento dell'Unità ai censimenti del centenario. Un secolo di vita della statistica italiana, text by R. Fracassi, Rome, n. d. [1961].

Izzo, L., *La finanza pubblica nel primo decennio dell'Unità*, Milan, 1962.

Introduzione allo studio della popolazione nel Mezzogiorno d'Italia nel secolo XIX, Naples, 1975.

Jaja, G., *Scritti di geografia e di storia della geografia concernenti l'Italia in onore di Giuseppe della Vedova*, Florence, 1908.

John, V., *Geschichte der Statistik*, Stuttgart, 1884.

Jordanova, L., "Earth science and environmental medicine: the synthesis of the late Enlightenment," in L. Jordanova and R. S. Porter, eds., *Images of the Earth. Essays in the History of the Environmental Sciences*, Aberdeen, Scotland, 1979.

Jordanova, L., ed., *Languages of Nature. Critical Essays on Science and Literature*, London, 1986.

Kaplan, S. L., and C. J. Koepp, eds., *Work in France. Representations, Meaning, Organization, and Practice*, Ithaca and London, 1986.

Keane, J. "Despotism and Democracy. The Origins and Development of the Distinction between Civil Society and the State 1750–1850," in Keane, ed., *Civil Society and the State. New European Perspectives*, London and New York, 1988, pp. 35–71.

Koepp, C. J., "The Alphabetical Order: Work in Diderot's *Encyclopédie*," in S. L. Kaplan and C. J. Koepp, eds., *Work in France. Representations, Meaning, Organization, and Practice*, Ithaca and London, 1986, pp. 229–257.

Koren, J., ed., *The History of Statistics. Their Development and Progress in Many Countries*, New York, 1908.

Krüger, L., L. J. Daston, and M. Heidelberger, eds., *The Probabilistic Revolution*, vol. I, *Ideas in History*, Cambridge, Mass. and London, 1987.

La Salvia, S., *Giornalismo lombardo: gli "Annali Universali di statistica" (1824–1844)*, Rome, 1977.

La Statistique en France à l'époque napoléonienne, Journée d'étude, Paris, 14 février 1980, Brussels, 1981.

LaCapra, D., "Rethinking Intellectual History and Reading Texts," in D. LaCapra, and S. L. Kaplan, eds., *Modern European Intellectual History: Reappraisals and New Perspectives*, , Ithaca, 1982, pp. 47–85.

Lanaro, S., *Nazione e lavoro. Saggio sulla cultura borghese in Italia 1870–1925*, Venice, 1979.

Latour, B., "Visualization and Cognition: Thinking with Eyes and Hands," *Knowledge and Society: Studies in the Sociology of Culture Past and Present* 6 (1986), pp. 1–40.

Lazarsfeld, P. F., "Notes on the History of Quantification in Sociology. Trends, Sources and Problems," *Isis* 52 (1961).

Le Bras, H., "La Statistique générale de la France," in P. Nora, ed., *Les Lieux de mémoire*, vol. II, *La Nation*, Paris, 1986, pp. 317–353.

Leclerc, G., *L'Observation de l'homme. Une histoire des enquêtes sociales*, Paris, 1982.

Lefebvre, H., *La Présence et l'absence. Contribution à une théorie des représentations*, Paris, 1980.

Levi, A., *Romagnosi*, Rome, 1935.

Levi, G., "Gli aritmetici politici e la demografia piemontese negli ultimi anni del '700," *Rivista storica italiana* 86 (1974), pp. 201–265.

Levra, U., "La 'statistica morale' del Regno di Sardegna tra la Restaurazione e gli anni Trenta: da Napoleone a Carlo Alberto," *Clio* 28 (1992), pp. 353–378.

Lombardo, E., "Il primo trattato italiano di statistica di Luca de Samuele Cagnazzi ed i suoi interessi demografici," in C. Corsini, ed., *Da osservazione sperimentale a spiegazione razionale. Per una storia della statistica in Italia*, (Pisa, 1989), pp. 33–48.

Lottin, J., *Quételet: statisticien et sociologue*, Louvain, 1912.

Lovejoy, A. O., *The Great Chain of Being. A Study of the History of an Idea*, Cambridge, Mass., 1936.

Lungonelli, M., "Sul servizio statistico del Regno d'Italia nel 1871 (Una lettera di Luigi Bodio a Luigi Luzzatti)," *Clio* 18 (1982), pp. 295–299.

Luraghi, R., *Agricoltura, industria e commercio in Piemonte dal 1848 al 1861*, Turin, 1967.

Lutz, G., "Geographie und Statistik im 18. Jahrundert. Zu Neugliederung und Inhalten von 'Fächern' im Bereich der historischen Wissenschaften," in M. Rassem, ed., *Statistik und Staatsbeschreibung in der Neuzeit*, Paderborn, 1980, pp. 249–263.

Lyttelton, A., "The National Question in Italy," in M. Teich and R. Porter, eds., *The National Question in Europe in Historical Context*, Cambridge, 1993, pp. 63–105.

Macchioro, A., *Studi di storia del pensiero economico e altri saggi*, Milan, 1970.

"La Raccolta Custodi di 'Scrittori Classici di Economia' fra la statistica e l'economia politica," in D. Rota, ed., *Pietro Custodi tra Rivoluzione e Restaurazione. Atti del Primo Convegno Nazionale*, Lecco, 1989, pp. 139–164.

"La 'philosophia naturalis' gioiana dell'economia," in *Melchiorre Gioia (1767–1829)*, pp. 269–302.

Maccioni Anguillesi, A., "Un geografo dell'Ottocento: Attilio Zuccagni Orlandini," *Rivista geografica italiana* 55 (1948), pp. 99–116.

Mack Smith, D., *Italy: A Modern History*, Ann Arbor, Mich., 1959.

A History of Sicily. Modern Sicily after 1713, London, 1968.

Macry, P., "La questione scolastica: controllo, conoscenza, consenso (1860–1872)," *Quaderni storici* 15 (1980), pp. 894–917.

"Notables, professions libérales, employés: la difficile identité des bourgeoisies italiennes dans la deuxième moitié du XIXe siècle," *Mélanges de l'Ecole Française de Rome* 97 (1985), pp. 341–359.

Mannori, L., *Uno stato per Romagnosi*, vol. 1, *Il progetto costituzionale*, Milan, 1984.

Marietti, P., *La Statistique générale en France*, Paris, 1949.

Marino, G. C., *La formazione dello spirito borghese in Italia*, Florence, 1974.

Marino, L., *I maestri della Germania. Göttingen 1770–1820*, Turin, 1975.

Martuscelli, S., ed., *La popolazione del Mezzogiorno nella statistica di Re Murat*, Naples, 1979.

Mayr, E., *The Growth of Biological Thought. Diversity, Evolution, and Inheritance*, Cambridge, Mass., and London, 1982.

Melchiorre Gioia (1767–1829). Politica, società, economia tra riforme e Restaurazione. Atti del Convegno di studi, Piacenza 5–7 Aprile 1990, Bollettino storico piacentino 85 (1990).

Meriggi, M., "Funzionari e carriere nella Lombardia della Restaurazione (1816–48)," *Società e storia* 7 (1980), pp. 61–96.

Il Regno Lombardo-Veneto, Turin, 1987.

Metz, K. H., "Paupers and Numbers: The Statistical Argument for Social Reform in Britain during the Period of Industrialization," in L. Krüger, L. J. Daston, and M. Heidelberger, eds., *The Probabilistic Revolution*, vol. 1, *Ideas in History*, Cambridge, Mass., and London, 1987, pp. 337–350.

Meyer, R., "The History and Development of Government Statistics in Austria," in J. Koren, ed., *The History of Statistics. Their Development and Progress in Many Countries*, New York, 1908.

Moe, N., "'Altro che Italia!' Il Sud dei piemontesi (1860–61)," *Meridiana. Rivista di storia e scienze sociali* 15 (1992), pp. 53–89.

"Representing the South in Postunification Italy ca. 1860–1880," Ph.D. dissertation, Johns Hopkins University, Baltimore, 1994.

Molfese, F., *Storia del brigantaggio dopo l'Unità*, Milan, 1964.

Monti, A., *L'idea federalistica nel Risorgimento italiano. Saggio storico*, Bari, 1922.

Mozzarelli, C. and S. Nespor, *Giuristi e scienze sociali nell'Italia liberale. Il dibattito sulla scienza dell'amministrazione e l'organizzazione dello stato*, Venice, 1981.

Mozzillo, A., *Viaggiatori stranieri nel Sud*, Milan, 1964.

Mulas, A., "Il regionalismo nell'opera di Cesare Lombroso e della sua scuola," *Archivio storico sardo* 32 (1981), pp. 311–347.

Natali, G., "Jacopo Graberg di Hemso," *Rivista d'Italia* 20 (1917), pp. 175–189.

Nordman, D., "L'Espace objet: le département," *Annales: économies, sociétés, civilisations* 45 (1990), pp. 445–462.

Nordman, D. and J. Revel, "La Formation de l'espace français," in A. Bourguière and J. Revel, eds., *Histoire de la France*, vol. I, *L'Espace français*, Paris, 1989, pp. 33–169.

Oesterreichischen Statistischen Zentralamt, *Von der Direktion der Administrativen Statistik zum Oesterreichischen Statistischen Zentralamt 1840–1990*, Vienna, 1990.

Oldrini, G., *La cultura filosofica napoletana dell'Ottocento*, Bari, 1973.

Ottaviani, M. G., "Note per una storia dell'insegnamento della statistica in Italia: la statistica nell'ordinamento didattico dell'istruzione superiore dalle origini al 1938," *Statistica* 47 (1987), pp. 619–647.

Ozouf-Marignier, M.-V., *La Formation des départements. La représentation du territoire français à la fin du 18e siècle*, Paris, 1989.

Pancaldi, G., ed., *I congressi degli scienziati italiani nell'età del positivismo*, Bologna, 1983.

Panseri, G., "Il medico: note su un intellettuale scientifico italiano nell'Ottocento," in *Storia d'Italia. Annali IV. Intellettuali e potere*, Turin, 1981.

Paoli, M. P. and R. Graglia, "Marco Lastri: aritmetica politica e statistica demografica nella Toscana del '700," *Annali della Fondazione Luigi Einaudi* 12 (1978), pp. 117–215.

Parisi Acquaviva, D., *Il pensiero economico classico in Italia (1750–1860)*, Milan, 1984.

Pasini, M., "La filosofia della statistica di Melchiorre Gioia," in G. Tarello, ed., *Materiali per una storia della cultura giuridica*, vol. V, Bologna, 1975, pp. 473–532.

Patriarca, S., "Statistical Nation Building and the Consolidation of Regions in Italy," *Social Science History* 18 (1994), pp. 359–376.

Pavone, C., *Amministrazione centrale e amministrazione periferica da Rattazzi a Ricasoli (1859–1866)*, Milan, 1964.

Pazzagli, C., "Statistica 'investigatrice' e scienze 'positive' nell'Italia dei primi decenni unitari," *Quaderni storici* 15 (1980), pp. 779–822.

Pellanda, A., "Angelo Messedaglia parlamentare e accademico nel contesto socio-politico italiano del secondo Ottocento," in A. Tagliaferri, ed., *I ceti dirigenti in Italia in età moderna e contemporanea*, Udine, 1984.

Perrot, J.-C., "The Golden Age of Regional Statistics," in J.-C. Perrot and S. J. Woolf, *State and Statistics in France 1789–1815*, London, 1984.

Perrot, J.-C. and S. J. Woolf, *State and Statistics in France 1789–1815*, London, 1984.

Perrot, M., "Premières mesures des faits sociaux: les débuts de la statistique criminelle en France (1780–1830)," in *Pour une histoire de la statistique*, Paris, 1977, pp. 125–136.

Petersen, W., "Politics and the Measurement of Ethnicity," in W. Alonso and P. Starr, eds., *The Politics of Numbers*, New York, 1987, pp. 187–233.

Petracchi, A., *Le origini dell'ordinamento comunale e provinciale italiano. Storia della legislazione piemontese sugli enti locali dalla fine dell'antico regime al chiudersi dell'età cavouriana (1770–1861)*, 3 vols., Venice, 1962.

Petrusewicz, M., "Agromania: Innovatori agrari nelle periferie europee dell'Ottocento," in *Storia dell'agricoltura italiana in età contemporanea*, vol. III, *Mercati e istituzioni*, Venice, 1991.

Pezzino, P., "Monarchia amministrativa ed *élites* locali: Naro nella prima metà dell'Ottocento," in P. Pezzino, *Il Paradiso abitato dai diavoli. Società, élites, istituzioni nel Mezzogiorno contemporaneo*, Milan, 1992.

Piccialuti-Caprioli, M., "Il patrimonio del povero. L'inchiesta sulle opere pie del 1861," *Quaderni storici* 15 (1980), pp. 918–935.

Pick, D., *Faces of Degeneration: Aspects of a European Disorder, c. 1848–c. 1918*, Cambridge, 1989.

Polsi, A., "La 'statistica dell'industria manifattrice' del 1862," *Quaderni storici* 15 (1980), pp. 894–917.

"Le amministrazioni locali postunitarie fra accentramento e autonomia: il caso del Comune di Pisa (1860–1885)," *Società e storia* 6 (1983), pp. 828–867.

Poovey, M., *Uneven Developments. The Ideological Work of Gender in Victorian England*, Chicago, 1988.

Porter, T. M., *The Rise of Statistical Thinking 1820–1900*, Princeton, 1986.

"Lawless Society: Social Science and the Reinterpretation of Statistics in Germany, 1850–1880," in L. Krüger, L. J. Daston and M. Heidelberger, eds., *The Probabilistic Revolution*, vol. I, *Ideas in History*, Cambridge, Mass., and London, 1987, pp. 351–375.

Pour une histoire de la statistique, Paris, 1977.

Prato, G., *La vita economica in Piemonte a mezzo il secolo XVIII*, Turin, 1908.

Fatti e dottrine economiche alla vigilia del 1848. L'Associazione agraria subalpina e Camillo Cavour, Turin, 1921.

Pucci, L., "Un assertore della 'forza industriale': Melchiorre Gioia tra Chaptal e List," in *Melchiorre Gioia, (1767–1829)*.

Quaini, M., "Appunti per una archeologia del 'colpo d'occhio.' Medici, soldati e pittori alle origini dell'osservazione sul terreno in Liguria," in L. Coveri and D. Moreno, eds., *Studi di etnografia e dialettologia ligure in memoria di Hugo Plomteux*, Genoa, 1983, pp. 107–125.

Ragionieri, E., *Politica e amministrazione nella storia dell'Italia unita*, Bari, 1967.

"La storia politica e sociale," in *Storia d'Italia*, vol. IV, *Dall'Unità a oggi*, Turin, 1976, pp. 1668–1743.

Ragone, G., "La letteratura e il consumo: un profilo dei generi e dei modelli nell'editoria italiana (1845–1925)," in *Letteratura italiana*, vol. II, *Produzione e consumo*, Turin, 1983, pp. 687–722.

Rainone, C., "Liberisti e Liberali: pensiero economico e pensiero politico in Italia avanti il 1861," in *Nuove questioni di storia del Risorgimento e dell'unità d'Italia*, vol. I, Milan, 1976, pp. 513–564.

Rak, M., "Il popolo del libro. L'editoria popolare italiana nella prima fase della società industriale (1790–1890)," in Rak, *La società letteraria. Scrittori e librai, stampatori e pubblico nell'Italia dell'industrialismo*, Venice, 1990, pp. 101–145.

Raponi, N., *Politica e amministrazione in Lombardia agli esordi dell'Unità*, Milan, 1967.

"La Società Storica Lombarda e i suoi soci (1873–1899)," in *Milano fin-de-siècle e il caso Bagatti Valsecchi. Memoria e progetto per la metropoli italiana*, Milan, 1991, pp. 33–46.

Riall, L., "Elite Resistance to State Formation: The Case of Italy," in M. Fulbrook, ed., *National Histories and European History*, London, 1993, pp. 46–68.

The Italian Risorgimento. State, Society and National Unification, London, 1994.

Riffaterre, M., "Descriptive Imagery," *Yale French Studies* 61 (1981), pp. 107–125.

Rigotti, F., *L'umana perfezione. Saggio sulla circolazione e diffusione dell'idea di progresso nell'Italia del primo Ottocento*, Naples, 1980.

Rodolico, F., *La Toscana descritta dai naturalisti del Settecento. Pagine di storia del pensiero scientifico*, Florence, 1945.

Romanelli, R., *L'Italia liberale (1861–1900)*, Bologna, 1979.

"La nuova Italia e la misurazione dei fatti sociali. Una premessa," *Quaderni storici* 15 (1980), pp. 765–778.

Il comando impossibile. Stato e società nell'Italia liberale, Bologna, 1988.

"La nazione e il campanile. Il dibattito intorno alle dimensioni dello scambio politico nell'Italia liberale," in F. Andreucci and A. Pescarolo, eds., *Gli spazi del potere. Aree, regioni, stati: le coordinate territoriali della storia contemporanea*, Florence, 1987, pp. 184–191.

Romani, R., "Il Messedaglia dimezzato," *Venetica. Rivista di storia delle Venezie* 1 (1984), pp. 109–130.

"Quale sapere sociale per il Risorgimento? Alcuni libri recenti sulla statistica napoleonica," *Società e storia* 54 (1991), pp. 893–911.

"Romagnosi, Messedaglia e la 'scuola lombardo-veneta': la costruzione di un sapere sociale," in R. Camurri, ed., *La scienza moderata. Fedele Lampertico e l'Italia liberale*, Milan, 1992, pp. 177–210.

L'economia politica del Risorgimento italiano, Turin, 1994.

Romano, S. F., *Storia della questione meridionale*, Palermo, 1945.

Romeo, R., *Cavour e il suo tempo (1810–42)*, Bari, 1969.

"Il Risorgimento: Realtà storica e tradizione 'morale'", in *Dal Piemonte sabaudo all'Italia liberale*, Bari, 1974.

Rotondi, C., ed., *I Lorena in Toscana*, Florence, 1989.

Ruffilli, R., *La questione regionale dall'unificazione alla dittatura*, Milan, 1971.

Rutto, G., "Riforme e patriottismi nell'Austria di Maria Teresa," in A. De Maddalena, E. Rotelli, and G. Barbarisi, eds., *Economia, istituzioni e cultura in Lombardia nell'età di Maria Teresa*, vol. II, *Cultura e società*, Bologna, 1982, pp. 903–923.

Saitta, A., *Alle origini del Risorgimento: I testi di un "celebre" concorso (1796)*, Rome, 1974.

Salvadori, M., *Il mito del buongoverno. La questione meridionale da Cavour a Gramsci*, Turin, 1960.

Salvemini, B., *Economia politica e arretratezza meridionale nell'età del Risorgimento. Luca de Samuele Cagnazzi e la diffusione dello smithianesimo nel Regno di Napoli*, Lecce, 1981.

Salvo, R., "Nota su un progetto di Saverio Scrofani per l'istituzione in Sicilia di una direzione di statistica e del censimento (1821–23)," *Nuovi quaderni del Meridione* 100 (1988), pp. 529–533.

"Alcune notizie sul concorso per i posti di 'commesso' alla Direzione centrale di statistica di Palermo (1832–33)," in P. F. Asso, P. Barucci, and M. Ganci, eds., *Francesco Ferrara e il suo tempo. Atti del Congresso Palermo 27–30 ottobre 1988*, Rome, 1990, pp. 491–510.

"Melchiorre Gioia nel dibattito politico-economico in Sicilia (1824–1831)," in *Melchiorre Gioia (1767–1829). Politica, società, economia tra riforme e Restaurazione. Atti del Convegno di studi, Piacenza 5–7 Aprile 1990, Bollettino storico piacentino* 85 (1990), pp. 343–375.

Sapori, A., *Luigi Serristori*, Florence, 1925.

Schiera, P., *Dall'arte di governo alle scienze dello stato. Il cameralismo e l'assolutismo tedesco*, Milan, 1968.

Schlanger, J. E., *Les Métaphores de l'organisme*, Paris, 1971.

Scirocco, A., "La *Statistica murattiana* del Regno di Napoli," in S. Martuscelli, ed., *La popolazione del Mezzogiorno nella statistica di Re Murat*, Naples, 1979.

L'Italia del Risorgimento 1800–1871, Bologna, 1990.

Scott, J. W., *Gender and the Politics of History*, New York, 1988.

"A Statistical Representation of Work. *La Statistique de l'industrie à Paris, 1847–48*," in Scott, *Gender and the Politics of History*, pp. 113–138.

Scrofani, S., *Memorie inedite*, ed. G. Giarrizzo, Palermo, 1970.

Seifert, A., "Staatenkunde: eine neue Disziplin und ihr wissenschaftstheoretischer Ort," in M. Rassem, ed., *Statistik und Staatsbeschreibung in der Neuzeit*, Paderborn, 1980, pp. 217–248.

Sella, D. and C. Capra, *Il Ducato di Milano dal 1535 al 1796*, Turin, 1984.

Sestan, E., "Origini delle società di storia patria e loro posizione nel campo della cultura e degli studi storici," *Annali dell'Istituto storico italo-germanico in Trento* 7 (1981), pp. 105–141.

Sestan, E., ed., *Opere di Giandomenico Romagnosi, Carlo Cattaneo, Giuseppe Ferrari*, Milan and Naples, 1957.

Sestini, A., "Le regioni italiane come base geografica della struttura dello stato," in *Atti del XIV Congresso geografico italiano*, Bologna, 1949.

Signorelli, *La statistica preunitaria tra "assunto civile" e funzione burocratica*, Catania, 1983.

Sirugo, F., *L'economia degli stati italiani prima dell'unificazione. 1. Stati sardi di Terraferma (1700–1860). Saggio bibliografico*, Milan, 1962.

Small, A., *The Cameralists. The Pioneers of German Social Polity*, New York, 1909.

Sofia, F., *Una scienza per l'amministrazione. Statistica e pubblici apparati tra età rivoluzionaria e restaurazione*, Rome, 1988.

"Melchiorre Gioia e la statistica," in *Melchiorre Gioia (1767–1829). Politica, società, economia tra riforme e Restaurazione. Atti del Convegno di studi, Piacenza 5–7 Aprile 1990, Bollettino storico piacentino* 85 (1990), pp. 249–268.

Soja, E. W., *Postmodern Geographies. The Reassertion of Space in Critical Social Theory*, London and New York, 1984.

Sola, G., "Sviluppi e scenari della sociologia italiana: 1861–1890. Percorsi di lettura da una bibliografia," in F. Barbano and G. Sola, *Sociologia e scienze sociali in Italia 1861–1890. Introduzioni critiche e repertorio bibliografico*, Milan, 1985.

Soldani, S. and G. Turi, eds., *Fare gli Italiani. Scuola e cultura nell'Italia contemporanea*, vol. I, *La nascita dello stato nazionale*, Bologna, 1993.

Sori, E., *Città e controllo sociale in Italia tra XVIII e XIX secolo*, Milan, 1982.

Sternberg, M., "Ordering the Unordered: Time, Space, and Descriptive Coherence," *Yale French Studies* 61 (1981), pp. 60–88.

Stigler, S. M., *The History of Statistics. The Measurement of Uncertainty before 1900*, Cambridge, Mass., and London, 1986.

Talamo, G., "Il problema delle diversità e degli squilibri regionali nella cultura politica italiana dal periodo dell'unificazione alla caduta della Destra," in *Gli squilibri regionali e l'articolazione dell'intervento pubblico. Atti del Convegno di studio svoltosi a Torino e Saint Vincent dal 3 al 7 settembre 1961*, Milan, 1962.

Tedeschi, P. L., "Francesco Ferrara e la politica della statistica (1835–1846)," *Archivio storico per la Sicilia orientale* 78 (1982), pp. 94–151.

Teti, V., *La razza maledetta. Origini del pregiudizio antimeridionale*, Rome, 1993.

Thompson, J. B., *Studies in the Theory of Ideology*, Berkeley and Los Angeles, 1984.

Toews, J. E., "Intellectual History after the Linguistic Turn: The Autonomy of Meaning and the Irreducibility of Experience," *American Historical Review* 92 (1987), pp. 879–907.

Tribe, K., *Governing Economy. The Reformation of German Economic Discourse 1750–1840*, Cambridge, 1988.

Venturi, F., *Settecento riformatore. Da Muratori a Beccaria*, Turin, 1969.

"L'Italia fuori d'Italia," in *Storia d'Italia*, vol. III, *Dal primo Settecento all'Unità*, Turin, 1973, pp. 987–1481.

Verdery, K., *National Ideology Under Socialism. Identity and Cultural Politics in Ceauçescu's Romania*, Berkeley, Los Angeles, and Oxford, 1991.

Villani, P., "Gruppi sociali e classe dirigente all'indomani dell'Unità," in *Storia d'Italia. Annali I. Dal feudalesimo al capitalismo*, Turin, 1978, pp. 881–978.

Westergaard, H., *Contributions to the History of Statistics*, London, 1932.

Winch, D., "Adam Smith's 'Enduring Particular Result': A Political and Cosmopolitan Perspective," in I. Hont and M. Ignatieff, eds., *Wealth and Virtue. The Shaping of Political Economy in the Scottish Enlightenment*, Cambridge, 1983, pp. 253–269.

Woolf, S. J., *Il Risorgimento italiano*, vol. II, *Dalla Restaurazione all'Unità*, Turin, 1981.

"Statistics and the Modern State," *Comparative Studies in Society and History* 31 (1989), pp. 588–603.

Zaninelli, S., *Un secolo di vita economica lombarda 1748–1848*, Milan, 1950.

"Una fonte per la storia dell'economia del Lombardo-Veneto nella prima metà del secolo XIX: le 'Tafeln zur Statistik der Oesterreichischen Monarchie,'" *Archivio economico dell'unificazione italiana* 12, (1963), pp. 1–11.

INDEX